D1398216

Understanding Digital Subscriber Line Technology

ISBN 0-13-780545-4

9 780137 805457

90000

Prentice Hall Communications Engineering and Emerging Technologies Series

Theodore S. Rappaport, Series Editor

RAPPAPORT, *Wireless Communication: Principles & Practice*

RAZAVI, *RF Microelectronics*

GARG & WILKES, *Principles and Applications of GSM*

STARR, CIOFFI, & SILVERMAN, *Understanding Digital Subscriber Line Technology*

FORTHCOMING

LIBERTI & RAPPAPORT, *Smart Antennas for Wireless CDMA: IS-95 and Wideband CDMA Applications*

TRANTER, KOSBAR, RAPPAPORT, & SHANMUGAN, *Simulation of Modern Communications Systems with Wireless Applications*

Understanding Digital Subscriber Line Technology

Thomas Starr
Senior MTS
Ameritech

John M. Cioffi
Professor of Electrical Engineering
Stanford University

Peter Silverman
Senior Architect – New Business Initiatives
3COM Corporation

Prentice Hall PTR, Upper Saddle River, NJ 07458
http://www.phptr.com

Library of Congress Catalog-in-Publication Data

Starr, Thomas J. J.
  Understanding digital subscriber line technology / Thomas J. J.
Starr, John M. Cioffi, Peter Silverman.
    p. cm.
  Includes bibliographical references and index.
  ISBN 0-13-780545-4
  1. Digital telephone systems.   2. Telephone switching systems.
Electronic.   I. Cioffi, John M.   II. Silverman, Peter.   III. Title.
TK6421.S85   1999
dc.387'8--dc21
                                          98-47586
                                          CIP

Editorial/production supervision: *Vanessa Moore*
Interior Formatting: *Aurelia Sharnhorst*
Cover design director: *Jerry Votta*
Manufacturing manager: *Alan Fischer*
Acquisitions editor: *Bernard Goodwin*
Series editor: *Theodore S. Rappaport*
Marketing manager: *Kaylie Smith*

Fairleigh Dickinson
University Library

MADISON, NEW JERSEY

© 1999 Prentice Hall PTR
Prentice-Hall, Inc.
Upper Saddle River, NJ 07458

Prentice Hall books are widely used by corporations and government agencies for training,
marketing, and resale. The publisher offers discounts on this book when ordered in bulk
quantities. For more information, contact Corporate Sales Department, Phone: 800-382-3419,
Fax: 201-236-7141, Email: corpsales@prenhall.com
or write: Prentice Hall PTR
          Corporate Sales Department
          One Lake Street
          Upper Saddle River, NJ 07458

All rights reserved. No part of this book may be reproduced, in any form or by any means,
without permission in writing from the publisher.

All Trademarks are the property of their respective owners.
All Figures used in Chapter 11 of this book are reproduced with permission of ATIS.

Printed in the United States of America
10   9   8   7   6   5   4   3   2   1

ISBN 0-13-780545-4

Prentice-Hall International (UK) Limited, *London*
Prentice-Hall of Australia Pty. Limited, *Sydney*
Prentice-Hall of Canada, Inc., *Toronto*
Prentice-Hall Hispanoamericana S.A., *Mexico*
Prentice-Hall of India Private Limited, *New Delhi*
Prentice-Hall of Japan, Inc., *Tokyo*
Simon & Schuster Asia Pte. Ltd., *Singapore*
Editora Prentice-Hall do Brasil, Ltda., *Rio de Janeiro*

Dedicated to our dear wives,

Marilynn Starr

Sharon Cioffi

Patricia Lane Silverman

Generic DSL Reference Model

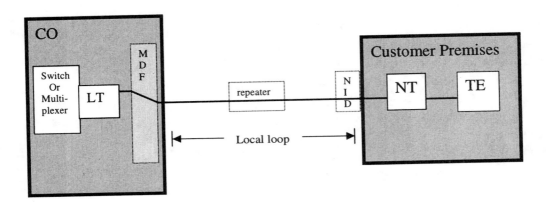

CO: Central Office – A building where local loops connect to transmission and switching equipment.

LT: Line Termination – DSL modem at the network end of the local loop.

MDF: Main Distributing Frame – A wire cross-connecting field used to any loop to any CO equipment.

Repeater – A signal regeneration device located near the midpoint of a cable. Used to enable transmission on long loops. Repeaters are not needed for the majority of loops.

Local loop – The telephone wire connecting the CO to the customer premises.

NID: Network Interface Device – The point of demarcation between the customer installation and the telephone company outside wire.

NT: Network Termination – DSL modem at the customer end of the local loop.

TE: Terminal Equipment – End-user equipment such as a personal computer or a telephone.

Contents

Preface   **xv**

Acknowledgments   **xvi**

About the Authors   **xvii**

Chapter 1   DSL Fundamentals   **1**
1.1   Alternatives to DSLs: Fiber, Wireless, and Coax   2
1.2   Worldwide Extent   2
1.3   Voice-Band Modems and DSLs   3
1.4   Transmission Modes   8
    1.4.1   Direction   *8*
    1.4.2   Timing   *9*
    1.4.3   Channels   *10*
    1.4.4   Single and Multipoint Topologies   *11*
1.5   DSL Terminology   12
1.6   Rate Versus Reach   12
1.7   Crosstalk   13
1.8   Enabling and Disabling Forces   16
1.9   Applications   17
1.10 Evolution of Digital Transmission   21

Chapter 2   Types of DSLs   **23**
2.1   DSL Design Margin   23
2.2   DSL Precursors   24

2.3   Basic Rate ISDN   25
      2.3.1   *ISDN Basic Rate Origins*   25
      2.3.2   *Basic Rate ISDN Capabilities and Applications*   26
      2.3.3   *Basic Rate ISDN Rate Transmission*   26
      2.3.4   *Extended-Range Basic Rate ISDN*   27
      2.3.5   *Digital Added Main Line*   29
      2.3.6   *IDSL*   30
2.4   HDSL   30
      2.4.1   *HDSL Origins*   30
      2.4.2   *HDSL Capabilities and Application*   31
      2.4.3   *HDSL Transmission*   32
      2.4.4   *Second-Generation HDSL*   35
2.5   ADSL   41
      2.5.1   *ADSL Definition and Reference Model*   41
      2.5.2   *ADSL Origins*   42
      2.5.3   *ADSL Capabilities and Application*   43
      2.5.4   *ADSL Transmission*   43
      2.5.5   *ADSL's Future*   46
2.6   VDSL   49
      2.6.1   *VDSL Definition and Reference Model*   49
      2.6.2   *VDSL Origins*   51
      2.6.3   *VDSL Capabilities and Applications*   51

Chapter 3   Twisted-Pair Transmission   53
3.1   Twisted-Wire-Pair Origins   53
3.2   Telephone Network and Loop Plant Characteristics   53
      3.2.1   *Feeder Plant*   54
      3.2.2   *Digital Loop Carrier*   54
      3.2.3   *Distribution Plant*   56
      3.2.4   *Wire Gauge*   56
      3.2.5   *Bridged Tap*   57
      3.2.6   *Loaded Loop*   58
      3.2.7   *Loop Length Distribution*   59
      3.2.8   *Customer Premises Configuration*   60
3.3   Line Powering   63
      3.3.1   *Activation / Deactivation*   63
3.4   Sealing Current   63

3.5   Transmission Line Characterization   64
     3.5.1   *"ABCD" Modeling*   *64*
     3.5.2   *Transmission Line RLCG Characterization*   *67*
     3.5.3   *Characterization of a Bridged-Tap Section*   *74*
     3.5.4   *Loaded Coils — Series Inductance*   *75*
     3.5.5   *Computation of Transfer Function*   *75*
     3.5.6   *Measurements for Computation of RLCG Parameters*   *78*
     3.5.7   *Balance — Metallic and Longitudinal*   *84*
3.6   Noises   85
     3.6.1   *Crosstalk Noise*   *86*
     3.6.2   *Radio Noise*   *92*
     3.6.3   *Impulse Noise*   *94*
3.7   Spectral Compatibility   97
     3.7.1   *Interference Between DSLs and Multiplexing*   *98*
     3.7.2   *Self-Interference*   *99*
     3.7.3   *Crosstalk FEXT and NEXT Power Spectral Density Models*   *100*
     3.7.4   *Emissions from DSLs*   *104*
3.8   More Two-Port Networks   105
     3.8.1   *Reciprocal and Lossless Two-Port Circuits*   *106*
     3.8.2   *Analog Filter Design and T(s)*   *107*
     3.8.3   *Lossless Realization of H(s)*   *113*
     3.8.4   *Frequency / Magnitude Scaling and Frequency Transformations*   *114*
     3.8.5   *Active Filters*   *116*
3.9   Three-Port Networks for DSLs   119
     3.9.1   *POTS Splitters*   *120*
     3.9.2   *Hybrid Circuits*   *128*
References   129

Chapter 4   Comparison with Other Media   **133**
4.1   Fiber-to-the-Home   133
4.2   Coax and Hybrid Fiber Coax   134
4.3   Wireless Alternatives   136
4.4   Satellite Services   137
References   137

Chapter 5   Transmission Duplexing Methods   **139**
5.1   Four-Wire Duplexing   139

5.2   Echo Cancellation   140
      5.2.1   Adaptive Echo Cancellation   142
5.3   Time-Division Duplexing   142
5.4   Frequency-Division Multiplexing   143
References   144

Chapter 6   Basic Digital Transmission Methods   147
6.1   Basic Modulation and Demodulation   147
      6.1.1   The Additive White Gaussian Noise Channel   150
      6.1.2   Margin, Gap, and Capacity   154
6.2   Baseband Codes   155
      6.2.1   The 2B1Q Line Code (ISDN and HDSL)   155
      6.2.2   Pulse Amplitude Modulation   159
      6.2.3   Binary Transmission with DC Notches   161
      6.2.4   4B3T Line Code   166
      6.2.5   4B5B Modulation   168
      6.2.6   Successive Transmission   169
6.3   Passband Codes   172
      6.3.1   Quadrature Amplitude Modulation   173
      6.3.2   Carrierless AMPM   174
      6.3.3   Other Quadrature Modulation Schemes   175
      6.3.4   Constellations for QAM/CAP and Relation to VSB   176
      6.3.5   Complex Baseband Equivalents   178
References   180

Chapter 7   Loop Impairments, Solutions, and DMT   183
7.1   Intersymbol Interference   183
      7.1.1   Quantifying ISI   184
      7.1.2   Equalization   186
      7.1.3   Transmit Equalization   196
      7.1.4   Partial-Response Detection   200
      7.1.5   Maximum-Likelihood Detection (Viterbi Algorithm)   202
7.2   Multichannel Line Codes   205
      7.2.1   Capacity of the AWGN Channel   205
      7.2.2   Basic Multichannel Transmission   206
      7.2.3   Loading Algorithms   208
      7.2.4   Channel Partitioning   217
      7.2.5   Equalization for Multichannel Partitioning   228

7.2.6   *ADSL T1.413 DMT*   *235*

7.2.7   *Clipping and Scaling (Peak-to-Average Issues)*   *236*

7.2.8   *Fast Fourier Transforms for DMT*   *242*

7.2.9   *Multiplexing Methods for Multicarrier Transmission*   *247*

7.2.10 *Narrowband Noise Rejection*   *251*

7.3   Trellis Coding   256

7.3.1   *Constellation Partitioning and Expansion*   *256*

7.3.2   *Enumeration of Popular Codes*   *262*

7.3.3   *Shaping Effects*   *262*

7.3.4   *Turbo Codes*   *263*

7.4   Error Control   264

7.4.1   *Basic Error Control*   *265*

7.4.2   *Reed-Solomon Codes*   *270*

7.4.3   *Interleaving Methods*   *274*

7.4.4   *Concatenated Coding and Multilayer Coding*   *277*

7.4.5   *ADSL Special Case*   *277*

7.4.6   *CRC Checks*   *281*

7.4.7   *Scramblers*   *285*

References   288

Chapter 8   Initialization, Timing and Performance   297

8.1   Initialization Methods   297

8.1.1   *Activation*   *297*

8.1.2   *Gain Estimation*   *299*

8.1.3   *Synchronization (Clock, Frame)*   *302*

8.1.4   *First Channel Identification*   *303*

8.1.5   *Channel Equalization*   *311*

8.1.6   *Secondary Channel Identification and Exchange*   *314*

8.2   Adaptation of Receiver and Transmitter   314

8.2.1   *Receiver Equalization Updating*   *315*

8.2.2   *Transmitter Adjustment*   *318*

8.3   Measurement of Performance   324

8.3.1   *Test Loops and Noise Generation*   *325*

8.3.2   *Measure of Performance*   *333*

8.4   Timing Recovery Methods   337

8.4.1   *Basic PLL Operation*   *337*

8.4.2   *Open-Loop Timing Recovery*   *341*

8.4.3   *Decision-Directed Timing Recovery*   *344*

8.4.4   *Pointers and Add / Delete Mechanisms*   *346*

8.4.5   *Frame Synchronization*   *348*

8.4.6   *Discrete-Time VCO Implementation*   *349*

References   352

**Chapter 9   Operations, Administration Maintenance,
         and Provisioning**   355

9.1   OAM&P Features   358

9.2   Loop Qualification   360

Chapter 10   DSL in the Context of the ISO Reference Model   **363**

10.1   The ISO Model   363

10.2   Theory and Reality   365

10.3   The Internet Protocol Suite   365

10.4   ATM in the Seven-Layer Model   366

         10.4.1   *Physical Layer Functions*   *367*

         10.4.2   *Link and Higher-Layer Functions*   *367*

Chapter 11   ADSL: The Bit Pump   369

11.1   ADSL System Reference Model   369

11.2   ATU-C Reference Model   370

11.3   ATU-R Reference Model   372

11.4   Specific Configurations to Support ATM   373

11.5   Framing   373

         11.5.1   *Superframe Structure*   *375*

         11.5.2   *Fast Data Buffer Frame Structure*   *375*

         11.5.3   *Interleaved Data Buffer Frame Structure*   *376*

11.6   Operations and Maintenance   377

11.7   Initialization   378

Reference   379

Chapter 12   ATM Transmission Convergence on ADSL   **381**

12.1   Functions of ATM Transmission Convergence   381

12.2   Transmission Convergence in an ADSL Environment   382

Reference   384

Chapter 13   Frame-Based Protocols over ADSL   **385**

13.1   PPP over a Frame-Based ADSL   385

         13.1.1   *RFC 1662 — PPP in HDLC-Like Framing*   *386*

13.1.2   *RFC 1661 — The Point-To-Point Protocol*   *387*
13.2   FUNI over ADSL   388
  13.2.1   *FUNI Frame Structure*   *388*
  13.2.2   *Encapsulation*   *389*
Reference   389

Chapter 14   ADSL in the Context of End-to-End Systems   391
14.1   An Overview of a Generic DSL Architecture   394
  14.1.1   *The Customer's Premises*   *394*
  14.1.2   *The DSL Loop*   *395*
  14.1.3   *Termination of DSL in the Carrier's Central Office
       or Remote Site*   *395*
  14.1.4   *The Carrier's Back-End Data Network*   *396*
  14.1.5   *The Interface to the Service Provider's Network*   *398*
14.2   Potential ADSL Services and the Service Requirements   398
14.3   Specific Architectures for Deploying ADSL in Different
    Business Models   399
14.4   Several ADSL Architectures   402
  14.4.1   *A Packet-Based Architecture for Small Deployments*   *402*
  14.4.2   *ATM Access Networks*   *403*
  14.4.3   *RFC 1483*   *405*
  14.4.4   *PPP over ATM*   *407*
  14.4.5   *Tunneled Gateway Architecture*   *408*
  14.4.6   *PPP Terminated Aggregation*   *409*
References   410

Chapter 15   Network Architecture and Regulation   411
15.1   Private Line   411
15.2   Circuit Switched   411
15.3   Packet Switched   412
15.4   ATM   413
15.5   Remote Terminal   414
15.6   Competitive Data Access Alternatives   414
15.7   Regulation   416

Chapter 16   Standards   419
16.1   ITU   420
16.2   Committee T1   421
16.3   ETSI   423

16.4   ADSL Forum   424
16.5   ATM Forum   424
16.6   DAVIC   425
16.7   IETF   425
16.8   EIA/TIA   426
16.9   IEEE   426
16.10   The Value of Standards and Participation in Their Development   427
16.11   Standards Process   428
        16.11.1   When to Develop a Standard   429
        16.11.2   Is a Standard Needed?   430
        16.11.3   Standard or Standards?   431

Appendix A   Glossary   433

Appendix B   Selected Standards and Specifications   443

Appendix C   Selected T1E1.4 Contributions and ADSL Forum
              Technical Reports (found on CD-ROM)   447

Index   465

Preface

Visionaries have spoken of a future where the common person has instantaneous access to data spread around the globe. Engaging in a live videoconference, or perhaps watching a personalized newscast are just two of examples of many. For this vision to become reality, a global broadband information infrastructure must be built that provides low-cost access to the consumers and sources of information. What connects to virtually every home and business in the industrialized world? Phone lines connect to 700 million sites today. Data rates of several kilobits per second are possible over phone lines using dial-up modems. This is enough to spark the appetite of the Internet surfer but is not nearly enough to satisfy the desire for immediate information on demand. Similarly, video and audio applications at dial-up modem data rates leave users demanding more.

Digital subscriber line (DSL) technology enables high-speed digital transmission on conventional telephone lines. A global broadband information infrastructure based on telephone lines is emerging, and it relies on DSL technology. The transformation of the telephone line access has begun; it is progressing with the addition of over one billion U.S. dollars worth of DSL equipment each year.

Accomplishing the impossible is an engineer's greatest reward. Digital subscriber line development has been most rewarding. In 1975, it was believed that 20 kb/s was the highest data rate that could be transmitted via telephone lines. Then, breakthrough concepts in digital transmission were enabled by enormous advances in very-large-scale integrated (VLSI) circuits and digital signal processing (DSP). Transceiver designs of breathtaking complexity (at that time) provided 144 kb/s basic rate ISDN (BRI) transport via most telephone lines. Experts then said that this was very near the capacity limit of telephone lines. This barrier was demolished by the 1.5 Mb/s high bit rate DSL (HDSL). The breakthrough cycle was repeated by 6 Mb/s asymmetric DSL (ADSL), and then 52 Mb/s very-high bit rate DSL (VDSL).

This book explains and details the key concepts for DSL technology and its applications. The reader will attain a strong familiarity with the crucial aspects and technical jargon of the DSL field. The scope encompasses applications, network architecture, network management, network operations, communications protocols, standards, regulatory issues, and the underlying technologies. This book was written to assist engineers and marketing managers — whether new to DSLs or experts in need of a convenient reference.

Background regarding voice-band transmission via telephone lines may be found in the excellent books by Witham Reeves on subscriber loops.

Acknowledgments

The authors thank Jim Loehndorf for assistance with the sections on data communications protocols, and Kim Maxwell for providing his input regarding voice-band modems and other sections.

The authors would also like to sincerely thank Dr. Kiho Kim, Richard Goodson, and Dr. Martin Pollakowski for their review of this material and their helpful comments and suggestions.

The second author, John Cioffi, especially would like to thank the following people (in alphabetical order) for their significant discussions and direct assistance on specific topics of this book: Mike Agah, John Bingham, Jacky Chow, Peter Chow, John Cook, Joice DeBolt, Kevin Foster, Mathias Friese, Richard Goodson, Werner Henkel, Atul Salvekar, Jose Tellado, Po Tong, Craig Valenti, Jean-Jacques Werner, and George Zimmerman. He further wishes to thank Dr. Joe Lechleider for enticing him into DSL in 1987, and thanks beyond measure the outstanding technical staff of Amati (1989–1997, now Texas Instruments), and the first to believe: his past and present students at Stanford.

Thanks also go out to Steve Blackwell and Kevin Schneider of Adtran, who kindly offered the use of their good summary of HDSL2 work in the T1E1.4 Working Group.

The first author, Tom Starr, has had the pleasure of chairing the T1E1.4 Working Group for over ten years. Thanks to the professionalism, dedication, and expertise of its members, T1E1.4 has done more than merely write the industry's DSL standards. Multidisciplinary collaboration has allowed T1E1.4 to set the industry's objectives and chart the course to meeting these objectives. There have been moments of agony and disappointment but, on the whole, serving as T1E1.4 chair has been rewarding. Thank you, members of T1E1.4, for being the world's foremost creators of DSL technology.

The views expressed in this book are those of the authors and do not necessarily reflect the views of their employers or the organizations in which the authors hold office.

Thomas Starr
John M. Cioffi
Peter Silverman

About the Authors

THOMAS STARR

Tom Starr is a Senior Member of Technical Staff in Ameritech's Network Architecture Planing Department. He is responsible for the development and management of new local access technologies for Ameritech's network. These technologies include ADSL, HDSL, VDSL, and ISDN.

Starr serves at Chairperson of ANSI accredited standards working group T1E1.4, which develops XDSL standards for the United States. He also serves on the board of directors for the ADSL Forum, which addresses end-to-end systems aspects related to ADSL-based services, and participates in the ITU SG15 Q4 group on XDSL international standards.

He previously worked 12 years at AT&T Bell Laboratories on ISDN and local telephone switching systems. Seven U.S. patents in the field of telecommunications have been issued to Starr. He holds a MS degree in computer science and a BS degree in computer engineering from the University of Illinois, Champaign-Urbana.

JOHN M. CIOFFI

John M. Cioffi received the BSEE in 1978 from Illinois (Champaign-Urbana) and the PhDEE degree in 1984 from Stanford. He worked as a modem designer at Bell Laboratories from 1978–1986. He has been on the faculty of Stanford since 1986, where he is now a tenured associate professor. He also founded Amati Communications Corp. in 1991, has served as an officer and director since 1991, and is presently Chief Technical Officer of Amati in addition to continuing his Stanford position.

Cioffi's specific interests are in the area of high-performance digital transmission and storage systems, where he has over 100 publications and 20 patents, all of which are licensed. He is a Fellow of the Institute of Electrical and Electronic

Engineers (IEEE), and received the 1991 *IEEE Communications Magazine* Best Paper Award as well as the 1995 T1 Outstanding Achievement Award of the American National Standards Institute. Cioffi was an NSF Presidential Investigator from 1987–1992 and has served in a number of editorial positions for IEEE magazines and conferences.

PETER SILVERMAN

Peter Silverman is a Senior Architect in the ADSL Development Group of the New Business Initiatives Division at 3COM. His responsibilities include formulating the architectural plans for 3COM's DSL product line to allow implementation, provisioning, and management of DSL services in a mulitvendor and mulitprovider environment. He has been active in the ADSL Forum, the T1E1.4, and ITU SG 15 Q4 DSL standards efforts.

Prior to joining 3COM in 1997, Silverman worked eight years as Senior Member of Technical Staff at Ameritech — including ADSL and Interactive Digital Cable Television. He began his career with eight years as a consultant for CAP Gemini America, developing digital telephone switching software for their clients. His education includes a BS in biology from the University of Chicago and graduate work in computer science at the University of Illinois, Chicago.

DSL Fundamentals

Digital subscriber line (DSL) technology provides transport of high-bit-rate digital information over telephone subscriber lines. Telephone lines, whose heritage dates back to Alexander Graham Bell's invention of the telephone in 1875, can now transport data at millions of bits per second. This is accomplished via sophisticated digital transmission techniques, which compensate for the many transmission impairments common to telephone lines. The digital transmission techniques involve complex algorithms that have recently become practical due to the enormous processing power of digital signal processors based on very-large-scale integrated (VLSI) circuits. Marketing alchemists claim that DSLs turn copper into gold.

DSL technology has added a new twist to the utility of telephone lines. Telephone lines, which were constructed to carry a single voice signal with a 3.4 kHz[1] bandwidth channel, can now convey nearly 100 digitally compressed voice signals, or a video signal with quality similar to broadcast television. High-speed digital transmission via telephone lines requires advanced signal processing to overcome transmission impairments due to signal attenuation, crosstalk noise from the signals present on other wires in the same cable, signal reflections, radio-frequency noise, and impulse noise.

The twisted-wire-pair infrastructure connects to virtually every home and workplace in the world, but DSLs have their limitations. Approximately 15% of telephone lines in the world will require upgrade activity to permit high-speed DSL operation. Corrective measures for long loops include installation of mid-span repeaters, installation of fiber-remoted multiplexers, and removal of load coils.

---

1.   kHz = kiloHertz, or frequency measured as thousands of cycles per second.

In this book we use the term *DSL* to refer to all types of digital subscriber line technologies, including ADSL, HDSL, basic rate ISDN DSL, VDSL, and IDSL. The term *XDSL* has also been used in the industry to refer to all types of DSLs.

1.1 Alternatives to DSLs: Fiber, Wireless, and Coax

The obsolescence of the twisted-wire-pair telephone line has been predicted many times. The popular belief of many telephone industry experts in the late 1980s was that most of the world's telephones would be directly connected via fiber optic lines in "a few years." Fiber optic lines are now common to major business sites. However, economics combined with the logistical challenges of cable construction will delay the arrival of the totally phonic world for many decades. In the early 1990s came the promise of universal wireless access. However, for the near term, limited radio spectrum bandwidth combined with the logistical challenges of placing radio hub sites where needed (terrestrial, or in earth orbit) restricts wireless transport to the subset of applications that require mobility and those that gain benefit by broadcasting the same information to a large number of sites. Coaxial cable can transport interactive data services and voice telephone service in addition to the traditional broadcast video service. However, the interactive data and voice services are best served via two-way cable facilities, which were available to about 10% of cable customers in 1997. Cable companies are upgrading their facilities, but only in selected areas. Coax-modem service is achieving some success in the areas with two-way coax facilities. Compared to the nearly 100% presence of phone lines, cable facilities pass by the vast majority of residences in the United States and a much lower proportion of businesses.

Fiber, wireless, and coax transport have proved valuable for many applications. There is no universal access technology that best serves all locations and all applications. However, now that DSL technology has enabled telephone lines to convey multimedia applications that were once thought to be the exclusive province of fiber, telephone lines are the most economic means to transport a broad range of communications services to millions of customers. The primary application weaknesses of DSLs are the lack of mobility and poor broadcast efficiency.

A preexisting infrastructure, such as telephone lines, enabled by the right technology is economically superior to deploying a new infrastructure. Even radio requires new infrastructure: sites for transceivers and network links to these sites. A new technology can be proven only where the existing infrastructure is incapable of supporting essential applications (such as mobile communications) or where legal/regulatory conditions alter the balance. In addition to being expensive, building new infrastructure takes a long time to obtain right-of-way, lay cable, obtain permits to place radio towers, or launch satellites.

Twisted pair telephone lines may be buried, but they are far from dead.

1.2 Worldwide Extent

Nearly every business and residence in the industrialized areas of the world is already connected to the global telephone network. The telephone industry has spent approximately one trillion dollars (U.S.) over the past century constructing the worldwide twisted pair outside plant facilities. Nearly 700 million telephone lines were in place in 1996. Telephone companies continue to spend millions of dollars each year constructing more twisted-wire-pair telephone lines. More than 900 million telephone lines are expected to exist by the year 2001.[2] The vast majority of these telephone lines will support the transport of approximately one million bits per second (Mb/s) when suitable high-speed DSL transceivers are connected to the customer and telephone company ends of the twisted wire pair. In most cases, no modifications are needed to the outside plant facilities. Many telephone lines will support data rates well above 1 Mb/s.

1.3 Voice-Band Modems and DSLs

Voice-band modems were introduced in the late 1950s for the purpose of sending data through the public switched telephone network (PSTN). See Figure 1.1. The word *modem* comes from *mo*dulator-*dem*odulator (see Chapter 6 for details on modulation and demodulation). Data transmitted via the PSTN must be modulated because the PSTN does not convey frequencies below approximately 200 Hz. Unmodulated data require transmission of frequencies approaching zero Hz. In effect, the modem transforms the frequency characteristics of the data to resemble the voice signals that the PSTN was designed to convey. The PSTN conveys signals in the 200 Hz to 3400 Hz frequency band. Thus, the modulated data appear to be a normal voice call to the PSTN. Facsimile (fax) machines contain a voice-band modem to transmit the digital representation of a page.

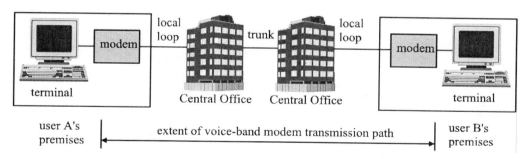

Figure 1.1   Voice-band modem reference model.

One of the first modems, AT&T's Bell 103, was used for full-duplex, asynchronous teletype transmission at 300 b/s using FSK (frequency shift keying). CCITT (now ITU) V.21 modems were similar, but not compatible with the Bell 103 modem. A few years later, the Bell 202 modem increased the bit rate to 1200 b/s using half-duplex FSK transmission. In late 1973,

2.   Kim Maxwell, ADSL Forum Contribution 96-049. "General Market Essays," March 25, 1996.

Vadic, Inc. introduced the VA3400, the first true full-duplex 1200 b/s modem using PSK (phase shift keying). A few years later the Bell 212 and then the CCITT V.22 also provided 1200 b/s full-duplex transmission using PSK. In 1981, V.22bis yielded 2400 b/s full-duplex. V.32 introduced trellis coding and took the bold step of echo-canceled transmission of information in both directions using the same frequency band. Modems preceding V.32 placed upstream transmission in a frequency band different from the downstream transmission (FDM). V.32 achieved 9600 b/s full-duplex transmission. Next came V.34, which used bandwidth optimization, constellation shaping, and channel-dependent precoding to enable 28.8 kb/s full-duplex. In 1995, 33.6 kb/s modems were introduced to the marketplace. V.34 modems utilize up to 3.6 kHz of bandwidth. This is technically a little more than the traditional 3.4 kHz telephony voice band. However, the V.34 modem can function on lines with less bandwidth by reducing the transmitted bit rate. By sending 33.6 kb/s in the 3.6 kHz voice band, V.34 modems send nearly 10 bits/s per Hz, a remarkable feat that approaches the theoretical limit for voice-band data transmission. History teaches us to be skeptical of "theoretical limits," which are sometimes exceeded by creative persons who break the rules by creating a new model. In late 1996, 56 kb/s "PCM" modems appeared, which were standardized by the V.90 ITU Recommendation in 1998. The PCM (pulse code modulation) modems are asymmetric since they support up to 56 kb/s downstream (toward the customer) and at most 33.6 kb/s in the upstream direction. In practice, PCM modems rarely achieve transmission rates above 50 kb/s because of limitations on transmitted power, tandem conversions, and line impairments such as load coils. Providing that there is a direct digital path (with no analog conversions) from the digital source to the PCM modem connected to the network end of the subscriber's line, the PCM modem transmission rate can exceed 33.6 kb/s by directly mapping the digital signal to the transmitted symbol without the impairment of quantizing noise.

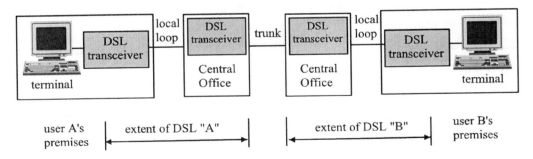

Note: this end-to-end reference model contains two DSLs, one for user A, and a second for user B.

Figure 1.2   DSL reference model.

The PCM modem network architecture departs from the universal PSTN capability of the prior generations of voice-band modems. The PCM modem at the network end must have a direct digital connection to the analog-to-digital conversion (CODEC) connected to the PCM modem user's telephone line. The PCM modem traverses the PSTN as a normal dial-up phone call. The PCM modem is like a DSL in that a direct digital connection from the network to the subscriber line interface is required, but differs from the DSL model (shown in Figure 1.2) since the PCM modem call is carried as a PSTN POTS (plain old telephone service) call. Architecturally, PCM modems reside in a middle ground between DSLs and traditional voice-band modems. PCM modems can utilize up to 4 kHz of bandwidth.

The fundamental limit of voice-band modems is the voice coder/decoder (CODEC) that is located at the local telephone switch or digital loop carrier (DLC) terminal. The CODEC converts the analog signals on the phone line to a 64 kb/s digital representation using pulse code modulation. A voice-band modem whose signal is carried in a PSTN voice call cannot exceed the 64 kb/s bit rate.[3]

With ITU Recommendations V.70 and V.61, voice-band modems can support simultaneous data and digitized voice via a PSTN call. V.70, using V.34 modulation and G.729 Annex A voice coding, can simultaneously convey 8 kb/s coded voice and approximately 20 kb/s data using a single PSTN call. Because G.729 Annex A provides for silence detection, a higher data rate can be achieved during periods of silence.

Data-compression techniques, such as specified in ITU Recommendation V.42, can achieve an effective data rate more than twice the modem rates listed above. However, highly random data (such as some binary files and digitized video) curtail the benefit of data compression. Data compression can also be applied to DSLs. For example, basic rate ISDN transport using two B channels can provide 128 kb/s uncompressed throughput and over 300 kb/s of effective throughput by compressing redundant types of data. When data compression is used, it is usually performed on the digital information prior to the DSL transceiver. The impact of transmission bit errors may be magnified by data compression.

The singular advantage of modems is their ubiquity. A modem may be connected to any phone line and immediately call to any of the millions of other phone lines that have a modem already attached. Modems cost less than DSL equipment and are easier to install. However, the data rates required by applications have outpaced the rates possible by voice-band modems. Other drawbacks of modems are the blocked calls resulting from overloaded local switches and modem pools (which were engineered for short duration calls), the inability to connect to multiple destinations simultaneously, and high error rates. These drawbacks of modems are solved by DSLs.

The fundamental difference between voice-band modems and DSLs is that voice-band modems operate over an end-to-end PSTN connection, whereas the DSL operates over a local loop. Figures 1.1 and 1.2 illustrate this difference.

---

3. At additional expense, rates above 64 kb/s can be achieved by inverse multiplexing multiple voice-band modem lines.

As shown in Figure 1.1, the voice-band modem transmission path may consist of the local loop for user A, a local Central Office, trunk facilities reaching thousands of miles in some cases, another Central Office, which serves another customer, and lastly a local loop serving user B. In contrast, the DSL transmission path consists of only one local loop from the user site to the nearby Central Office.

Another key distinction between voice-band modems and DSL is that the DSL keeps the information in the digital domain all the way from one user terminal to the other user terminal. In contrast, the voice-band modem sends the information through the PSTN as an analog representation of the user's digital information. With DSL, the signal is digitally regenerated at each step through the public network so that analog impairments do not accumulate at each step. Though the information is transported via a network consisting of many elements, the DSL transmission needs to address only the local loop portion.

The trunk facilities interconnect Central Offices with each other directly or via intermediate switching offices. The trunks are usually high-speed digital fiber optic transmission systems conveying information from many customers.

For those customers served via digital loop carrier (DLC), or other remote terminal systems, the DSL extends from the customer site to the DLC site. DLC and next-generation DLC (NGDLC) are used to serve customers who are too far to be economically served via a direct copper loop from the CO. The DLC remote terminal may be located in an outdoor cabinet in an underground vault (controlled environment vault, CEV), in a remote equipment hut, or sometimes in the utility room of a large multitenant office building. DLC systems multiplex 20 to 2000 customers onto a trunk to the CO. The DLC trunk typically is an optical fiber, but sometimes HDSL or T1 lines are used for smaller DLCs. See Section 4.2.2 for more discussion of DLC and NGDLC.

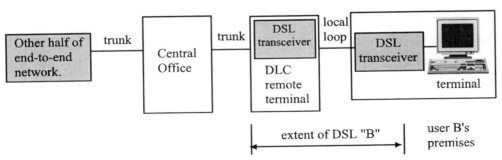

Figure 1.3 DSL reference model with digital loop carrier.

A DSL consists of a direct copper path from the user site to the nearest point of active network equipment. The one exception to this rule is the mid-span repeater, which is used to extend the reach of the DSL by placing a transceiver in the middle of the local loop. The DSL repeater is powered via DC power supplied from the CO via the same copper wire pair(s) that convey the data. Mid-span DSL repeaters are typically located in waterproof apparatus cases that house 4 to 20 repeaters. The apparatus cases may be located in an underground manhole, mounted on a utility pole, or hung from an overhead wire strand. The cost of the DSL repeater electronics is small compared to the cost of the environmentally hardened apparatus case and the labor required to splice the case into the cable.

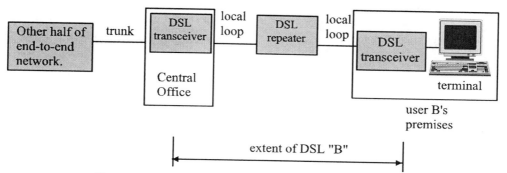

Figure 1.4 DSL reference model with mid-span repeater.

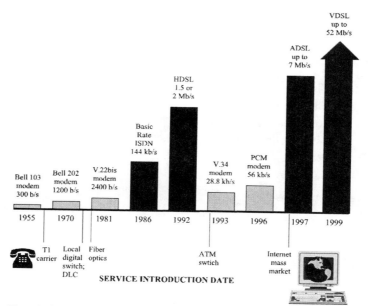

Figure 1.5 Progress in data-transmission bit rates (not to scale).

The voice-band modem is designed to operate over the combined limitations of the local loops at *both* ends of the network plus the limitation of the Central Offices. The Central Office usually contains PCM CODECs (Coder Decoders), which convert the analog signals on the local loop to a 64 kb/s digital signal (called a DS0) for transport via trunks. This transmission path is specified to provide a frequency band of 200 Hz to 3.4 kHz. The DSL is designed to operate over the limitations posed by only one subscriber loop. Typical subscriber loops have a bandwidth of hundreds of kilohertz. Thus, the potential performance of the DSL can exceed modems by a factor of 100 or more. Nonetheless, modems still have an important advantage in that they can operate over virtually any phone connection to anywhere in the world. Also, DSLs owe much to voice-band modems since many of the transmission techniques employed by DSLs originated in voice-band modems.

1.4 Transmission Modes

There are many types of transmission modes; their use depends on the application requirements and channel characteristics.

1.4.1 Direction

Simplex transmission is permanently one way from a source to a destination. Examples of simplex transmission include radio broadcast and security alarm circuits. Virtually all DSL applications require two-way transmission. Thus, simplex transmission is usually not used for DSLs. However, one could describe a T1 carrier as an example of simplex transmission. T1 lines are comprised of two simplex lines sending in opposite directions.

Half-duplex transmission periodically transmits from Station A to Station B and at other times in the opposite direction. Thus, at any point in time information is sent in one direction (simplex). Two-way transmission is achieved by the transceivers at the two ends of the line knowing when to "turn around" the line at the same time by exchanging the transmitter and receiver roles. In the early applications of half-duplex transmission, an entire message was sent before the line was turned around. Some DSL systems employ a variation of half-duplex called time compression multiplexing (TCM, also known as "ping pong" is discussed further in Chapter 5). TCM has reduced the turnaround period to an interval of a few milliseconds. Thus, TCM sends fixed-length blocks of a few thousand bits in alternate directions. Examples of traditional half-duplex transmission include telegraphy, teletype transmission, and two-way radio using the same frequency.

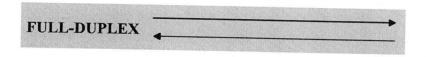

Full-duplex transmission sends information continuously in both directions on the same wire pair. Examples include traditional voice telephones, voice-band modems, basic rate ISDN (per ANSI T1.601), and HDSL. Simultaneous two-way transmission is accomplished by each transceiver subtracting the locally transmitted signal from its received signal. An echo-canceled hybrid (ECH) method is often used to permit both directions to use the same frequency band. The advantage of this approach is that the transmissions in both directions can reside in the lowest possible frequency band where signal loss and radio frequency interference is minimized.

An asymmetric version of full-duplex transmission is used by asymmetric digital subscriber line (ADSL). Information is sent in both directions simultaneously, but the data rate downstream (towards the customer) is much greater than the upstream (towards the network) data rate. This permits high downstream data rates on much longer lines by reducing the near-end crosstalk (NEXT) between ADSLs.

1.4.2 Timing

Synchronous transmission sends bits at a continuous rate. DSL receivers usually derive their timing from the periodicity of the received bit transitions. Synchronous and asynchronous transmission may apply for simplex, half-duplex, and full-duplex transmission. In general, DSLs use synchronous, not asynchronous transmission.

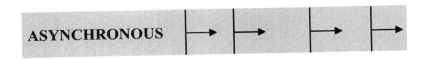

Asynchronous transmission sends units (characters or blocks) with a unique flag signal to mark the start of each unit. ATM (asynchronous transfer mode) is usually conveyed via *synchronous* transmission at the bit level; however, the start of each ATM cell may be at any idle bit. Thus, with ATM the cells are asynchronous, not the bits.

1.4.3 Channels

DSLs must convey more than one channel of information, where each channel is for different applications or services. ISDN has two B channels for voice/data, a D channel for signaling, and an embedded operations channel (eoc) for control and maintenance. HDSL has one large channel and an eoc. ADSL has data channels, an eoc, and a separate band for analog voice service.

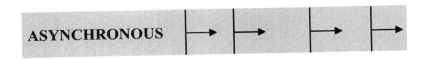

Time division multiplexing (TDM) is the most commonly used method to convey multiple channels of information. Information is organized into fixed-length frames with a fixed number of bits allocated to each channel. To reduce latency, the bits for a given channel may be divided into several small blocks, which are dispersed within a frame. Frames may be organized into superframes to accommodate low-bit-rate channels such as an embedded operations channel. In addition to sending multiple channels of information in the same direction, TDM may serve as a duplexing method. Information may be alternately sent upstream and downstream. This is called time compression multiplexing and virtually eliminates near-end crosstalk (NEXT), which limits the performance of echo-canceled hybrid transmission systems.

Frequency division multiplexing (FDM) places each channel in a separate frequency band. Thus, all channels are sent at the same time. One application of FDM is the use of one frequency band for upstream information and a different frequency band for downstream information. FDM duplexing also virtually eliminates NEXT. ADSL uses FDM by placing the analog voice in the lowest frequency band and the data in a higher frequency band. FDM designs involve a tradeoff between filter complexity and the amount of frequency spectrum wasted for guard bands.

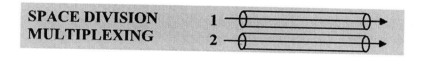

Space division multiplexing simply places each channel on a separate set of wires. This simplicity is appropriate for sending signals over distances measured in centimeters, but the cost of wires and the additional transceivers for each set of wires quickly becomes expensive. To minimize total cost, DSLs place all information on one pair of wires. HDSL uses two pairs of wires (for 1.5 Mb/s) and up to three pairs of wires (for 2 Mb/s) to achieve longer line distances.

1.4.4 Single and Multipoint Topologies

DSLs are point-to-point transmission systems. One transceiver is connected to each of the two ends of a pair of wires. One end may be located at a telephone company site such as a Central Office, and the other end may be located at a customer premises. Compared to multipoint systems, the point-to-point transmission enjoys simplicity, high reliability, and a higher security. The point-to-point topology provides dedicated bandwidth to each customer. With a suitable switching system at the central site, the throughput performance for each customer remains nearly constant as the number of network nodes grows.

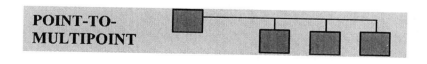

Point-to-multipoint systems consist of a central (master) station transceiver, which communicates with multiple directly connected terminals. The terminals do not directly communicate with each other. Coaxial television (CATV) systems employ point-to-multipoint transmission. Multipoint-to-multipoint permit the terminals to communicate directly among themselves. 10baseT local area networks (LANs) are multipoint-to-multipoint systems. The number of transceivers for a network of N terminals will be N+1 for a point-to-multipoint system, and 2N for a point-to-point system. In general, multipoint systems are more appropriate for shorter distances, and point-to-point systems are preferred for longer distances. At longer distances, the connection of multiple terminals results in greater signal loss and more difficult signal timing.

The discussion above applies for the physical level. At the logical level, where the information flow at higher protocols is considered, point-to-point and multipoint information flow may occur via any physical topology.

1.5   DSL Terminology

The following explanation of terms will be helpful in understanding DSLs. The oddest term is *kilofeet (kft)*, the traditional measure of the length of a telephone line measured in thousands of feet: 1 kft equals 306 meters. The diameter of a wire is measured in millimeters (mm), except in the United States where *the American Wire Gauge (AWG)* number represents 1/Nth of an inch (e.g., 24 AWG has a conductor diameter of 1/24 of an inch, which equals 0.5 mm). Signal power and signal loss are measured in the logarithmic units of *decibels (dB)*, named after Alexander Bell. A 3 dB increase in power equals a doubling of power, a 3 dB decrease in power equals one-half the power, a 6 dB increase in power equals four times the power, etc. The frequency of an electrical signal is measured in *kiloHertz (kHz,* thousands of cycles per second) or *megaHertz (MHz,* millions of cycles per second). Traditional circuit-switched analog voice telephone service is often called *POTS* for Plain Old Telephone Service. This has prompted some to jest by calling DSL-based services *PANS* (Positively Amazing New Services).

Other terms and acronyms are explained in the Glossary at the end of this book.

1.6   Rate Versus Reach

The strength (e.g., power) of an electrical signal decreases with distance traveled due to the resistance of the wires carrying the signal. Also, signal attenuation (e.g., loss) becomes greater at higher frequencies. Simply put, the amount of signal power dissipated in the wires grows with transmission speed and distance. DSL loop reach is limited by the signal becoming too weak to be accurately received.

Digital transmission engineers maximize the transmission distance by use of sophisticated modulation techniques that send a given data rate with a limited transmitted signal power in a certain frequency band. For a given transmission method, the maximum achievable transmission bit rate decreases as the line length increases. Thus, it is possible to achieve high data-transmission rate for short loops and relatively lower rates for longer loops. The achievable data rate also depends on other factors, including crosstalk (noise coupled from signal on other wire pairs in the same cable).

The 2.5 peak volt transmitted signal for basic rate ISDN (BRI) systems on a maximal length loop may encounter 42 dB of signal attenuation with a resulting minuscule received peak signal of 0.02 volts (20 mV). The BRI system performs a remarkable task: recovering a signal that amounts to only 1/125 of the transmitted signal. The corresponding values for DSL systems are provided below.

DSL Type (rate)	Transmit Peak (volts)	Maximum Signal Power Loss (decibels)	Minimum Received Peaks (volts)
BRI 2B1Q (144 kb/s)	2.5	42	0.02
HDSL 2B1Q (1.5 Mb/s)	2.5	35	0.045
ADSL DMT (1.5 Mb/s)	15*	45	0.085
VDSL SDMT (26 Mb/s)	3–4	30	0.09–0.12

*The peak ADSL voltage depends on the implementation of the transmitter; in some cases, the peak transmitted ADSL voltage could exceed 15 volts. For the commonly used 20 dBm transmit level, the average ADSL transmit signal level is 3.1 volts and an average received signal voltage of 0.02 volts for a maximum length loop.

Figure 1.6 shows the approximate achievable transmission rate as a function of line length. The lower curve is for symmetric transmission, and the upper curve shows the downstream rate for asymmetric transmission with an asymmetry ratio of 10:1. Thus, the upstream rate is assumed to be one-tenth of the asymmetric rate shown on this graph. Conventional crosstalk and a 6 dB margin is assumed. This graph shows the advantage of asymmetric transmission — much higher downstream transmission rates.

1.7 Crosstalk

A telephone cable contains up to several thousand separate wire pairs packed closely together. The electrical signals in a wire pair generates a small electromagnetic field, which surrounds the wire pair and induces an electrical signal into nearby wire pairs. The twisting of the wire pairs reduces this inductive coupling (also know as *crosstalk*), but some signal leakage remains. Crosstalk is most pronounced at the segment of cable near the interfering transmitters. The crosstalk resulting from other transmission systems in the same cable (and especially the same binder group within the cable) is a primary factor limiting the bit rates and loop reach achievable by DSLs. Management of pair-to-pair crosstalk requires care in the bandwidth and signal power of transmitters and out-of-band signal rejection by receivers. This is often referred to as *spectral compatibility* and is reminiscent of the management of radio frequency broadcasters.

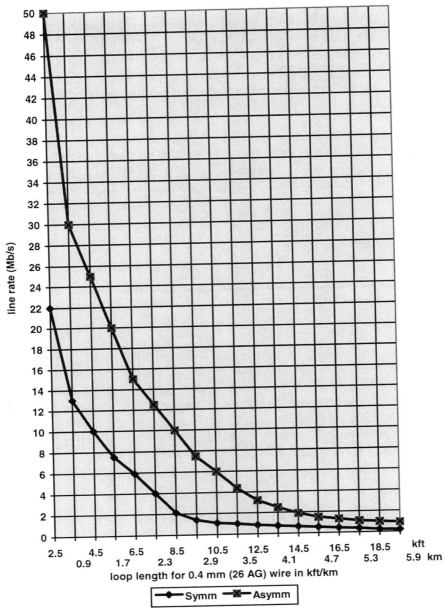

Figure 1.6 DSL data rate versus loop length.

Near-end crosstalk (also known as *NEXT*) is a major impairment for systems that share the same frequency band for upstream and downstream transmission (e.g., echo-canceled hybrid transmission). NEXT noise is seen by the receiver located at the same end of the cable as the transmitter that is the noise source.

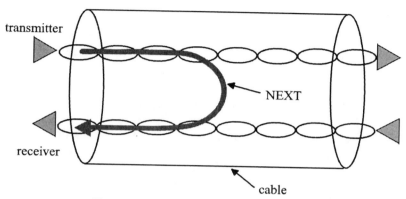

Figure 1.7 Near-end crosstalk (NEXT).

Transmission systems can avoid NEXT by using different frequency bands for upstream and downstream transmission. Frequency division multiplexed (FDM) systems avoid NEXT from like systems (also know as *self NEXT*). FDM systems still must cope with NEXT from other types of systems that transmit in the same frequency band, and another phenomenon known as FEXT.

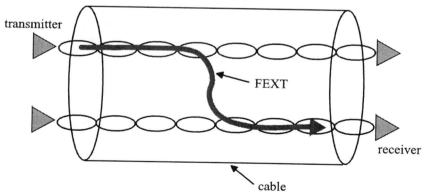

Figure 1.8 Far-end crosstalk (FEXT).

Far-end crosstalk (*FEXT*) is the noise detected by the receiver located at the far end of the cable from the transmitter that is the noise source. FEXT is less severe than NEXT because the FEXT noise is attenuated by traversing the full length of the cable.

A major advantage of fiber optic transmission is the lack of any crosstalk.

1.8 Enabling and Disabling Forces

In 1970, the telecommunications world consisted of voice- and character-oriented communications to mainframe computers. Voice was "king," and there was little need for DSLs. Then came millions of personal computers (PCs), multimedia (audio, still image, and video) applications, and eventually the Internet. In the early 1980s, the number of computers (including microprocessors in cars and appliances) exceeded the world population of humans, and in the mid-1990s the minutes of usage for digital applications (including fax) in the public network exceeded voice. Though Internet access is now a big application for DSLs, the deployment of DSLs began long before Internet became a household word. Voice transport remains important, even for DSLs. For example, HDSL is used for voice trunks to PBXs and wireless cell sites.

The advent of low-cost and powerful digital signal processors (DSPs) has enabled the use of algorithms that were once reserved for exotic aerospace and defense applications. The DSP breakthrough also enabled full-motion color video codecs (coder-decoder) at DSL rates using standard algorithms (MPEG1, MPEG2, JPEG, and H.261). High-quality video conferencing is supported at 384 kb/s, entertainment video is possible at 1.5 Mb/s, and high-definition TV is possible at 20 Mb/s. In 1985, the transport of high-quality video over most telephone lines was unimaginable; today it is common practice.

The DSL technology we take for granted today was very nearly prevented from being deployed by two powerful obstacles: fiber fever and regulatory uncertainty. During the 1980s many of the leading telecommunications policymakers believed that copper phone lines would soon be replaced by fiber optic lines directly to every customer. Two versions of fiber fever argument existed: (1) fiber-based transport would soon be so inexpensive that copper transport would be abandoned, and (2) DSL technology would prolong the use of copper and thereby delay the strategically vital deployment of fiber. As the debate raged it became clearer that fiber to the home (FTTH) would remain expensive and that the telephone companies could not tell their customers to wait several years until fiber arrived. The telephone companies focused their fiber deployments to where they were economically justified: to major business sites and to fiber-remoted multiplexers (digital loop carrier), which serve hundreds of customers. When asked if DSLs were an interim technology, Ray Smith (CEO, Bell Atlantic) replied, "ADSL is an interim technology, for the next 40 years."

The second threat to the introduction of DSLs was the uncertainty regarding who would own the DSL transceiver at the customer end of the line. The telephone companies felt that the customer-end transceiver should belong to the network to assure good performance, simplify trouble resolution, and to ease the upgrade to future technologies. Many regulators, on the other hand, insisted that the customer-end transceiver be owned by the customer to permit the customer the freedom to choose among many competitive providers of equipment. System development slowed because equipment designers did not know what features were required, equipment vendors did not know which sales channels to use, and telephone company engineers did not know who would be responsible for installation and maintenance. For basic rate ISDN in the United States, it was resolved that the customer-end transceiver would be owned by the customer. However, in many other countries the opposite resolution was reached. Either model can work, but only after the decision is firmly made.

ADSL encountered a different regulatory quagmire: would the telephone companies offer ADSL as a regulated or unregulated service. The regulatory environment in the United States made both alternatives equally unattractive, and worse yet was the prospect that the rules could change again at any time. Once again, the price paid for stimulating a competitive market was delaying the introduction of new services by nearly two years. With a market window of five years for some services, the chilling effect of regulation can easily cause innovators to become stodgy.

The success of a service (and its underlying technology) depends greatly on its price and its relation to available alternatives. Service price, in turn, depends greatly on the cost of equipment and labor costs for operation. Equipment and operating costs are reduced as the number of customers grows. Low-cost service is best achieved by establishing a service that addresses the most customers and minimizes additional infrastructure costs via use of existing facilities. For DSLs, additional transceiver circuitry that extends loop reach or enables additional applications may ultimately enable the reduction of service price by expanding the addressable market. A recurring theme in the field of DSLs is that the cost of additional capability in the transceiver yields greater savings from reduced operating costs, greater loop reach, or possible additional applications.

1.9 Applications

The first step in developing a technology or system is identification of the customers' needs and their implications on functional requirements. End-user demand for products and services is driven by saving money, making money, accomplishing necessary tasks, saving time, and sometimes the possession of a status symbol. These demands are satisfied by applications: hardware and software that perform certain tasks for the user. An *application* is a package of hardware, software, and, in some cases, a network service that provides a solution to specific end customer needs. A *service* performs certain tasks, or provides certain capabilities, for which the customer (or someone else) pays for on a recurring basis (e.g., per use, or per month). In many cases, a service may be used to support many applications. At the risk of circular reasoning, a *service* is something provided by a service provider, such as a telecommunications carrier company. The applications listed in the following tables apply to both public and private network services.

The end user may be charged for a service per use, per minute of use, per byte, or a flat rate per month. Services may be characterized in terms of guaranteed throughput and availability. Also, the service may be "free" to the end user, in which case another party actually pays for the service (e.g., 800 number phone service, advertiser-paid broadcast TV service).

Each application has specific characteristics, such as willingness to pay, sustained and peak data rates downstream and upstream, potential market size, connectivity, connection duration, constant-rate/bursty, error tolerance, maximum acceptable signal delay, maximum delay variation, privacy, and security (authentication of user). Real-time applications (such as streaming video) have stricter requirements on delay and bit rate than near-real time (web browsing) and non-real time (email). To be successful, all aspects of the system should be designed to assure that the characteristics of the key applications are satisfied. In other words, define the problem before attempting to solve it.

Applications are also partly defined by the market segment, terminal location, and terminal type. Market segments include consumer, business, education, government, and military. Each has its own characteristics for willingness-to-pay, privacy, security, reliability, and ease of use. Work-at-home (WAH) is a business-type application, which may be paid for by the end user or by the employer. Terminals may be located at the home, office, public, or mobile site. At first, one might dismiss the mobile application for DSL-based services. However, the DSL may be used to connect a radio base unit to the network with the final link to the terminal via radio. The end-user platform may consist of a PC, workstation, TV, information appliance, or public kiosk. For the same applications, *dumb terminals* and *network computers* require more communication bandwidth than a PC because much of the information must be downloaded from a distant server rather than the memory/disk-drive within the PC.

There have been many debates about the existence of a "killer application," a single application with such a large market demand that it alone justifies the creation of a service or system. Given the high complexity and cost of high-bit-rate public data networking, it is doubtful that a single application would suffice. Some will argue that Internet access is a killer application. However, Internet access is actually more rightly described as a service that supports many applications (electronic mail, web browsing, file transfer, database access, shop/bank at home, etc.). Even if there were a killer application, it certainly would not be the only important application. The power of a public data network is its ability to support many applications.

In the following tables, the first number provided for bit rate indicates the minimum satisfactory rate for some uses, and the second number is the rate that would satisfy nearly all uses in the near term. These are *throughput bit rates*, not *interface bit rates*. Popular interface bit rates of 10 Mb/s (Ethernet), 26 Mb/s (ATM Forum), and 100 Mb/s (Fast Ethernet) may be used as the local interface to convey the application throughput bit rates listed in the following tables. This information is based upon discussions within the ADSL Forum and T1E1.4.

ADSL is well suited to supporting many types of applications, with the notable exception of broadcast television and some high-performance business applications. ADSL's point-to-point nature provides advantages for reliability and security, however the dedicated ATU-C per customer makes it costly for broadcast use. Direct broadcast satellite (DBS) or terrestrial broadcast such as multichannel multipoint distribution system (MMDS) are excellent means for pro-

Table 1.1   Residential Consumer Applications[*]

Applications	Downstream Demand Rate (kb/s)	Upstream Rate (kb/s)	Willing to pay	Demand Potential
Voice telephony	16–64	16–64	M	H
Internet, online service access	14–3,000	14–384	H	M
Electronic mail	9–128	9–64	M	M
High definition TV (HDTV)	12,000–24,000	0	H?	M?
Broadcast video	1,500–6,000	0	L	H
Premium video channel	1,500–6,000	0	M	M
Pay-per-view video	1,500–6,000	1	M	M
Movies on demand	1,500–6,000	9	H	M?
Music on demand	384–1,500	9	L	M?
Video phone	128–1,500	128–1,500	H	M
Distance learning (live or stored video instruction	384–3,000	128–3,000	H	M
Database access (community, library, government information)	14–384	9	H	M
On-line directory; yellow pages	14–384	9	L	H
Software download	384–3,000	9	L	L
Shop-at-home (online catalog)	128–1,500	9–64	L	M
Video games (multiplayer and pay-per-use)	64–1,500	64–1,500	M	M
Vote-at-home	9	9	L	M
Bank-at-home	9	9	L	M
Radio from other cities	384–1,500	1	L	L
Gamble at home	9–128	9	L	L
Medical consultation at home	384–1,500	384–1,500	M	L
Security alarm	1	1	L	L
Utility meter reading	1	1	L	?
Energy management	1	1	L	?

[*]H = High; M = Medium; L = Low

Table 1.2 Business Application (Including Work-at-Home)[*]

Application	Downstream Rate (kb/s)	Upstream Rate (kb/s)	Willing to Pay	Demand Potential
Voice telephony	16–64	16–64	M	H
Facsimile	9–128	9–128	M	H
Point-of-sale, credit check	9–128	9–128	L	M
Public terminal, kiosk, teller	28–384	9	M	M
On-line directory; yellow pages	14–384	9	M	H
Internet access (public network)	28–3,000	14–384	H	H
Intranet/Extranet access (private network)	64–3,000	64–1,500	H	M
Electronic commerce (EDI)	28–384	28–384	H	L
Electronic mail	9–128	9–64	M	H
Remote access to LAN/host from home/remote office	128–6,000	64–1,500	H	M
Access to remote database, library	64–6,000	9–1,500	M	M
LAN interconnection	384–10,000	384–10,000	M	M
Collaborative design, screen sharing	128–1,500	128–1,500	M	L
Training/education	384–1,500	9–384	M	L
Business/financial news	128–1,500	9	M	L
Video phone (desktop)	384–1,500	384–1,500	H	L
Video conference (conference room)	384–3,000	384–3,000	H	L
Remote video surveillance	128–1,500	1	M	L
Software download	384–6,000	9	M	L
Supercomputing, CAD (Computer Aided Design)	6,000–45,000	6,000–45,000	H	L
Tele-radiology (x-ray, etc.)	1,500–12,000	128–12,000	H	L

[*]Downstream is towards the user; upstream is towards the network.

viding broadcast television service and could be used in combination with ADSL to support a full range of applications, including two-way Internet access and remote LAN access.

The following table indicates the transfer time for certain types of images without compression, except for the row labeled "facsimile" where typical compression is accounted for. In the last row, the picture size and subjective quality is indicated for digital compressed video. The transmission rates indicated in the column headings were used for the calculation of transfer rate. However, in many cases, the actual transfer rate may be less than the raw line transmission rate due to upper-layer protocol overhead, server delay, PC-processing delay, and backbone network traffic delay.

Table 1.3   Image Transfer Time

Image Type	Size	28.8 kb/s	128 kb/s	384 kb/s	1.5 Mb/s	26 Mb/s
Full screen of ASCII text	32 kb	1.1 s	0.25 s	nil	nil	nil
Compressed fax (1 page)	90 kb	3.1 s	0.7 s	0.23 s	nil	nil
Complex web page, **or** full-screen 8-bit color VGA	2.5Mbit	87 s	19.5 s	6.5 s	1.6 s	nil
Full screen 16-bit color XGA	13 Mb	7.3 min	1.6 min	33 s	8.2 s	0.5 s
Chest X-ray	64 Mb	37 min	8.3 min	2.8 min	41 s	2.5 s
Compressed digital video % of screen, picture quality	—	10%	25%	100%	100%	100%
		poor quality	fair quality	fair quality	VHS-like	HDTV

The file sizes in Table 1.3 are measured in kilobits or megabits. A VGA screen is 640 × 480 pixels. An XGA screen is 1024 × 768 pixels. Compressed video at 1.5 Mb/s has picture quality as good as an original videotape played on a VHS player.

Further information about DSL applications is provided in Sections 14.2 and 14.3.

1.10 Evolution of Digital Transmission

Digital technology was first introduced during the early 1960s to the interoffice trunks to solve the problem of long distance noise due to the accumulation of noise inherent to analog transmission. Each analog repeater in a long distance trunk amplified the noise as well as the signal. Despite the most advanced amplifier design, some additional noise is generated by each repeater. Digital transmission eliminated noise accumulation because an exact digital replica is recreated at each repeater. Digital repeatered transmission enabled virtually perfect transmission regardless of the distance.

Telephone Central Offices are connected to each other via trunks, each of which carries many voice circuits. In most cases, a hierarchical network architecture connects the local Central Offices to a tandem or toll office. By 1970, most of the analog trunks had been replaced by digital T1 trunks, each of which carried 24 voice circuits. As a result, the tandem and toll offices were surrounded by digital trunks. Switching system designers realized that it was inefficient for a tandem/toll office to convert the circuits from the digital trunks to analog, to be switched via a traditional analog switching matrix, and then back to digital again to another trunk. Thus, tandem/toll offices quickly converted to digital.

By the late 1970s, the entire interoffice network was digital. The local COs connected to digital trunks and analog lines, which were carrying a growing proportion of digital traffic from computers and fax machines. It was predicted that digital traffic would become greater than analog voice calls, and indeed by the mid-1990s it did. The entire electronics industry was moving to digital, the local switches (such as AT&T's 1A ESS) were already controlled by digital computers, and interoffice signaling (such as CCIS and SS7) was digital. Despite some resistance by the local switching community, the digital tide swept over the local switch with the introduction of digital telephone switches such as the Nortel DMS 100, AT&T 5ESS, Siemens EWSD, and Ericsson AXE. These "digital" switches initially terminated only analog lines, which were converted to 64 kb/s digital via CODECs at the switch line unit.

In 1985, ISDN extended the digital domain to the customer. For the first time, end-to-end digital service was available on a large scale. ISDN provides the customer with both circuit-switched and packet-switched digital service. Prior to this, digital data service (DDS) lines operating at rates in the range of 9.6 to 64 kb/s provided access to packet-switched data service. The DDS service was very limited due to its high price and availability only in a few selected areas. ISDN is primarily a circuit-switched network, with packet-switching suitable only for narrowband packet traffic.

Broadband ISDN (BISDN) with high-performance asynchronous transfer mode (ATM) switching was at first envisioned as connecting to all customers via direct fiber optic lines. BISDN would be limited to a privileged (and wealthy) few. HDSL and ADSL opened the world of switched broadband data service to a vastly larger market.

Types of DSLs

As the processing power of digital signal processors has grown, so have DSL bit rates. DSL technology began with 144 kb/s basic rate ISDN (BRI), and has evolved to 1.5 and 2.0 Mb/s versions of HDSL, 7 Mb/s ADSL, and now 52 Mb/s VDSL.

2.1 DSL Design Margin

DSLs are designed with a 6 dB SNR margin. This means that the DSL will provide 10^{-7} bit error rate (BER) when the crosstalk signal power is 6 dB greater than the defined "worst-case" crosstalk model. In many cases, the worst-case crosstalk model is a 50-pair binder group filled with 49 self crosstalkers. With pure Gaussian noise, a 6 dB SNR margin would result in a 10^{-24} BER. However, in the real world, noise is often non-Gaussian. Thus, for typical conditions, the 6 dB margin provides assurance that DSLs usually operate at a BER of better than 10^{-9} and that DSLs will provide reliable service even when the transmission environment is worse than normal.

The 6 dB value originated during work on the ANSI basic rate ISDN standard in T1D1.3 (the predecessor of T1E1.4) with a 1985 contribution from Richard McDonald of Bellcore. As described in T1E1.4/95-133, the 6 dB design margin still serves as an appropriate value. The design margin provides for cable variations (aging, splices, wet cable), additional noise in CO and customer premises wiring, other noise sources, imperfect transceiver designs, and manufacturing variations. The amount of design margin is a tradeoff between assuring reliable operation in all conditions and permitting the use of the technology on the longest possible loops.

More sophisticated transmission methods can achieve higher performance, but the need for design margin remains. However, systems that measure margin at start-up can provide the installer an instant indication if the loop has inadequate margin. The installer can then take cor-

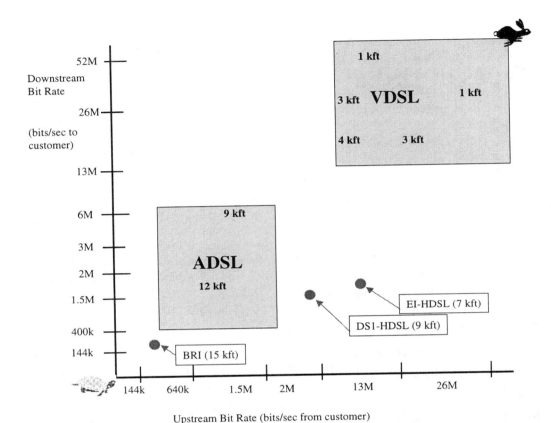

Figure 2.1   DSL bit rates with loop reach shown for 26 AWG wire (kilofeet — no repeater).

rective actions such as finding a better wire pair or removing bridged taps. An argument can be made that systems that provide a real-time indication of transmission margin could reasonably be used with a margin threshold of 5 dB. However, relaxing the design margin by one or two decibels translates into expanding the size of accessible loop population by only about 1% of loops.

2.2 DSL Precursors

One could argue that T1 trunks, E1 trunks, and DDS (digital data service) lines were the first DSLs. Although T1 (1.544 Mb/s Alternate Mark Inversion (AMI) used primarily in North America), and E1 (2.048 Mb/s HDB3) transmission systems were originally intended for use as trunks between Central Offices (COs), they later proved useful as high-speed links from COs to customer sites. T1 carrier was first used by AT&T in 1962. CO-to-CO trunks today are entirely fiber and microwave based. T1/E1 lines are not used today for their original purpose. T1/E1 lines are still used on subscriber lines, but they have their drawbacks. They are expensive and time consuming to install and are usually segregated into binder groups (wire pair bundles) separate from other types of transmission systems. A T1 line consists of four wires. Two wires convey information to the customer, and another two wires convey information from the customer. To reduce near-end crosstalk between the two directions of transmission, one cable binder group (a bundle of wires) carries only outbound T1 pairs, and a different binder group carries only inbound T1 pairs. T1 lines are designed with a maximum of 15 dB (e.g., 2 to 3 kft) of line loss at 772 kHz for the CO end section (CO-to-first repeater), a maximum of 36 dB (e.g., 3 to 6 kft) loss for repeater-to-repeater sections, and up to 22.5 dB of line loss from the last repeater to the customer premises. T1 lines must be unloaded and have no bridged taps. Distances of many miles may be covered by the use of many repeaters. T1 repeaters are powered via +/– 130 volt DC line power. For the purposes of this book, we shall consider T1/E1 and DDS not to be DSLs.

The AMI line code for T1 transmission is simple to implement but is inefficient by today's standards. AMI sends one bit per baud; a baud is one signal element. T1 transmission uses a high transmitted signal power, which generates high levels of crosstalk from 100 kHz to 2 MHz. Other DSLs, which use these same frequencies, can be affected if placed in a binder group with T1 lines. In extreme cases, T1 crosstalk can affect loops in other binder groups.

2.3 Basic Rate ISDN

2.3.1 ISDN Basic Rate Origins

In this book, we shall consider basic rate ISDN (BRI) to be the first in the family of DSLs. Integrated services digital network (ISDN) was first conceived in 1976 and was largely defined by Recommendations developed within the CCITT (now called the ITU, International Telecommunications Union). The ISDN vision was ambitious: a uniform global network for data communications and telephony. Development of the ISDN transmission, switching, signaling, and operations systems required a herculean effort reminiscent of the construction of a continental railway network, only to be followed by the invention of the airplane. The effort to develop ISDN spanned a decade, with efforts of thousands of people from hundreds of companies in more than 20 countries. We estimate that the development of ISDN cost over $50 billion; it is not known if this investment will be fully recovered. ISDN was focused on telephony services and lower-speed packet-switched data. This focus ultimately became a major weakness. ISDN networks were poorly suited for the high-speed packet switching and long holding-time sessions that characterize Internet access. Nonetheless, those who would claim the failure of ISDN must not forget the millions of happy ISDN customers.

ISDN service trials began in 1985. The first North American ISDN service was provided in 1986 by AT&T–Illinois Bell (now called Ameritech) in Oakbrook, Illinois. Early trial BRI systems employed TCM (ping-pong), or alternate mark inversion (AMI) transmission techniques. These early systems were simpler to implement, but 2B1Q (2 binary, 1 quaternary) transmission was selected for the standard transmission techniques for nearly all parts of the world except the Federal Republic of Germany and Austria, which use 4B3T (4 Binary, 3 Ternary), and Japan, which uses a ping-pong AMI transmission method. The loop reach of the 2B1Q and 4B3T systems is greater than the prestandard systems, which quickly faded from use.

The total number of BRI lines in service worldwide grew from 1.7 million in 1994 to nearly 6 million by the end of 1996. The approximate number of ISDN lines for the countries with the largest ISDN deployments are provided below. The 1994 information is based on ITU statistics. The 1996 values are based on information provided by experts from the respective countries. The U.S. 1996 number is from FCC statistics. ISDN deployment is growing at 30% to 50% per year in many countries.

Table 2.1 Basic Rate ISDN Lines in Service

Country	1994 BRI lines	1996 BRI lines
Germany	428,000	2,000,000
United States	352,000	843,115
Japan	320,000[*]	1,000,000
France	240,000[*]	1,400,000
United Kingdom	75,000[*]	200,000

*Extrapolated values.

The deployment of ISDN in Germany was accelerated by government mandate, whereas other countries have followed a market demand deployment model. ISDN service was available in 1996 to about 90% of telephone customers in the countries listed in Table 2.1.

2.3.2 Basic Rate ISDN Capabilities and Applications

BRI transports a total of 160 kb/s of symmetric digital information over loops up to approximately 18 kft (5.5 km, or up to 42 dB of loss at 40 kHz). This is channelized as two 64 k/s B channels, one 16 kb/s D channel, and 16 kb/s for framing and line control. The B channels may be circuit switched or packet switched. The D channel carries signaling and user data packets. An embedded operations channel (eoc) and indicator bits are contained within the 8 kb/s of overhead. The eoc conveys messages used to diagnose the line and the transceivers. The indicator bits identify block errors so that the transmission performance of the line may be measured.

2.3.3 Basic ISDN Rate Transmission

BRI modulates data using one four-level pulse (a quat) to represent two binary bits, hence 2 Binary one Quarternary (2B1Q). Data is sent in both directions simultaneously using echo-canceled hybrid (ECH) transmission. The simple 2B1Q baseband transmission technique sends 160 kb/s using 80 kHz of bandwidth, yielding a modest bandwidth efficiency of 2 bits/s per Hz. Adaptive equalization automatically compensates for attenuation across the transmission band.

BRI can work on a loop with bridged taps, providing the total loss is less than 42 dB at 40 kHz. BRI loops must be unloaded.

2.3.4 Extended-Range Basic Rate ISDN

Loops beyond direct BRI reach of 5.5 km (18 kft) from the CO may be served via alternative methods: BRITE, mid-span repeater, and extended-range BRI.

2.3.4.1 BRITE

Basic rate ISDN transmission extension (BRITE) (see Figure 2.2) uses digital channel banks (e.g., D4 and D5 type multiplexers, which time division multiplex 24 DS0 channels into one 1.544 Mb/s line) and digital loop carriers (DLC) as a means to extend ISDN service to areas served by these channel banks. Special ISDN channel units use three DS0s in the channel bank to transport BRI. Due to the additional channel units, the BRITE configuration has a relatively high cost per line. However, when using preexisting SLC or channel bank equipment, the low start-up cost of BRITE is ideal for serving very small numbers of lines in remote areas.

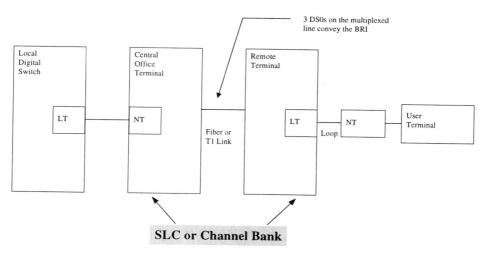

Figure 2.2 BRITE.

2.3.4.2 Mid-Span Repeater

The loop reach may be nearly doubled by placing a repeater in the middle of the loop. See Figure 2.3. Since the repeater is essentially a back-to-back NT and LT, the loop is divided in a tandem pair of DSLs. Each of the two loops may have up to 42 dB of loss at 40 kHz, which corresponds to a total reach of approximately 30 kft (2×15). Repeaters are typically located in a multislot repeater apparatus case located in a manhole or mounted on a pole. Since a manhole with available space may not be located at the exact midpoint of the loop, the repeater often is located somewhat off-center. As a result, the attainable loop reach of a repeated line may be slightly less than twice the unrepeated reach. Loading coils must be removed from the loop for BRI operations with or without repeaters.

Mid-span repeaters are typically powered via a DC voltage (usually –130 volts DC) in the United States supplied from a CO power feed circuit. For yet longer reach, a second repeater may be employed. The two-repeater configuration is rarely used due to power feeding and administrative complexities. The cost of a repeated line is dominated by the labor for loop design, the apparatus case, and installation of the apparatus case (including cable splicing). The cost of the repeater electronics is relatively small in comparison.

The repeated configuration and the BRITE configuration have twice the signal transfer delay (2.5 ms one way) of the direct DSL configuration (1.25 ms).

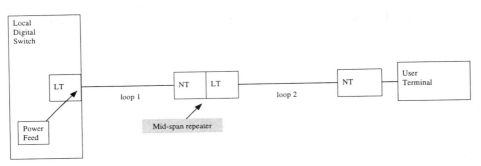

Figure 2.3   Mid-span repeater configuration.

2.3.4.3 Extended-Range BRI

Transmission techniques have advanced since the creation of the BRI standard (ANSI T1.601). Techniques, such as trellis coding, permit 160 kb/s to be transmitted over loops up to 8.5 km (28 kft) without the need for mid-span repeaters. For backward compatibility, the extended-range BRI systems present the standard ANSI T1.601 interface to the LT in the CO switch and also to the customer's NT1. See Figure 2.4. Normally, a conversion unit is located in a miscellaneous equipment bay in the CO, and the other conversion unit is located in an enclosure located at the outside of the customer premises. However, locating the remote conversion unit in a mid-span location may extend the loop reach further. As a result, a total reach of approximately 43 kft (15 + 28) may be attained. Furthermore, the network-side converter may also be placed remotely, provided that local power is available at the site.

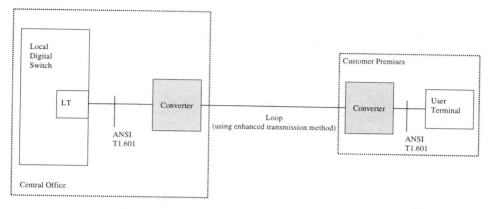

Figure 2.4 Extended-range ISDN configuration.

2.3.5   Digital Added Main Line

BRI transceivers are also used for non-ISDN applications — most notably digital added main line (DAML). DAML systems permits one loop to convey two voice telephone circuits. See Figure 2.5. Voice coder/decoders (CODECs) at each end of the DAML system convert the 64 kb/s BRI B channel to an analog telephone interface. Thus, the traditional voice telephony interface is provided to the CO switch and the customer's phones. DAML systems are used to provide additional telephone service to sites in an area having a shortage of spare wire pairs between the CO and the customers. The DAML unit at the customer's premises is usually powered from CO power fed via the loop. DAML systems using BRI technology have a maximum loop reach of 5.5 km (18 kft). HDSL-based DAML systems can convey more than two voice circuits via one pair of wires.

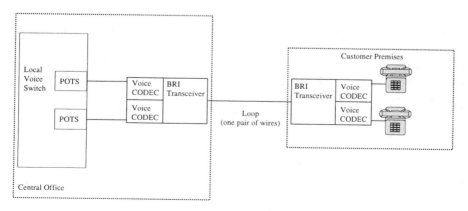

Figure 2.5 Digital added mail line.

2.3.6 IDSL

Another non-ISDN application of BRI transceivers is IDSL (ISDN DSL). The BRI symmetric channels (128 kb/s or 144 kb/s) are concatenated to form one channel for transmission of packet data between a router and the customer's computer. Most forms of IDSL will work with a conventional ISDN NT at the customer end of the line. Thus, with IDSL the ISDN local switch is replaced by a packet router. This configuration is used for Internet access.

2.4 HDSL

2.4.1 HDSL Origins

The early concept definition of HDSL (high-bit-rate digital subscriber line) took place in late 1986 at AT&T Bell Laboratories and Bellcore. HDSL transceiver designs were essentially up-scale basic rate ISDN designs. Laboratory prototype HDSL systems appeared in 1989. The first HDSL was placed into service in March 1992 by Bell Canada using equipment manufactured by Tellabs Operations Inc. in Lisle, Illinois. Nearly every major telephone company in the world now uses HDSL. In 1997, approximately 450,000 HDSL lines were in service worldwide, with approximately 350,000 lines of these lines in North America. HDSL deployment is growing at more than 150,000 lines per year. In October 1998, the ITU approved Recommendation G.991.1 for first generation HDSL; this is based closely on the ETSI Technical Specification TM-03036. The ITU has started work on a second generation HDSL (HDSL2) Recommendation that will be called G.991.2.

The need for HDSL became evident as T1 and E1 transmission systems ceased to be used for their original purpose as interoffice trunks and saw rapid growth as private lines from Central Office to customer premises. T1/E1 transmission systems operated over the existing telephone wires, but at a large cost for special engineering, loop conditioning (removal of bridged taps and loading coils), and splicing for apparatus cases to hold the repeaters that were required every 3,000 to 5,000 feet. The transmission methods used for T1/E1 lines placed high levels of transmit signal power at frequencies from 100 kHz to above 2 MHz; this required the segregation of T1/E1 lines into binder groups separate from many other services. In addition to being expensive to install and maintain, T1/E1 lines often took many weeks from service order to service turn-up. What was needed was a *plug-and-play* transmission system that could quickly and easily provide 1.5 or 2 Mb/s transport over most subscriber lines, thus HDSL.

HDSL's benefits are largely due to the elimination of mid-span repeaters. Each repeater site must be custom-engineered to assure that each section of the line remains within the limits for signal loss. The repeated signals can cause severe crosstalk; thus special care must be taken in the design of repeatered facilities to avoid excessive crosstalk to other transmission systems. The repeater is placed in an environmentally hardened apparatus case in a manhole or on a pole. The apparatus case must be spliced into the cable. The apparatus case costs far more than the repeaters it holds. A repeater failure results in a field service visit. Repeaters are usually line powered; this requires a special line feed power supply at the CO. Most of the power fed by the CO power supply is wasted due to loop resistance and power supply inefficiencies.

HDSL is also preferred over traditional T1 carriers because HDSL provides more extensive diagnostic features (including SNR measurement) and HDSL causes less crosstalk to other transmission systems because its transmit signal is confined to narrower bandwidth than the traditional T1 carrier.

2.4.2 HDSL Capabilities and Application

HDSL provides two-way 1.544 or 2.048 Mb/s transport over telephone lines up to 3.7 km (12 kft) of 0.5 mm (24 AWG) twisted pair without a mid-span repeater and up to nearly twice this distance with one mid-span repeater. More than 95% of HDSL lines have no repeater. As a rule, no line conditioning or binder group segregation is required for HDSL. HDSL provides reliable transmission over all carrier serving area (CSA) lines with a typical bit error rate of 10^{-9} to 10^{-10}. DS1 (1.544 Mb/s) HDSL systems use two pairs of wires, with each pair conveying 768 kb/s of payload (784 kb/s net) in both directions. Thus, the term *dual duplex* is used to describe HDSL transmission. See Figure 2.6. E1 (2.048 Mb/s) HDSL systems have the option of using two or three wire pairs, with each wire pair using full-duplex transmission. The three-pair 2.048 Mb/s HDSL uses the very same 784 kb/s transceivers as the 1.544 Mb/s systems. HDSL loops may have bridged taps, but no loading coils.

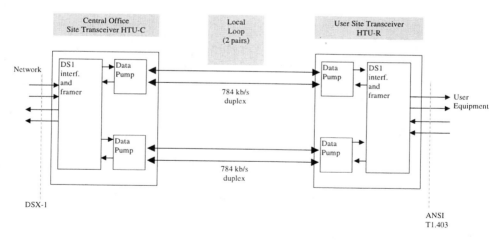

Figure 2.6 HDSL Dual duplex transmission system.

Despite early descriptions of HDSL as a "repeaterless technology," HDSL repeaters are commonly used for lines beyond HDSL's nonrepeated reach of 2.75 to 3.7 km (9 to 12 kft). For 24 AWG wire, up to 7.3 km (24 kft) can be reached with one repeater and up to 11 km (36 kft) with two repeaters. The actual reach can be less where it is not possible to place the repeater at the precise midpoint. Early two-repeater HDSL systems powered the first repeater via line power from the CO, and the second repeater was powered from the customer site. Power feeding from the customer site poses maintenance and administrative drawbacks. With recent reductions in transceiver power consumption, it has become possible to line power two tandem HDSL repeaters from the CO power source.

Primary rate (1.544 or 2.048 Mb/s) private line circuits from a user to the network is the leading HDSL application. HDSL is a popular means to connect private branch exchange (PBX) and packet/ATM data equipment to the public network. HDSL links are used to link wireless radio sites into the landline network. HDSL is used to connect small digital loop carrier (DLC) sites to the CO. During its first few years of use, the high cost of HDSL equipment limited the use of HDSL to situations where there was no economical site to place a repeater apparatus case. By the end of 1994, the price of HDSL equipment had reached the point where HDSL was economically preferred over traditional T1/E1 transmission equipment for nearly all new installations. Traditional T1/E1 equipment is still used for very short lines (less than 3 kft) that require no repeater and for very long lines (more than 30 kft) that would require more than two HDSL repeaters.

The annual maintenance costs of HDSL lines are lower than for T1/E1 lines because HDSL lines have fewer repeaters to fail, superior transmission robustness, and improved diagnostic capabilities. However, existing T1/E1 lines are rarely replaced by new HDSL lines due to the cost of installing the new line.

Although HDSL is most used by local exchange carriers (telephone companies), there is some use of HDSL in private networks to provide high-speed links between buildings within a campus.

2.4.3 HDSL Transmission

Echo-canceled hybrid dual-duplex 2B1Q transmission is used for nearly all HDSL systems worldwide, with some discrete multitone (DMT) and carrierless AM/PM (CAP) systems used in parts of Europe. For 1.544 Mb/s transport, dual-duplex transmission uses each pair of wires to convey one-half of the two-way payload (768 kb/s) plus framing and embedded operations channel (eoc) overhead of 16 kb/s for a total of 784 kb/s transmission. Two pairs of wires make up the total 1.544 Mb/s HDSL transmission system. Since the same overhead information is conveyed on both wire pairs, the receiver selects one wire pair for the overhead information. Usually, the receiver selects the wire pair with the better signal-to-noise ratio (SNR).

Several alternatives were considered for the original HDSL systems: single duplex, dual simplex, and dual duplex.

Single duplex provides the benefits of using only one pair of wires and requiring only one transmitter-receiver pair at each end of the line. See Figure 2.7. The two directions of transmission may be separated by frequency division multiplexing (FDM) or by echo-canceled hybrid (ECH) transmission. However, transmitting the full payload rate over most loops was beyond the

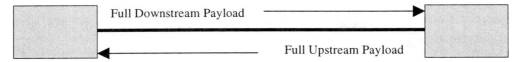

Figure 2.7   Single-duplex HDSL.

abilities of the technology in the early 1990s. Furthermore, the large bandwidth needed presented concerns for spectral compatibility with other types of transmission systems. The single-pair 1.544 Mb/s HDSL systems (sometimes called SDSL) developed in the early 1990s had a loop reach of less than 6 kft on 26 AWG wire; this short reach greatly limited their utility. Only with the most advanced technology available in the late 1990s does it appear that single-duplex 1.544 Mb/s transport may become practical for full carrier serving area (CSA) loop reach. HDSL2, described in Section 2.4.4, employs single-duplex transmission.

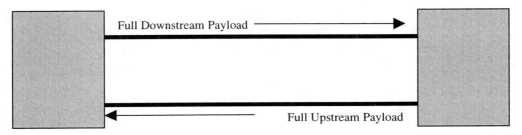

Figure 2.8   Dual-simplex HDSL.

Dual-simplex transmission uses two pairs of wires, with one wire pair carrying the full payload in one direction and the second wire pair carrying the full rate transmission in the opposite direction. See Figure 2.8. This provides a very simple means for the separation of the signals in the two different directions of transmission. Traditional T1 carrier uses dual-simplex transmission. Dual-simplex transmission has the disadvantage of transmitting a signal with a wide frequency bandwidth, which is subject to great loss and crosstalk at the higher frequencies. Due to crosstalk, the signals sent on the two pairs of wires are not fully segregated. Thus, the dual-simplex transceivers may be simpler, but the resulting performance is inferior to dual duplex.

Dual-duplex transmission improves the achievable loop reach and spectral compatibility by sending only one-half of the total information on each wire pair. See Figure 2.9. HDSL further reduces the bandwidth of the transmitted signal by using ECH transmission to send the two directions of transmission in the same frequency band. The dual-duplex HDSL transmitted signal power is progressively less for frequencies above 196 kHz. As a result, the signal crosstalk and attenuation is reduced. Another advantage of dual-duplex transmission is that using one pair of wires can easily provide a half-rate transmission system.

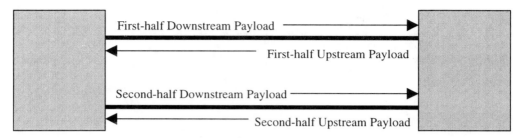

Figure 2.9   Dual-duplex HDSL.

One-pair, fractional-rate HDSL systems are used for the transport of fractional-rate private-line services of 768 kb/s and below and also for small loop carrier systems supporting 12 or fewer voice channels. Fractional-rate HDSL plug-ins for the D4 channel bank permit up to 12 DS0s of HDSL transport information to be multiplexed with information from other channel units in the same D4 channel bank.

Identical maintenance information (indicator bits and eoc) is conveyed on each wire pair of the dual-duplex HDSL system. This redundant transport of the overhead permits the use of the same transceiver components for one-, two-, and three-wire pair HDSL systems. Furthermore, the redundant overhead information ensures reliable operation of the maintenance functions even if transmission has failed or is impaired on one of the loops.

2.4.3.1 Timing

HDSL framing contains positions for stuff quats (quaternary symbols representing two bits). The stuff quats are added to frames as needed to synchronize the T1/E1 payload bit rate to the HDSL line transmission rate. To permit effective echo-canceled operation, the upstream and downstream HDSL symbol rates must be exactly the same. There are some situations where the upstream T1/E1 payload bit rate may slightly differ from the downstream payload bit rate. The stuff quats, together with a little buffering, permit the payload rate to differ slightly from the HDSL line rate. Many public network T1/E1 circuits are loop timed, which means that the upstream timing is derived from the downstream bit clock. Loop-timed circuits do not require stuff quats. However, this feature is provided on all HDSLs in the event that a circuit is not loop timed.

2.4.3.2 Delay (Latency)

Traditional T1 transmission systems have an end-to-end signal transfer delay of less than 100 microseconds. Due to the digital signal processing, HDSL circuits typically have approximately 400 microseconds of signal transfer delay as measured one way between the DSX-1 interface and the T1.403 interface. The additional delay found in HDSL systems rarely poses a problem, but there have been a few cases where upper-layer protocol handshakes have timed-out due to the total end-to-end delay. For this reason, HDSL systems are designed to assure that the one-way signal transfer delay for a nonrepeatered HDSL line is less than 500 microseconds.

HDSL lines with one mid-span repeater have twice this delay. Other network elements including SONET terminals and digital cross-connect systems (DCS) may have delays in excess of 500 microseconds. Thus, end systems should allow for a few milliseconds of network delay regardless of the presence of HDSL.

2.4.3.3 Bit Error Rate

HDSL systems, like BRI and ADSL, are designed to assure better than 10^{-7} bit error rate (BER) on worst-case loops with crosstalk noise power 6 dB greater than the theoretical worst-case crosstalk model. This design criteria is based on the engineering judgment and agreements among leading experts in the T1E1.4 standards working group. A decade of field experience has proved this design criteria to be a good compromise between over-engineering (underuse due to overly conservative design) and under-engineering (poor reliability due to the lack of robustness).

Nonetheless, there are two prevalent misconceptions regarding the BER design of HDSLs and other DSLs. The first misconception is that most HDSLs operate at 10^{-7} BER. The 10^{-7} BER value is for a *worst-worst*-case situation, which is rarely seen in the field. Approximately 99% of HDSLs in the field operate at a BER better than 10^{-9}. When errors do occur, they tend to appear in short bursts. This bursty characteristic is more benign than random bit errors. The second misconception is that HDSLs are grossly over-engineered. Considering the design with 6 dB of margin beyond a worst-worst-case model, it is easy to see why some people have this opinion. However, the seemingly overconservative design is justified for two reasons. HDSLs are required to operate reliably all the time for all qualified loops. Unlike voice-band modems used on switched circuits, one can not "hang up" and dial up again in hopes of obtaining a better connection. Furthermore, the real-world environment contains many impairments that can consume the 6 dB design margin (e.g., water in cable, bad splices, poor-quality inside wire, or a line longer than indicated in cable records).

2.4.4 Second-Generation HDSL

Standards development for a second-generation HDSL technology (HDSL2) began in 1995 to provide the same bit rate and loop reach as first-generation HDSL but using one pair of wires instead of two. This reduction of wire pairs is important because many LECs have a shortage of spare wire pairs in some areas. HDSL2 uses more sophisticated modulation and powerful coding techniques. Carefully selected offset frequency placement of the up- and downstream directions is used for HDSL2 to help combat crosstalk. The newer versions of HDSL borrow many ideas from ADSL. A rate-adaptive version of HDSL is likely to appear. There has also been consideration of HDSL placed in a frequency band above baseband analog voice or above basic rate ISDN. The term SDSL (symmetric, or single-pair DSL) has also been used to describe later versions of HDSL.

2.4.4.1 Performance Requirements

Although several suggestions for line codes were made to T1E1.4 following a request in 1995 (T1E1.4/95-044), progress was slow until detailed requirements were established. These requirements, specified primarily by the operating companies, were first proposed in March 1996 (T1E1.4/96-094 and T1E1.4/96-095) and revised since that time (T1E1.4/97-180, 180R1, 181, 469). They currently are as follows:

Loop Reach: CSA coverage (same as two-pair ANSI HDSL):

9000 ft (2.7 km) of 26 AWG (0.4 mm)

12000 ft (3.6 km) of 24 AWG (0.5 mm)

Bridged taps limited to 2.5 kft total, 2 kft per tap

Cable parameters as specified in T1.601

Impairments/Performance: minimum of 5 dB of performance margin with 1% worst-case crosstalk from the following interfering services:

49-disturber HDSL

39-disturber HDSL2

39-disturber EC-ADSL

49-disturber FDM-ADSL

25-disturber T1

24 T1 + 24 HDSL2

24 FDM-ADSL + 24 HDSL

Spectral Compatibility: To all existing services, no more impairment than the services tolerate today, with the following exceptions: shall not degrade HDSL by more that 2 dB (T1E1.4/97-434, 440R1) and ADSL by more than 1 dB (T1E1.4/97-444). These services include the following customer interface specifications: T1.413 (ADSL), TR-28 (HDSL), ANSI T1.403 (DS1), and T1.601 (ISDN-BRA).

Latency: The maximum latency for HDSL2 is to be no more than for HDSL (500 μs).

2.4.4.2 Impairments

The impairments were selected as typical of severe-case crosstalk combinations that HDSL2 may encounter. Of the CSA test loops in ANSI TR-28, it was found that CSA 4 represented the limiting case. Near-end crosstalk coupling is modeled using the Unger model, as specified in T1E1.4/96-036, and far-end crosstalk coupling is modeled as specified in ANSI T1.413 Annex B. Models for T1.601, TR-28, and T1.403 transmitters were taken from T1.413 Annex B. A variety of models for the echo-canceled (EC) and frequency division multiplexed

(FDM) version of ADSL were used. Most of the latest work incorporated modified versions of the PSDs from Annex B.4 and B.5 of T1.413. Most of the variation dealt with split points for the FDM, roll-off of the upstream PSD, and low frequency roll-off of the downstream EC PSD. It was generally accepted that the Sinc term from B.4 and B.5 should not be used. The mixed crosstalk cases were added to the requirements (T1E1.4/97-180, 181) after it was found that they were worse than homogeneous crosstalk for non–self-NEXT limited modulation techniques.

Impulsive noise has not been considered to be a significant impairment in the T1E1.4 deliberations. Also, all calculations regarding spectral compatibility are with respect to other ANSI DSLs. No calculations/measurements have been published with respect to their ETSI or ITU counterparts.

2.4.4.3 Spectral Compatibility

Determining spectral compatibility between new and existing service proved to be a significant challenge. For ISDN-BRA it was easily shown that the proposed line codes were definitely less of an impairment than self-NEXT. The other listed services were not so easy. For T1.403, (DS1/T1), the initial technique involved measuring the total amount of NEXT power present at the T1 receiver. This was compared to the power from T1 self-crosstalk to see if a problem resulted. In several contributions, the crosstalk was weighted by a measured (T1E1.4/97-071) or calculated T1 receive filter. Later, it was noticed that spectral compatibility with T1 was eased since the first segment from the CO only has 15 dB of loss and not the 30 dB that the other segments must operate over.

With ADSL, spectral compatibility was defined by ideal margin calculations. It was found that slight changes in the assumed noise floor, transmit PSDs, and minimum carrier number (for the FDM case) could have a significant impact on the performance estimate. Most calculations have found that the agreed PSD would degrade ADSL (T1.413) margins by less than 1 dB for the worst-case standard interferer combination.

With HDSL, the initial compatibility work was done using theoretical calculations, but later testing (T1E1.4/97-339) showed that for some modulation formats this was insufficient. (This is addressed more completely in the next section.)

2.4.4.4 Modulation Format

Early on, both symmetric echo-canceled transmission (SET) and frequency division multiplexed transmission (FDM) approaches were considered. SET proved to have a self-crosstalk limitation 2 to 3 dB short of the requirements. In contrast, FDM transmission is not limited by self-NEXT, but by (ingress) crosstalk from other services. It is also limited by (egress) crosstalk into other services due to the higher transmit frequencies involved with transmitting a symmetric payload in this manner. The ingress and egress crosstalk make the FDM solution even less desirable than SET. A "staggered FDM" scheme (T1E1.4/96-340) was proposed in an attempt to limit these undesirable effects.

In T1E1.4/97-073, partially overlapped echo-canceled transmission (POET) was proposed. POET involves overlapping, but not identical, spectra in the two transmit directions.

These spectra are carefully shaped to provide maximum performance in the presence of self- and foreign crosstalk while causing minimal degradation of other services due to POET crosstalk into other services. Various versions of this approach were proposed in the standards process, all incorporating the same basic concept (POET-PAM (97-073), OverCAPped (97-179), OPTIS (97-237,320), MONET (97-307,412)).

One characteristic that all these POET modulation schemes exhibit is the effect of heterogeneous crosstalk on performance. For SET, performance in homogeneous and heterogeneous crosstalk is quite similar. However, with POET modulation it is possible to have performance in the presence of heterogeneous crosstalk that is significantly worse than performance in the presence of homogeneous crosstalk. The actual performance of these systems also varies with the symbol rate and modulation type. With digital oversampled transceivers, it is possible to decouple the transmit PSD from the actual symbol rate. (This uses principles similar to those used in a traditional CAP transceiver.) This property was first exploited in a CAP version of POET (T1E1.4/97-170), but ultimately it was found that with the impairment set for HDSL2, PAM modulation reaps even larger benefits from this decoupling (T1E1.4/97-237). For each unique crosstalk PSD, there is a particular symbol rate that gives maximum performance. For ease of implementation, a single symbol rate that offers performance at near the optimal level over a wide variety of crosstalk PSDs is desirable.

Most of the later modulation scheme proposals have PSDs where some of the upper frequencies are boosted above the nominal value. These "boosted" portions of the PSD are also above the level of any other DSL that operates at those frequencies. This boost was first introduced in T1E1.4/97-170 and incorporated in a pronounced way in T1E1.4/97-273. After this concept was introduced, it was discovered that, when transmitting such signals, theoretical calculations alone were not sufficient to predict spectral compatibility with existing services. Testing of deployed HDSL systems (T1E1.4/97-339) revealed a significant difference between theoretical calculations and measured performance in the presence of OPTIS crosstalk. As a result, modifications were made to the proposed HDSL2 PSD to reduce this degradation (T1E1.4/97-435). Final measurements after modification showed this degradation to be 2 dB or less (T1E1.4/97-434, 440R1).

The current agreed-upon modulation format incorporates the key elements proposed in T1E1.4/97-257:

- The upstream and downstream transmitters will each have a unique spectral shape.
- The upstream and downstream transmitter spectra will be partially overlapped in frequency.
- The shape of the transmit spectrum will be decoupled from the symbol rate to allow for flexible use of excess bandwidth.
- The transmit modulation used will be pulse amplitude modulation (PAM).
- Coded modulation will be used.

The result (T1E1.4/97-435) is a POET system using a modification of the OPTIS PSD. This modulation format uses PAM with 3 information bits per symbol and a 16-level coded constellation. A symbol rate of one-third the payload rate in both the NT-to-LT and LT-to-NT directions was chosen as a good compromise symbol rate. Advantage is achieved through the use of excess bandwidth in the LT-to-NT direction and a high degree of spectral shaping in both directions. The transmit power is approximately 16.5 dBm in each direction. This modulation technique has been shown (via optimal DFE calculations) to have a minimum theoretical uncoded margin on the worst-case required loop of 1.0 dB. Realized performance near the theoretical values is only possible through the use of a fractionally spaced equalizer.

The spectral shaping employed in the agreed PSDs was designed specifically for this application: to maximize the folded SNR at the HDSL2 receiver (with a symbol rate of 517.33 kHz) in the presence of the specific crosstalk mixes listed in the requirements, while simultaneously minimizing the impact of egress crosstalk into the ANSI DSLs. See Figure 2.10. Not only are maximizing HDSL2 performance while minimizing egress crosstalk conflicting goals, simply trying to simultaneously minimize egress crosstalk into two different DSLs may result in a conflict. For instance, to minimize impact into ANSI HDSL, it is desirable to use lower and wider PSDs, whereas to minimize crosstalk into ANSI T1.413 ADSL, a narrower and higher PSD is preferable. Since this optimization was done specifically for the set of ANSI DSLs in the requirements, one should not extrapolate that a frequency scaled version of the same shaping filter would be the best solution for the ITU version of HDSL2. Further work on this issue will be required before conclusions can be drawn.

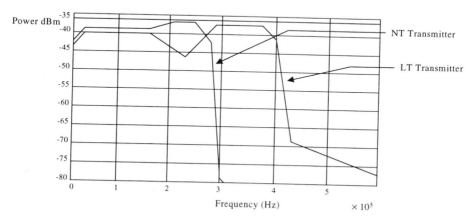

Figure 2.10 Nominal transmit PSDs for the HDSL2 modulation format.

2.4.4.5 Trellis Code Structure

To meet the difficult requirements, coded modulation must be used to increase the crosstalk limited performance of HDSL2 over that possible using only uncoded modulation. With the latency limitation, interleaved concatenated coding and turbocoding techniques proved to be impossible. This left traditional trellis-coded modulation in combination with channel equalizing precoding (such as Tomlinson-Harashima precoding). Although multidimension and multilevel coding approaches were examined, a simple one-dimensional trellis-coded PAM approach has proved to be the best at achieving high coding gains with minimal latency (T1E1.4/97-337).

With the agreed modulation format, 4.0 dB of coding gain is needed to achieve the requirements. For one-dimensional codes with Viterbi decoding, 32 states are needed to achieve over 4 dB of realized BER coding gain. However, the 5.0 dB margin budget must include some non–coding-related implementation loss which must be made up for in coding gain. (This implementation loss, can also affect the realized coding gain.) Thus, a variable amount of coding gain may be needed, based on other losses in the design of the system. The agreement includes a programmable rate ½ one-dimensional trellis (T1E1.4/97-443). See Figure 2.11. This structure allows receivers the flexibility to trade off complexity of the trellis-decoder with complexity in the remainder of the transceiver. This programmable structure also allows for alternative decoding techniques to be used (e.g., sequential decoding), which require substantially different codes than those used for Viterbi decoding.

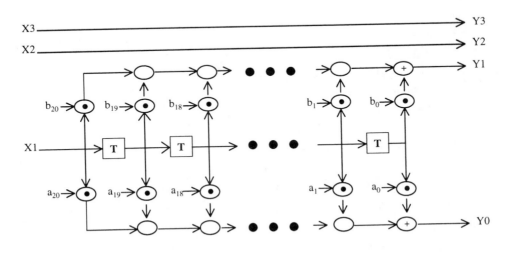

Figure 2.11 Block diagram of the programmable encoder (from T1E1.4/97-443).

2.4.4.6 Complexity Differences Relative to HDSL

Since the requirements for ANSI HDSL2 were quite challenging, a significant complexity increase as compared to HDSL is required to meet them. In this section, we briefly examine the complexity differences.

- The transmit power of HDSL2 is 3 dB higher than that of HDSL. Furthermore, the use of precoding and the spectral shaping together cause the peak-to-rms ratio to be larger than that of 2B1Q-based HDSL. Higher peak voltage levels will increase the power consumption of the line driving circuitry.

- The channel equalizing precoder has a function that is similar to the feedback filter of the decision feedback equalizer used in HDSL. However, the data in the precoder are many bits (12–16) wide instead of the 2 bits for 2B1Q DFE. This increases the complexity. Furthermore, the presence of the precoder increases the complexity of the echo canceler in the same manner.

- To meet the performance requirements through the use of Viterbi decoding, it appears that trellis codes on the order of 512 states will be necessary. The Viterbi decoder for such a code represents an enormous increase in both processing power and memory compared to uncoded HDSL systems.

- To obtain adequate performance, HDSL2 requires a fractionally spaced equalizer and echo canceler, both which are of significantly higher complexity than the baud-spaced equivalents typically employed in HDSL transceivers.

2.5 ADSL

2.5.1 ADSL Definition and Reference Model

Asymmetric digital subscriber line (ADSL) is a local loop transmission technology that simultaneously transports the following via one pair of wires:

- Downstream (towards customer) bit rates of up to about 9 Mb/s
- Upstream (towards network) bit rates of up to 1 Mb/s
- Plain old telephone service (POTS, i.e., analog voice)

The bit rate towards the customer is much greater than from the customer, hence the term *asymmetric*. Analog voice is transmitted at baseband frequencies and combined with the pass-band data transmission via a low-pass filter (LPF) that is commonly called a "splitter." In addition to the splitters, the ADSL consists of an ADSL transmission unit at the Central Office side (ATU-C), a local loop, and an ADSL transmission unit at the remote side (ATU-R).

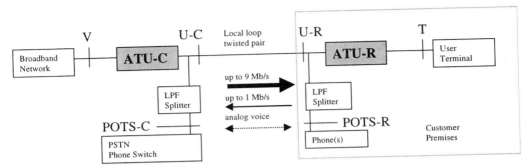

Figure 2.12   ADSL reference model.

2.5.2   ADSL Origins

The early conceptual definition of ADSL began in 1989, primarily by the work of J. W. Lechleider and others at Bellcore. Early ADSL development began at Stanford University and AT&T Bell Labs in 1990. ADSL prototypes arrived at telephone companies and Bellcore Laboratories in 1992, and early ADSL products moved into field technology trials in 1995. ADSL drew from the earlier work on voice-band modems, ISDN, and HDSL.

In October 1998, the ITU gave preliminary approval to ("determination" in ITU language) a set of ADSL Recommendations. Recommendation G.922.1 specifies full-rate ADSL. This is nearly identical to ANSI T1.413 Issue 2 with two major exceptions:

1. the T1.413 tone-based initialization sequence is replaced by a message-based process described in G.994.1, and

2. a special mode has been added to improve performance in the presence of crosstalk from TCM-type ISDN used in Japan.

Recommendation G.992.2 (previously known as G.lite) specifies ADSL for use without a POTS splitter. G.922.2 is based on G.992.1 with the following major differences:

1. added provisions for power saving modes at the ATU-C and ATU-R,

2. the addition of a fast retrain mechanism to permit rapid recovery from on/off-hook events,

3. the number of tones is reduced from 256 to 128, and

4. the number of bits per tone is reduced from 15 to 8.

Recommendation G.994.1 (previously known as G.hs) specifies a message-based initialization handshake to allow multimode DSL transceivers to negotiate a common operating mode. Recommendation G. 995.1 provides an overview of the family of DSL recommendations. Recommendation G.996.1 specifies methods for measuring the performance of DSL equipment.

Recommendation G.997.1 specifies the physical layer operations, administration, and maintenance provisions for ADSL. This includes the ADSL embedded operations, channel (eoc), and Management Information Bases (MIBs).

2.5.3 ADSL Capabilities and Application

2.5.3.1 2.5.3.1ADSL1, ADSL2, and ADSL3

The ADSL concept evolved during the early 1990s. At first, ADSL was considered at a fixed rate 1.5 Mb/s downstream and 16 kb/s upstream for video dial tone (VDT) MPEG-1 applications. Some members of the industry refer to this as ADSL1. Later, it became clear that some applications would require higher speeds and that more advanced transmission techniques would enable the higher speeds. Three Mb/s downstream and 16 kb/s upstream ("ADSL2") was briefly considered to enable two simultaneous MPEG-1 streams. In 1993, interest shifted to ADSL3 with 6 Mb/s downstream and at least 64 kb/s upstream to support MPEG2 video. The Issue 1 ANSI T1.413 ADSL standard grew out of the ADSL3 concept. The terms ADSL1, ADSL2, and ADSL3 have seen little use after approval of the ANSI T1.413 standard.

2.5.3.2 RADSL

Rate-adaptive digital subscriber line (RADSL) is a term that applies to ADSL systems capable of automatically determining the transport capacity of the individual local loop and then operating at the highest rate suitable for that local loop. The ANSI T1.413 standard provides the capability for rate-adaptive operation. The rate adaptation occurs upon line start-up, with an adequate signal quality margin to assure that the start-up line rate can be maintained during nominal changes in the line transmission characteristics. Thus, RADSL will automatically provide higher bit rates on loops with better transmission characteristics (less loss or less noise). RADSL implementations have supported maximum downstream rates in the range of 7 to 10 Mb/s and maximum upstream rates in the range of 512 to 900 kb/s. On long loops (5.5 km/18 kft or more), RADSL may operate at rates of about 512 k/s downstream and 128 kb/s upstream.

RADSL borrowed the concept of rate adaptation from voice-band modems. RADSL provides the benefit of one version of equipment that assures the highest possible transmission rate for each local loop and also permits operation on long loops at lower rates.

2.5.4 ADSL Transmission

The ADSL concept contains two fundamental parts: (1) near-end crosstalk is reduced by having the upstream bit rate and bandwidth much less than the downstream bit rate, and (2) simultaneous transport of POTS and data by transmitting data in a frequency band above voice telephony. Two-way transmission of multimegabit rates is not on most telephone lines due to the combined effect of loop loss and crosstalk. As shown in Figure 2.13, received signal power diminishes in proportion with frequency, and received crosstalk noise increases with frequency. Thus, two-way transmission is not possible at frequencies where the crosstalk noise overwhelms the received signal.

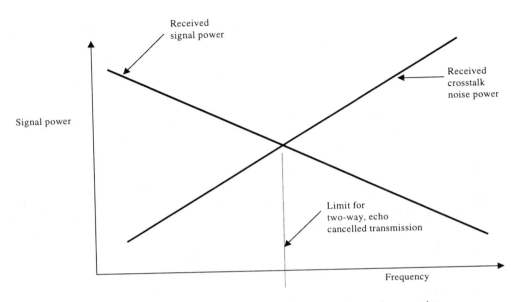

Received
signal power

Received
crosstalk
noise power

Signal power

Limit for
two-way, echo
cancelled transmission

Frequency

Figure 2.13 Two-way transmission is limited to lower frequencies.
This figure is based upon one created by Kim Maxwell of Independent Editions.

ADSL performs two-way transmission where possible: below the two-way cut-off frequency. The upper frequencies that are unsuitable for two-way transmission are used for one-way transmission. This permits downstream transmission rates far in excess of those possible for two-way transmission.

Many ADSL systems use a frequency division multiplexed (FDM) transmission technique, which places upstream transmission in a frequency band separate from the downstream band to prevent self-crosstalk. The guard band is necessary to facilitate filters that prevent POTS noise from interfering with the digital transmission. See Figure 2.14.

Some ADSL systems use an ECH transmission technique where the upstream frequency band resides within the downstream band. See Figure 2.15. By overlapping the bands, the total transmitted bandwidth may be reduced. However, the ECH is subject to self-crosstalk, and its implementation involves more complex digital signal processing. There is some debate as to whether the digital complexity is offset by simplification of the analog front end.

Due to the lack of self-crosstalk at the CO end, FDM ADSL offers much better upstream performance than ECH ADSL. However, the wider downstream bandwidth of ECH ADSL permits better downstream performance, especially for shorter loops.

The performance of symmetric DSL is primarily limited by self–near-end-crosstalk (self-NEXT). ADSL overcomes self NEXT at the customer end simply by reducing the source of the self-NEXT. By reducing the upstream bit rate, the upstream channel may be positioned to mini-

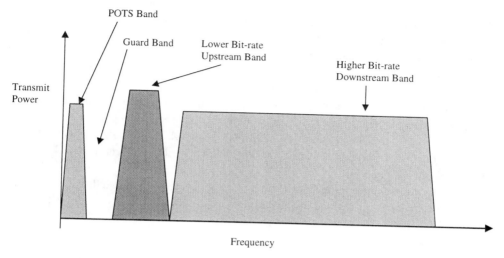

Figure 2.14   FDM ADSL.

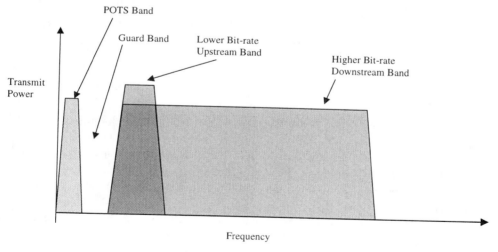

Figure 2.15   ECH ADSL.

mize crosstalk into the downstream transmission. For ADSL, reception of the upstream channel is made easier by placing it at lower frequencies where loop loss is less and crosstalk noise is less.

ADSL systems employ advanced digital transmission techniques to enhance performance. The modulation and frequency placement of the transmitted signal dynamically adapts to achieve optimal performance from the characteristics unique to the subscriber line being used. Trellis codes are used to reduce the effect of steady-state wide-band noise. Adaptive equalizers protect against narrowband noise such as radio frequency interference (RFI). Forward error control (FEC) codes and interleaving protect against noise impulses. Interleaving protects against error bursts by shuffling blocks of data so that a long burst of errors results in a few (correctable) errors in each block, rather than a large (uncorrectable) number of errors in one block. An interleaving depth of 20 ms will protect against noise bursts up to 500 μs in duration. Surveys of loop impulse events suggest that a vast majority of impulses are less than 500 μs in duration. However, this degree of interleaving causes an additional transport delay of 20 ms, which can slow down the throughput of protocols such as TCP/IP, which require packets to be acknowledged before more data are sent.

ADSL loops may have bridged taps, but no loading coils. An in-depth discussion of ADSL transmission techniques is provided in Chapter 6.

2.5.5 ADSL's Future

ADSL will be integrated into fiber-fed digital loop carrier (DLC) systems to address those loops that are not served directly from a CO. ADSL is well suited to providing high bit rates over the DLC-fed loops, which are rarely longer than 3.7 km (12 kft). Despite an industry standard for ADSL (ANSI T1.413), early ADSL systems did not interoperate. Efforts by ADSL manufacturers and standards committees are expected to achieve multivendor interoperability for future ADSL systems. In addition to the physical layer, full interoperability requires compatibility at all layers of the protocol stack.

It has become clear that ADSL is the access technology that asynchronous transfer mode (ATM) needed to open the door to home and small office. Before ADSL, ATM appeared to be restricted to only those who could afford the premium prices of links at 45 Mb/s and above: big business and the backbone network. Work is underway to deal with ATM transport over ADSL's unique characteristics: error rates, latency, asymmetry, and dynamic rate change.

For a time it appeared that the focus for ADSL evolution was high speeds such as 10 Mb/s downstream and 1.5 Mb/s upstream. However, this direction has faded due to overlap with VDSL, spectral compatibility concerns, and doubts regarding the need for these speeds. Instead, the focus now is on improved loop reach at more modest rates near 1 Mb/s, and lower-cost, lower-power, and reduced crosstalk implementations. ADSL systems are being developed to convey multiple digital derived voice circuits in addition to high-speed data.

2.5.5.1 ADSL + ISDN

Some vendors are introducing a version of ADSL where the upstream and downstream frequency band have been placed above the 0 to 80 kHz ANSI T1.601 basic rate ISDN transmission band. For BRI using the 4B3T line code, the BRI frequency band is 0 to 120 kHz. This substantially reduces the ADSL bit rates but does permit simultaneous ISDN and ADSL service on one loop. The ADSL + ISDN configuration is unlikely to provide the full 5.5 km (18 kft) reach normally provided by ISDN. ADSL + ISDN is of particular interest in Germany and France, where ISDN service is particularly widespread. This configuration may also be used to provide two voice circuits and moderately high data rates.

2.5.5.2 Splitterless ADSL

The customer premises installation of ADSL service can require new or modified inside telecommunications wiring. For a conventional ADSL configuration, the ADSL terminates at the network interface device (NID), where a low-pass filter (the *splitter*) extracts the voice-band signals that are attached to the red and green inside wires to the telephones, and the wideband signals are connected to the yellow and black inside wires to the customer's ADSL modem. This requires the installation of the splitter and also requires the use of the yellow and black inside wires, which are not found in some premises or may already be used for second-line voice service. Furthermore, in some cases, substandard inside wire has been used, which will impair ADSL (and even ISDN) operation. As a result, new inside wire will often be required from the NID to the customer's ADSL modem.

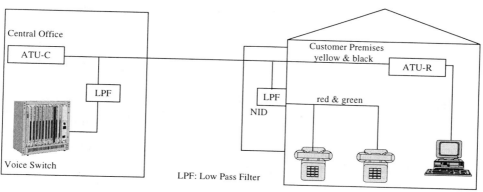

Figure 2.16 Conventional ADSL configuration with splitter.

The most common ADSL POTS splitter configuration (shown in Figure 2.16) places a low-pass filter (LPF) for the voice wiring at or near the NID and a high-pass filter (HPF) located within the ATU-R. Alternatively, the splitter (LPF and HPF) may be integrated within the ATU-R. The splitter within the ATU-R has the disadvantages of the possible loss of POTS service when the ATU-R is removed and possibly excessive crosstalk when using existing premises wiring.

The splitterless ADSL concept eliminates the splitter filter at the customer end of the line. Many other terms have been used to describe this concept: *ADSL Lite, Consumer DSL (CDSL)*, or *Universal ADSL (UADSL)*. Splitterless ADSL is defined in ITU Recommendation G.992.2. The ADSL modem and phones are all directly connected to the existing red and green wires within the premises. Simultaneous data and voice operation are supported. ADSL installation is easily performed by plugging the ADSL modem into any phone jack in the premises. Neither new inside wiring nor splitter installation is needed. See Figure 2.17.

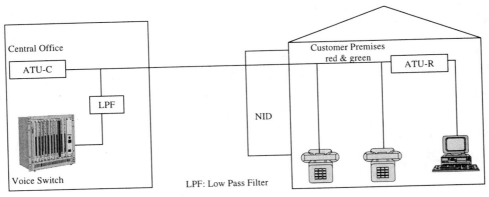

Figure 2.17   ADSL configuration without splitter at customer premises.

The concept is wonderful; the practical implementation is being investigated and will present compromises. The ADSL splitter has two purposes: (1) the splitter attenuates POTS signaling noise, which could corrupt ADSL data transmission, and (2) the splitter attenuates the ADSL signals to prevent audible noise on the telephones. Due to the nonlinear impedance of some telephones, ADSL-transmitted energy at frequencies well above the audible band can be modulated into the voice band. Shifting the ADSL transmission bands to higher frequencies can reduce these problems. However, this will reduce the ADSL data rates and the loop reach. Brief error bursts are likely to occur when the phone rings, and surely during the *ring trip* transient when a ringing phone is taken off-hook. An objectionable hissing sound may be heard in phones with poor wideband characteristics.

One solution to these problems is to place a low-pass filter in series with each phone. See Figure 2.18. This filter would be inexpensive and have modular connectors so that an untrained customer could install it in seconds. Installation for a wall phone would not be so easy, but at worst the customer could buy a new "ADSL-compatible" wall phone with a low-pass filter built inside the phone. This configuration should prevent POTS noise from impairing ADSL transmission and ADSL noise from being heard on the telephones. No new inside wire is needed, and no splitter installation is needed at the NID. The customers could plug their ADSL modem into any telephone wall jack. The ADSL data rate would be somewhat less than the conventional ADSL configuration. There would be some reduction in data rates due to other noise and the loading effect of many filters and wiring stubs. The ITU Recommendation G.992.2 ("G.lite") has less performance than the full-rate Recommendation G.992.1 ("G.DMT"), due to a reduced number of DMT tones, and fewer bits-per-tone. Reduced voice-band transmission quality may also result from many low-pass filters being placed in parallel. Another concern is trouble resulting from customers who forget to place a LPF in the line to one of their phones.

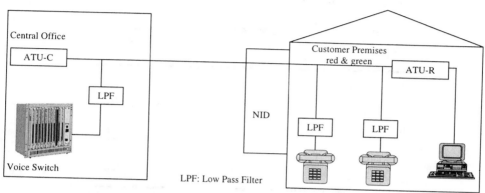

Figure 2.18 ADSL configuration with low-pass filter by each phone.

If the technical and operational hurdles are overcome, splitterless ADSL may ultimately become the dominant type of ADSL. For the near term, most ADSLs are being installed with a splitter at both ends of the line. The use of a splitter at the customer site may see continued use for the installation of service to customers who require higher-bit-rate service. Some ADSL service providers have suggested that their ADSL service should work with both the splitterless and splittered customer configurations, while using the same type of ATU-C at the CO. The progress of splitterless ADSL was accellerated by the technical and marketing activities of the Universal ADSL Working Group (UAWG), a collection of leading telephone companies and computer companies.

2.6 VDSL

Very-high-bit-rate digital subscriber line (VDSL) is an extension of ADSL technology to higher rates, up to 52 Mb/s downstream. At such high bit rates, the loops must be so short that optical fiber will be used for all but the last few thousand feet.

2.6.1 VDSL Definition and Reference Model

Most DSLs are primarily intended for use on loops from a CO to a customer premises and secondarily for use from fiber-fed distribution multiplexers. The opposite is true for VDSL. VDSL will primarily be used for loops fed from an optical network unit (ONU), which is typically located less than a kilometer from the customer. Few VDSL loops will be served directly from a CO.

Optical fiber connects the ONU to the CO. VDSL transmission over a twisted wire pair is used for the few thousand feet from the ONU to the customer premises. See Figure 2.19. VDSL requirements developed by the T1E1.4 standards working group specified the following objectives for rates and distances from the ONU to the customer site:

Downstream rate (Mb/s)	Upstream rate (Mb/s)	Distance (kft - m)
52	6.4	1,000 – 300
26	3.2	2,500 – 800
26	26	1,000 – 300
13	13	1,800 – 600
13	1.6	3,750 – 1,200

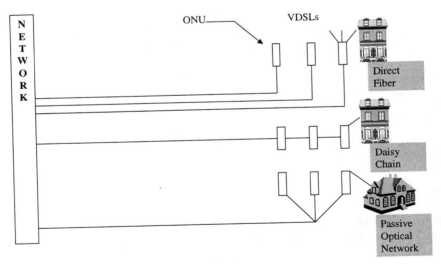

Figure 2.19   VDSL architectures.

The fiber from the network to the ONUs may be connected directly to the ONU, daisy chained, or via a passive optical splitter.

2.6.2   VDSL Origins

Discussion of the VDSL concept began in standards committees in late 1994, with the definition of VDSL system requirements in ETSI TM6 and T1E1.4. Several proposals are currently be studied by these groups.

2.6.3   VDSL Capabilities and Applications

VDSL, as part of a full-service network (FSN), is intended to support all applications simultaneously: voice, data, and video. Ultimately, VDSL would support high-definition television (HDTV) and high-performance computing applications. Symmetric application of VDSL will provide two-way data rates up to 26 Mb/s that will be attractive for business sites where fiber-to-the-building is not justified.

The DAVIC VDSL type specification employs carrierless amplitude phase (CAP) modulation for rates of 13, 25.92, and 51 Mb/s downstream and 1.6 Mb/s upstream via an unshielded twisted wire pair. The DAVIC VDSL specification is based on a passive NT architecture, which permits direct connection of multiple VDSL transceivers at the customer end of the line. See Figure 2.20. Typically, the passive NT architecture requires the ONU to be less than 100 meters from the customer VDSL units, thus making it more suitable for fiber-to-the-pedestal and in-premises applications.

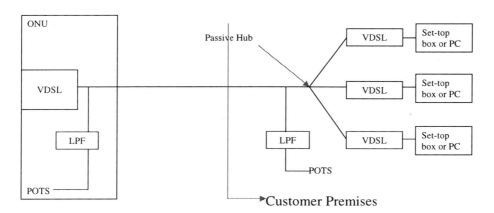

Figure 2.20   VDSL passive hub architecture.

The VDSL Active hub architecture shown in Figure 2.21 permits a greater (rate) × (reach) product by using a point-to-point configuration for loop transmission. The active hub consists of a single VDSL transceiver, and a separate short-haul link within the premises to each terminal (shown), or a short-haul bus within the premises (not shown).

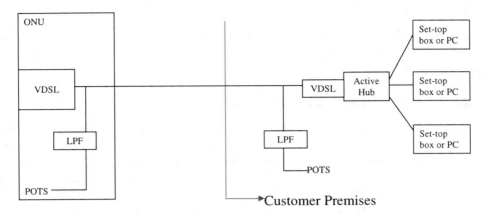

Figure 2.21   VDSL active hub architecture.

Twisted-Pair Transmission

3.1 Twisted-Wire-Pair Origins

Telephone service began in 1877 when Alexander Graham Bell connected telephones via a single iron wire with the earth as a return path for the electrical circuit. This avoided the cost of the second conductor, but signal transmission proved to be unreliable due to corrosion of the grounding connection and poor conduction during long periods of dry weather. Customers were sometimes told to pour buckets of water on their ground rods. These problems were solved later by the use of a pair of bare wires, which were strung in parallel a few inches apart. This provided a reliable return path for the current. However, the phenomenon of crosstalk was quickly discovered when voice-band signals from one wire pair exhibited electromagnetic coupling to nearby wire pairs. The signals on one telephone line would be faintly heard on other lines. It was discovered that the crosstalk could be reduced by periodically swapping the positions of the left and right conductors. Transmission performance and ease of installation were both improved. Bell invented the twisted-wire pair, in 1881 with a pair of individually insulated conductors twisted together. With a sufficiently short space between twists, the electromagnetic coupling of energy over a small segment of wire is canceled by the out-of-phase energy coupled on the next segment of wire. Modern telephone cables are designed with slightly different twist rates for each wire pair to assure minimal crosstalk. Copper conductors are used to minimize signal attenuation due to electrical resistance. Aluminum conductors were installed in portions of Europe for a short time but discontinued due to the higher resistance and wire-splice troubles.

3.2 Telephone Network and Loop Plant Characteristics

The twisted-wire-pair infrastructure (known as the loop plant and shown in Figure 3.1), connecting customers to the telephone company, was designed to provide economical and reliable plain old telephone service (POTS). A loop plant intended for operation of DSLs and POTS would have been designed quite differently. Local loop design practices have changed relatively little over the past 20 years. The primary changes have been the use of longer-life cables and some reduction in loop lengths via the use of digital loop carrier (DLC). However, DSLs must cope with the huge embedded base of loop plant, some of which is 75 years old.

The term *loop* refers to the twisted-pair telephone line from a CO (Central Office) to the customer. This term originates from current flow through a looped circuit from the CO on one wire and returning on another wire.

For the first 100 years, the telephone network had one goal: universal voice service. High reliability was a top priority, and low cost was a close second priority. Voice was carried as an analog signal of 3.4 kHz bandwidth. Services beyond voice began to gain significance in the 1970s.

3.2.1   Feeder Plant

Larger COs may serve over 100,000 telephone lines; they all terminate at the main distributing frame (MDF) in the CO. Feeder plant cables lead from the CO to the serving area interface (SAI), shown in Figure 3.1, which serves 1,500 to 3,000 lines.

The loop plant consists of twisted wire pairs that are contained within a protective cable sheath. In some parts of Europe and Asia the wires are twisted in four-wire units called quads. Quad wire has the disadvantage of high crosstalk coupling between the four wires within a quad. Within the Central Office, cables from switching and transmission equipment lead to the MDF, which is a large wire cross-connect frame where jumper wires connect the CO equipment cables (at the horizontal side of the MDF) to the outside cables (at the vertical side of the MDF). The MDF permits any subscriber line to be connected to any port of any CO equipment. Cables leaving the Central Office are normally contained in underground conduit with up to 10,000 wire pairs per cable and are called feeder cables, E-side, or F1 plant. The feeder cables extend from the CO to a wiring junction and interconnection point, which is known by many names: serving area interface (SAI), serving area concept box (SAC box), crossbox, flexibility point, primary crossconnection point (PCP). The SAI contains a small wire jumper panel that permits the feeder cable pairs to be connected to any of several distribution cables. The SAI is at most 3000 feet from the customer premises and typically serves 1,500 to 3,000 living units. The SAI contains only a wiring crossconnect field; it has no active electronics. The loops emanating from the SAI to the customer are sometimes called the "distribution plant" (see Section 3.2.3).

3.2.2   Digital Loop Carrier

The digital loop carrier (DLC) was introduced in 1972 in the United States as an electronic multiplexing device that resides at the SAI point to multiplex up to 96 lines into a few T1-carrier feeder lines to the CO. The DLC replaces the large number of copper pairs in the feeder with a multiplexer in the serving area. Later, a fiber-fed next-generation digital loop carrier (NGDLC) terminated up to 2,000 customer lines. Approximately 15% of subscriber lines in the United

States are served via DLC, though the proportion varies greatly by region. Loops served by DLC follow the carrier serving area (CSA) design rules, which stipulate a maximum CSA loop length of 3.7 km (12 kft) for loops entirely comprised of 24 AWG (American Wire Gauge) (see Ref. [3]), and a maximum of 2.75 km (9 kft) for loops entirely comprised of 26 AWG. Loops with a mixture of wire gauges are restricted to a length that corresponds to the proportional length of each type of wire. This corresponds to a maximum loop resistance of 850 ohms. The cumulative length of bridged taps may not exceed 762 m (2.5 kft). The maximum loop length is reduced by the cumulative bridged tap on the loop.

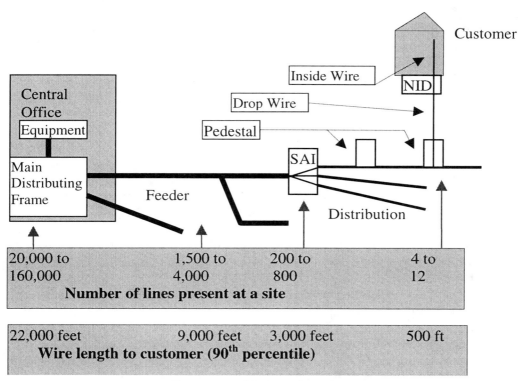

Figure 3.1 Telephone loop plant.

DLC does not eliminate the copper loops to each customer site, but DLC does make the loops shorter. The relatively short loops served by DLCs are ideal for use with BRI (basic rate interface), HDSL (high-bit-rate digital subscriber lines), and ADSL (asymmetric DSL). Since DSL transmission operates only over a path of continuous copper, the DLC remote terminal must be equipped with the correct type of DSL channel unit. Also, the DLC must have sufficient bandwidth on the link to the CO (an important limitation for copper-fed DLCs).

Outside the United States, DLC saw little use until the 1990s, when substantial DLC growth began. Figure 3.1 shows the typical U.S. loop plant design. The line counts represent the number of wire pairs present. The line count is 1.2 to 4 times the number of living units in the covered area.

3.2.3   Distribution Plant

Distribution cables (also known as D-sides) contain 25 to 1000 pairs. For residential and small business areas, the distribution cables lead to the drop wire that serves each customer. The distribution cable connects to the drop wires via a wiring pedestal (also known as a distribution terminal), which typically serves four to six living units. Typical drop wires contain two or three pairs of 22 AWG, although larger numbers are found in some areas. Much of the drop wire installed prior to 1992 was not twisted ("flat wire"). In the United States, the drop wire connects to the inside wire via the network interface device (NID). The NID contains an overvoltage protector and test access jack, which serves as the point of demarcation between the telephone company network and the customer premises. About 50% of residences in the United States have a NID; it is typically attached to the outside of the premises. For many other countries, the point of demarcation is inside the premises at the customer side of the network terminating equipment (NTE). The NTE may be the DSL transceiver at the customer end of the line. Inside wire is often two twisted pairs of 24 AWG, although a wide range of wiring practices may be found inside the customer premises.

The feeder and distribution cables are bundled into **binder groups** of 25, 50, or 100 pairs. The pairs within a binder group remain adjacent to each other for the length of the cable. As a result, the crosstalk of pairs within a binder group is somewhat greater than crosstalk between pairs in different binder groups. Despite the administrative complexities, telephone companies will sometimes segregate certain services (such as T1 carrier) into separate binder groups.

Cables connecting to a CO may have up to 10,000 pairs. As one follows the cable plant from the CO to the customer site, the cables branch. As a result, fewer customer lines are accessible at points closer to the customer site. The number of wire pairs per cable becomes progressively smaller at successive splice points approaching the customer site. Feeder and distribution cable pair counts were traditionally sized to meet the service demand forecast for 20 years from the construction date. More recently, cable designs have been based on a shorter service capacity life. Also, the demand for more than one line per living unit has grown far beyond what was expected. As a result, there is a strong need to conserve wire pairs. This is addressed by ADSL's ability to convey POTS and data on one wire pair, and digital added main line (DAML) systems, which convey two or more POTS channels over one wire pair.

3.2.4   Wire Gauge

Most of the loop plant in the United States follows a practice called 1300 ohm resistance design. As a rule, the first 10,000 feet of cable from the CO is 26 American Wire Gauge (AWG, a measure of wire diameter). Beyond this point, successively heavier gauge wire is used to avoid excessive loop resistance. Overall, the loop plant consists of the same amounts of 26 AWG and 24 AWG and approximately equal amounts of overhead (aerial) and buried wire. Very long loops will have some 22 or 19 AWG wire. Wire is commonly obtained on 500-foot spools. As a result, a typical loop may have 22 splices. Modern splices use a compression device to assure a solid connection without labor-intensive soldering. Older splices, where two wires are twisted to form a joint, can loosen and corrosion at the junction can develop high resistance and even act as a diode due to a layer of copper oxide between the wires. This phenomenon is reduced by sealing current (see Section 3.3).

Signal reflections can result from the impedance change due to splicing one wire gauge to a different wire gauge. Longer loops may have several changes in wire gauge. The degree of transmission impairment caused by a gauge change is debated. Most experts believe that DSLs with echo cancelers tolerate gauge changes well, and the effects of a gauge change are small enough to be ignored.

Outside of the United States, wire diameters are measured in millimeters, with the commonly used gauges corresponding approximately to the American Wire Gauge diameters. The following table provides the resistance of a loop at a temperature of 70°F. Loop resistance varies with temperature, for example a 26 AWG loop has a resistance of 373 ohms/mile at 0°F, and 489 ohms/mile at 120°F. Loop resistance is the total of the round-trip circuit for one conductor out and one conductor back.

AWG	Metric Size (mm)	Loop Resistance (ohms/mile)
28	0.32	685
26	0.4	441
24	0.5	277
22	0.63	174

3.2.5 Bridged Tap

In some countries, there is a common practice of splicing a branching connection (called a *bridged tap*) onto a cable, as in Figure 3.2. Thus, a bridged tap is a length of wire pair that is connected to a loop at one end and is unterminated at the other end. Approximately 80% of loops in the United States have bridged taps; sometimes several bridged taps exist on a loop. Bridged taps may be located near either end or at an intermediate point. One reason for a bridged tap is that it permits all the pairs in a cable to be used or reused to serve any customer along the cable route. Most countries in Europe claim to have no bridged taps, but there have been reports of exceptions. The reflection of signals from the unterminated bridged taps results in signal loss and distortion. The adaptive equalizer and echo canceler found in many DSLs partly reduce the transmission impairment caused by bridged taps. The worst-case bridged tap is a heavy-gauge tap of a length that equals one-quarter of the wavelength of any used transmission frequencies, causing an additional loss of 3 to 6 dB. The reflections from a quarter-wavelength tap are 180 degrees out of phase from the primary signal frequency and thus partly cancel the signal. DSLs can tolerate multiple bridged taps, provided that the combined signal loss due to loop length and bridged taps is within the system's loss budget. The effect of bridge taps can actually be evident at several frequencies (see Section 3.5).

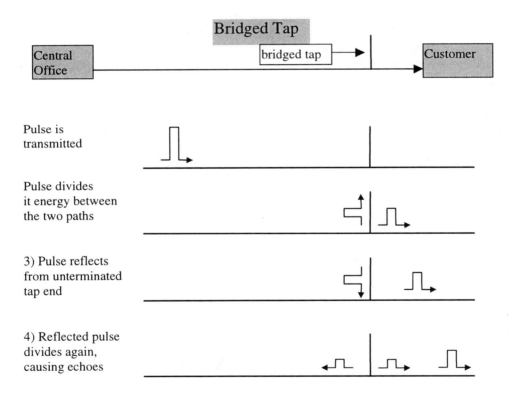

Figure 3.2   Effect of bridged tap on pulse propagation.

3.2.6   Loaded Loop

For loops beyond 5.5 km (18 kft), the signal loss at frequencies above 1 kHz is excessive, making voice transmission unacceptable. Series inductors (typically 88 mH) placed at 1.8 km (6 kft) intervals result in flatter frequency response across the voice band at the expense of much greater loss at frequencies above the voice band. As a result, DSLs will not operate on loaded loops. Figure 3.3 illustrates the effect of loading on frequency response. Depending on the region, 10% to 15% of loops in the United States have loading coils. In the 1970s, prior to the massive deployment of digital loop carrier, 20% of loops were loaded. In rare cases, loading coils are found on loops shorter than 5.5 km (18 kft). To permit DSL operation, loading coils may be removed. However, an expensive effort is required to find and remove the loading coils. In Europe, loops beyond 5.5 km (18 kft) are rarely found, so loading coils are not used.

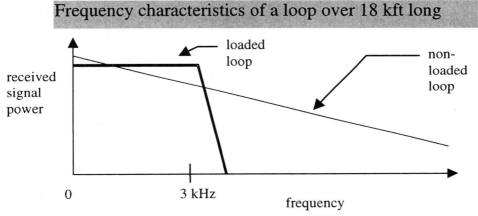

Figure 3.3 Loaded loop frequency response.

3.2.7   Loop Length Distribution

Central Offices are positioned as close as possible to customers. One-half of customers in the United States and the United Kingdom are served by loops shorter than two kilometers (6.6 kft). The graph in Figure 3.4 represents the distribution of all loops by country. Business loops tend to be shorter, and residential loops tend to be longer. The United States loop plant distribution tends to be longer than other countries. Note that the values are averages for each country and that there is a considerable variation of loop statistics by Central Office. For example, there are Central Offices that have no loops longer than 2.5 km (8 kft), other COs where most loops are longer than 4.6 km (15 kft), and some COs where the majority of loops are served by digital loop carrier. Very few copper loops exceed 30.5 km (100 kft).

Is the average loop length getting shorter? DLCs continue to be deployed widely to reduce the effective length of loops. However, this is offset by the growth pattern for new buildings, which tend to be at the edge of towns, far from the CO. Some new growth areas are served by DLCs or remote switch modules (RSMs). Overall, the distribution of loop lengths is changing very slowly towards shorter loops.

Local Loop Length Distribution

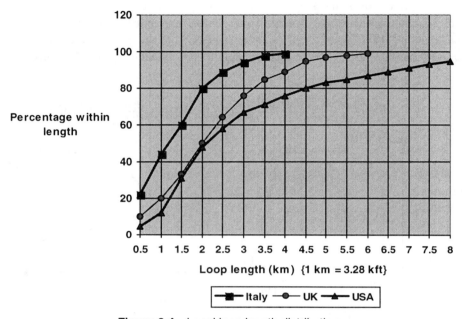

Figure 3.4   Local loop length distribution.
Source: *IEEE Communications,* March 1991.

3.2.8   Customer Premises Configuration

After its long and torturous journey from the CO to the customer premises, the DSL signal may find its greatest hurdle: the wiring within the customer's premises. For both residential and business premises, only one rule applies: *anything goes.*

The number of wire pairs may be anywhere from one to eight. Some of the wire pairs may not be connected at some wall jacks. The type of wire may be high-quality Category 5 unshielded twisted pair (UTP), shielded wire, quad wire, or flat (untwisted) wire. Quad wire consists of four insulated conductors twisted as a unit of four wires. Quad wire has high crosstalk between the wires within the quad. Flat wire is highly susceptible to the many types of electrical noise within the premises: light dimmers, electric motors, and radio transmitters. As in Figure 3.5, the wiring topology may be a star (also know as *home run*), a bus, a ring, or a combination of these configurations.

In the United States, it is not unusual for a home to have more than six phones connected to a line. In Europe, one or two phones per line is most common.

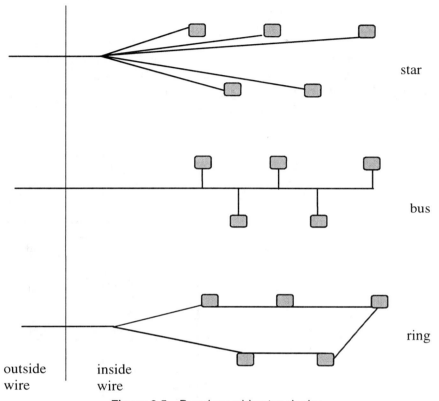

Figure 3.5 Premises wiring topologies.

The transmission impairments presented by low-quality premises wiring may be minimized by placing the DSL transceiver as close as possible to the point of entry. However, the point of entry is often an inconvenient location because it is not near the point of use, and local power may not be accessible at the point of entry. In some cases, it may be necessary to find an alternative media to convey the signals within the premises: new Category 5 UTP wire, coax cable, or wireless transmission.

Many newer buildings follow industry standards for inside wiring, but many existing buildings do not. Per TIA/EIA-568A, the standard premises wiring practice for telecommunications uses 24 AWG UTP Category 3 or 5 wire using a star topology. As discussed in T1E1.4/97-169, the pair-to-pair crosstalk and loss for UTP-3 is worse than typical outside plant telephone cables, whereas UTP-5 is as good or better than the outside plant telephone cable characteristics. Type D inside wire (DIW), which exists in older office buildings, has very poor crosstalk characteristics above 1 MHz.

DSLs are point-to-point transmission systems; thus the line connects to only one device at the customer end of the line. However, a user may require multiple PCs, phones, and other devices to communicate via a shared DSL, as in Figure 3.6. The DSL transmission unit at the customer site must perform a local fan-out function to enable the connection of multiple customer-end terminals (PCs, phones, etc.). For example, the DSL transmission unit may reside within a PC-NIC card, with the PC providing a gateway function that routes the traffic to other PCs connected via a LAN (e.g., 10baseT) connection.

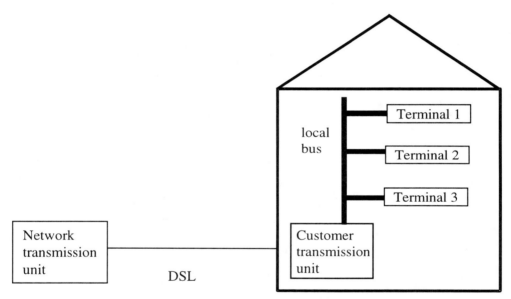

Figure 3.6   Customer premises local bus architecture.

A potential alternative to the conventional LAN method described previously could avoid the installation of new inside wire. A "phone LAN" adapter at each PC would connect to the in-home phone wiring using signals modulated in a frequency band well above 1.1 MHz to avoid the ADSL transmission band. A combined ADSL modem and "phone LAN" adapter (possibly located within a PC) would also connect to the home phone wiring. The ADSL modem would communicate with the network via the standard ADSL signals (in the 30 kHz to 1.1 MHz band), while simultaneously communicating with the other PCs in the home via the same inside wire via the LAN band (above 1.1 MHz). Ideally, this could be performed in addition to the traditional POTS in the frequency band below 4 kHz band. To avoid interference from the simultaneous use of the POTS band, ADSL band, and LAN band on the same inside wire, isolation filters might be needed at the phones and at the NID.

3.3 Line Powering

Mid-span repeaters must be line powered since no power source is available at repeater sites. The customer-end units for HDSL and some ISDN systems are line powered to assure a reliable power source and to reduce installation costs for equipment at the customer premises. Line power is usually provided from the CO as a negative DC voltage applied to one wire and ground to the other wire of a pair. Positive voltages are often avoided in line power feeding to prevent damage to copper conductors and associated apparatus due to electrolysis where there is moisture between the conductors and ground. In the United States, Bellcore GR-1089-CORE Class A3 states that up to 140 volts DC with respect to ground (positive and/or negative) may be applied to the line provided that the voltage is inaccessible for contact by the general public and untrained employees. Equipment with these voltages must have suitable safety warning labels, physical protection, and other safety features described in Bellcore GR-1089-CORE. As described in T1E1.4/96-110, power feeding with voltages up to 200 volts DC with respect to ground are allowed in Class A3 if the current to ground is limited to less than 10 mA. The design of the current limiting circuit must assure that the voltage is quickly turned off in the event of a current fault. However, the current limiting circuit should not be unnecessarily triggered by normal events such as induced noise.

In some places, a positive 130 volts is applied to one conductor and a negative 130 volts is applied to the other conductor. Bipolar power feeding can deliver far more power than unipolar feeding and may be used in places where moisture is not present in the cable. For example +/– 130 volt line power feeding is widely used for T1 repeater powering on pressurized cable. Pressurized cable is often found in the feeder plant but is less common in the distribution plant. High-pressure dry air is pumped into the cable continuously to force moisture out of the cable. Newer cables contain a gel filling agent within the cable to prevent the intrusion of moisture.

3.3.1   Activation/Deactivation

Feeding line power from the CO to thousands of lines, with over half of the power lost to loop resistance, can add up to an expensive CO power budget. Outside of North America, where ISDN CPE is often line powered, the ISDN CPE automatically enters a low-power mode when there is no communications activity. Deactivating the idle customer-end equipment saves considerable power. The deactivated equipment must quickly activate when communications activity resumes. The ability to enter a low-power mode and then quickly reactivate adds complexity to the transceivers. Some applications send messages periodically; this detracts from the potential power savings.

3.4 Sealing Current

Sealing current (also known as wetting current) is an electric current applied to a loop for the purpose of preventing transmission degradation due to the oxidation of wire splices. The oxide layer between wires that are not tightly bonded cause enough resistance to cause substantial signal loss. Furthermore, the nonlinear nature of the oxidized junction can cause signal distortion. In telephony parlance, a "wet" loop carries DC current, whereas a "dry" loop carries no DC current. There is no need for sealing current on POTS loops because the high ringing voltages and currents will break down oxidation on wire splices. DSLs that do not provide line powering can be subject to splice oxidation. ANSI T1.601 specifies the optional use of a current of 1 to 20 mA for the purpose of preventing splice oxidation. The current may be applied continuously or applied periodically for a short duration. Some studies have suggested that sealing current may not be helpful. Thus, ANSI T1.601 does not require the use of sealing current.

3.5 Transmission Line Characterization

This section directly addresses the transmission characteristics of twisted-pair phone lines.

Category 3 twisted-pair phone lines can be well modeled for transmission at frequencies up to at least $f < 30\ MHz$ by using what is known as **two-port modeling** or **"ABCD" theory.** ABCD theory will also be used to model the three-port networks[1] discussed in Section 3.9. Such ABCD theory is well covered in basic electromagnetic texts (see Ref. [1]) but is often not in a form convenient for use in DSLs. Werner presented essential results of such translation to DSLs in Ref. [2], and this section essentially repeats that effort but provides more detail along with updates based on various studies in standards groups that have led to DSL characterization to 30 MHz on Category 3 wiring.[2] Category 5 wiring appears to follow ABCD modeling to 150 MHz.

Section 3.5.1 describes ABCD modeling in general whereas Section 3.5.2 concentrates on the case of twisted-pair transmission lines. Section 3.5.2 also introduces the important concept of **return loss,** a measure of reflected energy from a two-port circuit. Sections 3.5.3 and 3.5.4 consider special cases of bridge taps and loading coils. Section 3.5.5 shows how to compute the transfer characteristics of a subscriber loop consisting of many sections and to relate transfer functions to the often-measured **insertion loss** of a transmission line or two-port. Section 3.5.7 shows how to measure RLCG parameters for loop characterization and lists models for several popular twisted-pair types. Section 3.5.8 concludes with a discussion of balance and the related metallic and logitudinal signal components on a twisted pair.

3.5.1 "ABCD" Modeling

Figure 3.7 shows a general two-port linear circuit. There is a voltage at each port and a current into, or out of, the upper path of each port. The figure and resulting equations use Fourier Transforms of voltage and current, and thus all quantities are functions of frequency in general.

---

1.   Examples include hybrid circuits, POTS splitters, and bridge taps.
2.   Category 3 twisted pair is actually for computer networks and specified by the Electronics Industries Association (EIA) [3], but is generally believed similar in characteristics and twist to phone wiring, having a twist every 2 to 6 inches (cables of phone lines vary the twist to reduce crosstalk). Category 5 is a higher grade of twisted pair with more twisting and about 20 dB less crosstalk that is often used in local area computer networks or premises wiring.

The voltages and currents will depend on the source (port 1) and load (port 2) impedances and voltage source(s), but nevertheless always satisfy the matrix relationship:

$$\begin{bmatrix} V_1 \\ I_1 \end{bmatrix} = \begin{bmatrix} A & B \\ C & D \end{bmatrix} \cdot \begin{bmatrix} V_2 \\ I_2 \end{bmatrix} = \Phi \cdot \begin{bmatrix} V_2 \\ I_2 \end{bmatrix} \qquad \text{or} \qquad \begin{array}{l} V_1 = AV_2 + BI_2 \\ I_1 = CV_2 + DI_2 \end{array} \tag{3.1}$$

where Φ is a 2×2 matrix (nonsingular in all but trivial situations not of interest) of the four frequency-dependent parameters, $A, B, C,$ and D, which all depend only on the network and not on external connections. The quantities have circuit definitions as in Table 3.1:

Table 3.1 ABCD Definitions

A	open-load voltage ratio, $\left.\dfrac{V_1}{V_2}\right|_{I_2=0}$
B	shorted-load impedance, $\left.\dfrac{V_1}{I_2}\right|_{V_2=0}$
C	open-load admittance, $\left.\dfrac{I_1}{V_2}\right|_{I_2=0}$
D	shorted-load current ratio, $\left.\dfrac{I_1}{I_2}\right|_{V_2=0}$

The transformation is reversed by Φ^{-1} so that

$$\begin{bmatrix} V_2 \\ I_2 \end{bmatrix} = \frac{1}{AD - BC} \begin{bmatrix} D & -B \\ -C & A \end{bmatrix} \cdot \begin{bmatrix} V_1 \\ I_1 \end{bmatrix} = \Phi^{-1} \cdot \begin{bmatrix} V_1 \\ I_1 \end{bmatrix} \tag{3.2}$$

When $\Phi = I$, i.e., an identity, the network is a trivial connection of the upper path and lower path across the network, essentially meaning there is no network. A relationship of interest is the ratio

$$T(f) = \frac{V_2}{V_1} = \frac{V_2}{A \cdot V_2 + B \cdot I_2} = \frac{1}{A + B\frac{I_2}{V_2}}$$

(3.3)

where the frequency dependence is shown explicitly for $T(f)$, but not for the other voltages to simplify notation. This ratio depends on the load impedance attached at port 2, or the ratio $Z_L = Z_2 = V_2/I_2$

$$T(f) = \frac{1}{A + B\Big/Z_L} = \frac{Z_L}{A \cdot Z_L + B}$$

(3.4)

$T(f)$ can be related to a **transfer function** $H(f)$ between an input voltage supply V_s (with finite internal impedance Z_s) to the output voltage $V_L = V_2$ (across a load $Z_L = Z_2$).

$$\frac{V_L(f)}{V_s(f)} = H(f) = \frac{V_L(f)}{V_2(f)} \cdot \frac{V_2(f)}{V_s(f)} = \frac{Z_1}{Z_1 + Z_s} \cdot T(f)$$

(3.5)

where $Z_1 = V_1/I_1$ is the input impedance of the terminated two-port. The transfer function generally depends on the load and source impedances, Z_L and Z_s. Sometimes this dependency may be tacitly obscure in that expressions are written that assume $Z_s = Z_L = Z_0$ or that do not explicitly show this dependence. Z_1 can be computed as in Equation (3.7) and is the ratio of input voltage to current when load Z_L is attached at the output.

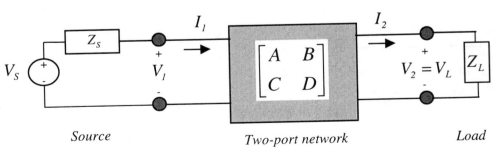

Figure 3.7 Two-port network model.

A cascade of two-ports has a two-port matrix that is the product, in order, of the two ports

$$\begin{bmatrix} V_1 \\ I_1 \end{bmatrix} = \Phi_1 \cdot \Phi_2 \cdots \Phi_{N-1} \cdot \begin{bmatrix} V_N \\ I_N \end{bmatrix} = \Phi \cdot \begin{bmatrix} V_N \\ I_N \end{bmatrix}$$

(3.6)

allowing for the calculation of transfer functions, and insertion losses of more complicated networks as long as a two-port model can be found for each subsection in the cascade. The inverse is found by reversing the order and taking the product of the inverse matrices. The input impedance of the two-port is

$$Z_1 = \frac{V_1}{I_1} = \frac{A + B/Z_L}{C + D/Z_L} = \frac{AZ_L + B}{CZ_L + D}$$

(3.7)

Two-port networks are very useful in the analysis of twisted-pair transmission lines as will be shown in the next several sections. In these sections, the transmission line is modeled as a cascade of two ports that are characterized by resistance, inductance, capacitance, and conductance per unit length, and by the length of the transmission-line segment.

3.5.2 Transmission Line *RLCG* Characterization

The two-port characterization of a transmission line derives from the per-unit-length two-port model shown in Figure 3.8. The *R*, *L*, *C*, and *G* parameters represent resistance, inductance, capacitance, and conductance per unit length of the transmission line.

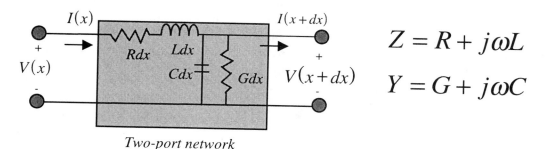

Two-port network

Figure 3.8 Incremental section of twisted-pair transmission line.

A segment of transmission line can be viewed as a cascade of such sections that are infinitesimally small in length. At any point *x*, the two-port voltages and currents relate through the differential equations

$$-\frac{dV}{dx} = (R + j\omega L) \cdot I$$

$$-\frac{dI}{dx} = (C + j\omega C) \cdot V \tag{3.8}$$

at any given frequency $\omega = 2\pi f$. V and I are **phasor** quantities representing peak amplitudes of sinusoids at frequency f (or amplitudes of the complex exponential $e^{j2\pi ft}$). The R, L, C, and G parameters themselves can vary with frequency, but are presumed constant with respect to length at any given frequency in the analysis to follow. This set of differential equations is equivalent to the pair of second-order differential equations with solutions:

$$\frac{d^2 V}{dx^2} = \gamma^2 \cdot V$$

$$\frac{d^2 I}{dx^2} = \gamma^2 \cdot I \tag{3.9}$$

where

$$\gamma = \alpha + j\beta = \sqrt{(R + j\omega L) \cdot (G + j\omega C)} = \sqrt{Z \cdot Y} \tag{3.10}$$

is the frequency-dependent **propagation constant** for the twisted pair and characterizes the segment of transmission line. The **impedance per unit length,** Z, and the **admittance per unit length,** Y, are also defined in Figure 3.8. The solutions to the differential equations is a sum of positive- and negative-going waves that vary with position as $e^{\pm \gamma x}$, with the sign of the exponent depending on direction (positive to the right and negative to the left). The propagation constant has a real part called the **attenuation constant,** α, and an imaginary part called the **phase constant,** β. When the attenuation constant is zero, the line is lossless $(R = G = 0)$. The attenuation constant is very important for twisted-pair DSLs, as they are not lossless unlike many other transmission lines, and thus $\alpha \neq 0$. As can be inferred from equations to come, the attenuation of a twisted-pair is approximated by 8.668α dB per unit length at the frequency of interest. The phase constant is related to speed of propagation on the twisted pair. At each frequency $\omega = 2\pi f$, a sinusoid propagates on the twisted pair with phase given by

$$\theta(\omega, x) = \omega t - \beta x \tag{3.11}$$

and has envelope amplitude attenuated as $e^{-\alpha x}$. The **wavelength** is the length (at fixed frequency and time) over which the sinusoid undergoes a full cycle and is thus given by

$$\lambda = \frac{2\pi}{\beta}$$

$$(3.12)$$

Remembering that β is tacitly a function of frequency, different frequencies thus have different wavelengths. The sinusoidal wave at frequency ω appears to propagate along the twisted pair at **phase velocity:**

$$v_p = \frac{\omega}{\beta}$$

$$(3.13)$$

and the **phase delay** per unit length at this same frequency is $\tau_p = 1/v_p = \beta/\omega$. When β is a linear function of frequency, the channel is said to have **linear phase** and the phase velocity and delay are constant over all frequencies. An example is the case where $R = G = 0$, and then $\beta = \omega\sqrt{LC}$ and (when L and C are constant with respect to frequency) means that all frequencies move at the same phase velocity $v_p = 1/\sqrt{LC}$. Such a transmission line is said to be **dispersionless.** In practical DSLs, this situation never occurs and different frequencies travel with different velocities, leading to **dispersion** of signal energy (and to the intersymbol interference discussed in Chapter 7). For a dispersive transmission line, it is of interest to investigate the relative speed at which a group of frequencies centered around ω propagates. To understand this concept of "group" or "envelope" velocity, suppose one investigates the differing speeds of the two frequencies $\omega \pm \Delta\omega$ where the offset or difference is small and the corresponding values of $\beta \pm \Delta\beta$. The resultant sum waveform is:

$$A\cos[(\omega + \Delta w)t - (\beta + \Delta\beta)x] + A\cos[(\omega - \Delta w)t - (\beta + -\Delta\beta)x]$$
$$= 2A\cos[\Delta w \cdot t - \Delta\beta \cdot x] \cdot \cos[\omega t - \beta x]$$

$$(3.14)$$

The right side of Equation (3.14) is an "envelope-modulated" sinusoid, a product of two sinusoids. When the phase velocity is constant and there is no dispersion, the phase velocity of the first term on the right in Equation (3.14) is the same as that of the second term, and the phase velocity equals the group velocity. However, when phase velocity is not constant, the first term moves at a different (often much slower) speed given by $\Delta\omega/\Delta\beta$. This slower speed is the **group velocity** and in general is computed by the inverse of the **group delay:**

$$\tau_g = \frac{d\beta}{d\omega} \qquad \text{or} \qquad v_g = \frac{1}{\tau_g} = \frac{d\omega}{d\beta}$$

$$(3.15)$$

Group delay in essence measures the spread in delay between the fastest and slowest moving frequencies in the immediate vicinity of ω. The greater the group delay, the greater the dispersion in the transmission line. The solution to the set of differential equations in Equation (3.9) is easily modeled as the sum of two opposite-direction voltage/current waves:

$$V(x) = V_0^+ \cdot e^{-\gamma x} + V_0^- \cdot e^{\gamma x}$$
$$I(x) = I_0^+ \cdot e^{-\gamma x} + I_0^- \cdot e^{\gamma x} \tag{3.16}$$

By insertion of either of these solutions into the appropriate first-order voltage/current differential equations in Equation (3.8), the ratio of the positive-going voltage to the positive-going current, as well as the (negative of the) ratio of the negative-going voltage to the negative-going current, is equal to a constant **characteristic impedance** of the transmission line:

$$Z_0 = \frac{V_0^+}{I_0^+} = -\frac{V_0^-}{I_0^-} = \sqrt{\frac{R + j\omega L}{G + j\omega C}} = \sqrt{\frac{Z}{Y}} \tag{3.17}$$

One easily verifies that the R, L, C, and G parameters are equal to:

$$R = \Re\{\gamma \cdot Z_0\}$$
$$L = \frac{1}{\omega}\Im\{\gamma \cdot Z_0\}$$
$$C = \frac{1}{\omega}\Im\left\{\frac{\gamma}{Z_0}\right\}$$
$$G = \Re\left\{\frac{\gamma}{Z_0}\right\} \tag{3.18}$$

For twisted-pair transmission and DSLs, it is rare that any of these four parameters are zero and so simplifications in textbooks or other developments that lead to so-called "lossless transmission lines" or "dispersionless" transmission are not of interest for DSLs. Furthermore, these parameters are frequency dependent for transmission lines and are best determined by measurement as in Section 3.5.7.

A segment of transmission line of length d has the solution $V_L = V_d$ and $I_L = I_d$ and thus:

$$V_L = V(d) = V_0^+ \cdot e^{-\gamma l} + V_0^- \cdot e^{\gamma l}$$
$$I_L = I(d) = I_0^+ \cdot e^{-\gamma l} + I_0^- \cdot e^{\gamma l} \tag{3.19}$$

Since the two voltage waves in each direction are related to the same-direction current waves by the common ratio Z_0, one can solve the above two equations for V_0^+ and V_0^- to get:

$$V_0^+ = \tfrac{1}{2}\left(V_L + I_L \cdot Z_0\right) \cdot e^{\gamma l}$$
$$V_0^- = \tfrac{1}{2}\left(V_L - I_L \cdot Z_0\right) \cdot e^{-\gamma l}$$

(3.20)

By substituting these constants back into the solution in general and evaluating for the voltage and currents at $x = 0$ in terms of those at $x = d$, one obtains the following two-port representation

$$\begin{bmatrix} V(0) \\ I(0) \end{bmatrix} = \begin{bmatrix} \cosh(\gamma d) & Z_0 \cdot \sinh(\gamma d) \\ \dfrac{1}{Z_0} \cdot \sinh(\gamma d) & \cosh(\gamma d) \end{bmatrix} \cdot \begin{bmatrix} V(d) \\ I(d) \end{bmatrix}$$

(3.21)

The ABCD entries can be read from the matrix or, equivalently, can be computed from the *R, L, C, G* values through in relations for γ in Equation (3.10) and for Z_0 in Equation (3.17). Then, for a given length of transmission line *d*, the engineer may model that transmission line as a single "lumped" two-port, replacing the distributed model in Figure 3.8. Knowing the load impedance so that $V(d)/I(d) = Z_L$, $T(f)$ then becomes

$$T = \frac{1}{\cosh(\gamma d) + \left(Z_0 \big/ Z_L\right) \cdot \sinh(\gamma d)}$$

(3.22)

The input impedance of the two-port is $V(0)/I(0)$, or

$$Z_1 = Z_0 \cdot \frac{Z_L + Z_0 \cdot \tanh(\gamma d)}{Z_0 + Z_L \cdot \tanh(\gamma d)} = \frac{Z_L + Z_0 \cdot \tanh(\gamma d)}{1 + \dfrac{Z_L}{Z_0}\tanh(\gamma d)}$$

(3.23)

The input impedance of a very long line reduces to $Z_1 = Z_0$, since $\tanh(\gamma d) \to 1$ for large *d*, or when $Z_0 = Z_L$.

The transfer function in any case becomes

$$H = \frac{Z_1}{Z_1 + Z_S} T = \frac{Z_0 \cdot \operatorname{sech}(\gamma d)}{Z_S \cdot \left[\frac{Z_0}{Z_L} + \tanh(\gamma d)\right] + Z_0 \cdot \left[1 + \frac{Z_0}{Z_L} \cdot \tanh(\gamma d)\right]}$$

(3.24)

Thus, this type of model applies to the upper example in Figure 3.10. Note also the two-port models that characterize the source and load. Thus, the cascade of two-ports studied in general in Equation (3.6) can be directly applied. If several transmission line segments with different R, L, C, and G were cascaded, then each would have its own two-port model. This situation corresponds to connection of twisted pairs (splicing) with different gauges.

3.5.2.1 Power for Transmission Lines

A sinusoid at any frequency on a transmission line represented by the peak phasor voltage V and (peak) phasor current I has average (rms) power

$$P(f) = \tfrac{1}{2}\Re\{VI^*\}$$

(3.25)

Figure 3.9 shows a simple circuit having input current I and voltage V across a load with impedance $Z_L = R_L + jX_L$. From basic circuit theory, a sinusoidal current with peak amplitude $|I|$ delivers power

$$P(f) = \tfrac{1}{2}|I|^2 R_L = \tfrac{1}{2}\left|\frac{V}{Z_L}\right|^2 R_L = \tfrac{1}{2}\Re\{VI^*\}$$

(3.26)

thus providing interpretation for the relation in Equation (3.25).

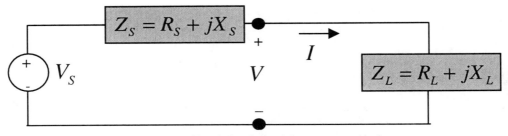

Figure 3.9 Simple load circuit for power analysis.

Maximum power is transferred from the power supply to the load when the source impedance is the conjugate of the load impedance shown in Figure 3.9, $Z_{S,opt} = Z_L^* = R_L - jX_L$. This corresponds to one-half the total power of the source being dissipated in the load. An example of the use of this maximum-power-transfer result is when one investigates the termination of a twisted-pair transmission line. To transfer maximum power from the line to the load, the load impedance should be designed to be the conjugate of the line impedance viewed going back into the line. When the line is long, this impedance will be the characteristic impedance of the line itself, meaning the best loading is

$$Z_{L,opt} \cong Z_0^*$$

(3.27)

meaning half the power in the line is transferred to the load (with the other half dissipated within the line itself). Similarly, the optimum driving impedance is the conjugate of the line impedance, which again for long lines is the characteristic impedance, so

$$Z_{S,opt} = Z_{L,opt} \cong Z_0^*$$

(3.28)

again, half the source power will be delivered to the line. For a lossless transmission line, the half of the source power delivered to the line is the same half of power delivered to the load. At higher frequencies, all transmission lines appear as purely resistive Z_0, and so the best load and source impedances become resistive and equal to the (real) characteristic impedance of the line. A purely resistive Z_0, however, should not be confused with a lossless transmission line (the latter has purely resistive Z_0 and $\alpha = 0$). Twisted pairs have resistive Z_0 at high frequencies, but large α.

The condition for maximum power transfer is not the same condition for elimination of reflections (see Section 3.5.2.2) unless the line has purely real characteristic impedance.

3.5.2.2 Reflection Coefficients and Return Loss

When the load impedance is equal to the characteristic impedance (*and not the conjugate of the characteristic impedance*), there is no negative-going wave, and $V_0^- = 0$ in Equations (3.18) and (3.19). There is then no reflected wave, and all of the above relationships simplify somewhat. In practice, such matching is not likely to occur, and the solution for the differential equation at $x = d$ has a general ratio of positive-going wave to negative-going wave of

$$\rho = \frac{V_0^- \cdot e^{-\gamma d}}{V_0^+ \cdot e^{\gamma d}} = \frac{Z_L - Z_0}{Z_L + Z_0}$$

(3.29)

This **reflection coefficient** is clearly zero when the transmission line is "matched" or terminated in its own impedance, $Z_L = Z_0$. This situation prevents "bouncing" of signals on a transmission line and thus reduces the dispersion (relative delay) of signals on the line. In this case of $Z_L = Z_0$, the input impedance is then also $Z_1 = Z_0$. When the transmission line impedance is approximately real, then the situation of no bouncing corresponds also to maximum energy transfer in Section 3.5.2.1 from the line into the matched load. However, when (as usual for twisted pairs) the line characteristic impedance is complex, then maximum energy transfer occurs when the load is the conjugate of the characteristic impedance, and thus elimination of bouncing does not guarantee maximum energy transfer for lossy lines. On many lines as the frequency increases, the R and G terms become negligible and so for these frequencies, maximum energy transfer and elimination of bouncing occur when the load impedance is matched to $Z_L = Z_0 \approx \sqrt{L/C}$.

A similar **source reflection coefficient** can be written as

$$\rho_S = \frac{Z_S - Z_0}{Z_S + Z_0}$$

(3.30)

This source reflection coefficient measures the reflected positive-going wave amplitude with respect to a negative-going wave that flows into the source impedance. Note that the source impedance that leads to maximum power transfer into the line $Z_S = Z_1^*$ again is not necessarily the same as that leading to no reflection at the source end. A wave launched from a source will traverse the loop with phase and group velocities, will be reflected at one end, reflected again at the source end, and so on. This series of reflections leads to a transient on the loop, unless the loop is terminated in a load impedance equal to the characteristic impedance of the line. Again when the line can be approximated over the used frequency range as lossless, and thus having real characteristic impedance, then the maximum energy transfer and reduction of bouncing objectives coincide.

The **return loss** of a transmission line or a two-port is the inverse ratio of reflected power to incident power on the load (or next section of circuitry). This return loss is simply the square of the reflection coefficient, thus

$$\text{return loss} = 10\log_{10}\left(\frac{1}{\rho}\right)^2 \text{ dB}$$

(3.31)

3.5.3   Characterization of a Bridged-Tap Section

For modeling of loops, a bridge tap can be viewed as a three-port section, but one of the ports appears as a load impedance to the line between the two sections on each side of the bridged tap. Such a situation can be modeled by the two-port with ABCD matrix shown in the example of Figure 3.10, i.e.,

$$\Phi_2 = \begin{bmatrix} 1 & 0 \\ 1/Z_{bt} & 1 \end{bmatrix}$$

(3.32)

Where the impedance of the bridged-tap section is computed according to Equation (3.7) with ABCD as in Equation (3.21) for the bridged-tap two-port line model.

The impedance of the tap section Z_t can also be computed according to the formula in Equation (3.22) for the input impedance of a section of transmission line terminated with an open circuit, which simplifies to

$$Z_t = Z_{0t} \cdot \frac{\cosh(\gamma d)}{\sinh(\gamma d)}$$

(3.33)

if the tap was not terminated in an open circuit, then the general formula in Equation (3.23) for the input impedance of the section can also be used. The input impedance for a lossless tap section simplifies to $jZ_{0t}/\tan(\beta d)$, which oscillates between a short circuit and an open circuit as $\beta d = (k + \frac{1}{2})\pi$ and $\beta d = k\pi$, respectively.

Circuits with bridged taps on bridged taps have an impedance that is calculated by working backwards from all open taps to the line, each modeled as the two-tap section's impedances in parallel with the connecting sections of the line. The resultant impedance then becomes a termination (load) impedance for the next section working backwards toward the main transmission pair of interest. While perhaps tedious, the calculation process is straightforward and recursive.

3.5.4   Load Coils — Series Inductance

A load coil is a series (high-Q) inductance placed between two sections of twisted pair in a phone line, typically 88 mH, as discussed earlier. The frequency at which $\omega = 1/\sqrt{LC}$, where C is the line capacitance, is roughly the frequency that will be boosted. The two-port model for such a circuit is

$$\Phi_{coil} = \begin{bmatrix} 1 & j\omega L_{coil} \\ 0 & 1 \end{bmatrix}$$

(3.34)

3.5.5   Computation of Transfer Function

The computation of the transfer function for twisted-pair transmission lines with multiple sections then simply becomes a process of multiplying in cascade the corresponding two-port ABCD matrices for each section. Some examples are provided in Figure 3.10, with the corresponding two-port matrices below each example. The matrices are multiplied left to right in the natural order of appearance in the figure. That is, the overall two port is just

$$\Phi = \Phi_0 \cdot \Phi_1 \cdot \ldots \cdot \Phi_N$$

(3.35)

where the source voltage divider is modeled by the two-port

$$\Phi_0 = \begin{bmatrix} 1 & Z_S \\ 0 & 1 \end{bmatrix}$$

(3.36)

The final output voltage and current are related by the usual $V_L = I_L \cdot Z_L$, which allows the transfer function to be computed from the ratio from Equation (3.24):

$$H = \frac{V_L}{V_S}$$

(3.37)

or in terms of the ABCD entries for Φ in Equation (3.35),

$$H = V_L/V_S = Z_L/(A \cdot Z_L + B + C \cdot Z_S \cdot Z_L + D \cdot Z_S)$$

In the upper example of Figure 3.10 a simple section of twisted pair with characteristic impedance Z_0 and propagation constant γ is modeled by the cascade of a two-port matrix description Φ_1 for a length d and the source two-port matrix Φ_0. This upper example is a straightforward application of the two-port theory. The lower example additionally has a bridged-tap section with Z_{02} and γ_2 of length d_2 and a second section of the transmission line with yet a third characteristic impedance, Z_{03}, and propagation constant, γ_{03}. The two sections of transmission line are modeled as usual, where the impedance and propagation constant can be computed for each frequency from the known R, L, C, G parameters for each section. The bridged-tap section is modeled as a parallel (shunt) impedance that is computed according to the formula for an open-ended transmission line of length d_2 (if the tap were terminated, the impedance shown need only be replaced by the more general expression for the inverse of the input impedance of that section). The overall two-port matrix is simply the product of the four two-port matrices shown.

A variety of simplifications are sometimes studied, assuming each section is very long and so appears to be terminated in its own characteristic impedance, leading to expressions for the transfer function and input impedance in various situations. While sometimes useful for interpretation, with modern day signal processing analysis tools (i.e., matlab), it is often easier to compute the transfer function without simplifying assumptions and then analyze the corresponding results.

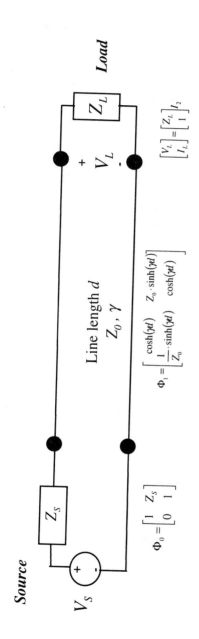

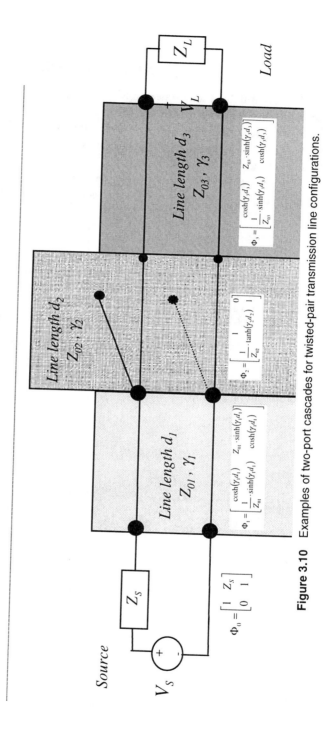

Figure 3.10 Examples of two-port cascades for twisted-pair transmission line configurations.

77

3.5.6 Measurements for Computation of *RLCG* Parameters

The four transmission-line parameters are often expressed as the resistance R in ohms/km, the inductance L in Henrys/km, the capacitance C in Farads/km, and the conductance G in Siemens/km. These parameters are best found for twisted pairs through measurement, and fortunately, standards groups and phone companies publish the characteristics for use by transmission designers.

The *RLCG* parameters in this section were provided by the following recommended measurement and curve-fitting procedure (extracted from a recent VDSL system requirements contribution of ANSI [4]).

3.5.6.1 Measurement Procedure

Looking towards load, the open-circuit impedance, Z_{OC}, and short-circuit impedance, Z_{SC}, for a length, l, of a twisted-pair transmission line are measured versus frequency. An $l = 10$ m length is used for measurements below $f = 2$ MHz and an $l = 1$ m length is used for measurements between $f = 2$ MHz and $f = 30$ MHz. The characteristic impedance and propagation constant are computed from the measured impedance according to characteristic impedance:

$$Z_0 = \sqrt{Z_{OC} \cdot Z_{SC}}$$

(3.38)

and propagation constant:

$$\gamma = \frac{1}{l} \tanh^{-1}\left(\sqrt{\frac{Z_{SC}}{Z_{OC}}} \right)$$

(3.39)

From the characteristic impedance and propagation constant, R, L, C, and G can be computed as:

$$R = \Re(\gamma Z_0)$$

(3.40)

$$L = \frac{1}{\omega} \Im(\gamma Z_0)$$

(3.41)

$$C = \frac{1}{\omega} \Im\left(\frac{\gamma}{Z_0} \right)$$

(3.42)

$$G = \Re\left(\frac{\gamma}{Z_0} \right)$$

(3.43)

3.5.6.2 Curve-Fitting

Because of error in practical measurements of the impedance, the measured *RLCG* values may not follow smooth curves with frequency, so parametrized (smooth) models of *RLCG* are then fit to the measured values. The models are:

$$R(f) = \frac{1}{\frac{1}{\sqrt[4]{r_{OC}^4 + a_c \cdot f^2}} + \frac{1}{\sqrt[4]{r_{OS}^4 + a_s \cdot f^2}}}$$

(3.44)

where r_{Oc} is the copper DC resistance and r_{Os} is (any) steel DC resistance, while a_c and a_s are constants characterizing the rise of resistance with frequency in the "skin effect."

$$L(f) = \frac{l_0 + l_\infty \left(f/f_m\right)^b}{1 + \left(f/f_m\right)^b}$$

(3.45)

where l_0 and l_∞ are the low-frequency and high-frequency inductance, respectively and b is a parameter chosen to characterize the transition between low and high frequencies in the measured inductance values.

$$C(f) = c_\infty + c_0 \cdot f^{-c_e}$$

(3.46)

where c_∞ is the "contact" capacitance and c_0 and c_e are constants chosen to fit the measurements.

$$G(F) = g_0 \cdot f^{+g_e}$$

(3.47)

where g_0 and g_e are constants chosen to fit the measurements.

Several types of twisted pair have been characterized in terms of these measure parameters.

Table 3.2   26-Gauge AWG Twisted Pair

Resistance (value)	r_{0c} 286.17578 Ω/km	r_{0s} ∞ Ω/km	a_c 0.14769620	a_s 0.0
Inductance (value)	l_o 675.36888 µH/km	l_∞ 488.95186 µH/km	b 0.92930728	f_m 806.33863 kHz
Capacitance (value)	c_∞ 49 nF/km	c_0 0.0 nF/km	c_e 0.0	
Conductance (value)	g_0 43 nS/km	g_e 0.70		

Some example values corresponding to Table 3.2 are

Frequency (Hz)	Resistance (Ω/km)	Inductance (H/km)	Capacitance (F/km)	Conductance (S/km)
5000	286.21516	673.7277e-6	49.e-9	16.701192e-6
10000	286.3332	672.26817e-6	49.e-9	27.131166e-6
20000	286.8039	669.55152e-6	49.e-9	44.074709e-6
50000	290.03566	662.28605e-6	49.e-9	83.70424e-6
100000	300.77488	651.94136e-6	49.e-9	135.97794e-6
1.e6	626.85069	572.86886e-6	49.e-9	681.50407e-6
10.e6	1.9606119e3	505.33352e-6	49.e-9	3.4156114e-3
10.5e6	2.0090081e3	504.66857e-6	49.e-9	3.5342801e-3
30.e6	3.3955368e3	495.20494e-6	49.e-9	7.3697598e-3

Table 3.3   24-Gauge AWG

Resistance (value)	r_{0c} 174.55888 ohms/km	r_{0s} ∞ ohms/km	a_c 0.053073481	a_s 0.0
Inductance (value)	l_o 617.29539 µH/km	l_∞ 478.97099 µH/km	b 1.1529766	f_m 553.760 kHz
Capacitance (value)	c_∞ 50 nF/km	c_0 0.0 nF/km	c_e 0.0	
Conductance (value)	g_0 234.87476 fS/km	g_e 1.38		

Frequency (Hz)	Resistance (Ω/km)	Inductance (H/km)	Capacitance (F/km)	Conductance (S/km)
5000	174.62121	616.69018e-6	50.e-9	29.882364e-9
10000	174.8078	615.95674e-6	50.e-9	77.774343e-9
20000	175.54826	614.35345e-6	50.e-9	202.42201e-9
50000	180.48643	609.15855e-6	50.e-9	716.82799e-9
100000	195.44702	600.41634e-6	50.e-9	1.8656765e-6
1.e6	482.06141	525.43983e-6	50.e-9	44.754463e-6
10.e6	1.5178833e3	483.72215e-6	50.e-9	1.0735848e-3
30.e6	2.6289488e3	480.34357e-6	50.e-9	4.8894913e-3

Table 3.4   Drop-Wire 10" (British Reinforced 0.5 mm copper PVC-insulated conductors, PVC-insulated steel strength member, Polyethylene sheath)

Resistance	r_{0c}	r_{0s}	a_c	a_s
(value)	180.93 ohms/km	∞ ohms/km	0.0497223	0
Inductance	l_o	l_∞	b	f_m
(value)	728.87 µH/km	543.43 µH/km	0.75577086	718888 Hz
Capacitance	c_∞	c_0	c_e	
(value)	51 nF/km	63.8 nF/km	0.11584622	
Conductance	g_0	g_e		
(value)	89 nMho/km	0.856		

Frequency (Hz)	Resistance (Ω/km)	Inductance (H/km)	Capacitance (F/km)	Conductance (S/km)
5000	180.98245	724.62777e-6	74.722723e-9	130.65969e-6
10000	181.13951	721.81902e-6	72.886768e-9	236.50605e-6
20000	181.76372	717.26896e-6	71.192474e-9	428.0977e-6
50000	185.96294	707.04768e-6	69.151688e-9	938.00385e-6
100000	199.01927	694.78496e-6	67.745589e-9	1.6978732e-3
1.5e6	579.72026	611.02577e-6	63.21713e-9	17.246997e-3
10.e6	1.493348e3	565.7413e-6	60.79255e-9	87.504681e-3
30.e6	2.5864318e3	553.86667e-6	59.613734e-9	224.11821e-3

Table 3.5   Flat Pair (No Twists)

Resistance	r_{0c}	r_{0s}	a_c	a_s
(value)	41.16 ohms/km	∞ ohms/km	0.001218	0
Inductance	l_o	l_∞	b	f_m
(value)	1000 µH/km	911 µH/km	1.195	174.2 kHz
Capacitance	c_∞	c_0	c_e	
(value)	22.68 nF/km	31.78 nF/km	0.1109	
Conductance	g_0	g_e		
(value)	53 nMho/km	0.88		

Frequency (Hz)	Resistance (Ω/km)	Inductance (H/km)	Capacitance (F/km)	Conductance (S/km)
5000	41.268736	998.73982e-6	35.041871e-9	95.360709e-6
10000	41.589888	997.16583e-6	34.127572e-9	175.49949e-6
20000	42.805363	993.76481e-6	33.280903e-9	322.98493e-6
50000	49.316246	983.62766e-6	32.257008e-9	723.38496e-6
100000	62.284991	969.66713e-6	31.548702e-9	1.3312998e-3
1.e6	186.92411	920.40732e-6	29.550852e-9	10.098942e-3
10.e6	590.76171	911.20963e-6	28.003118e-9	76.608308e-3
30.e6	1.023223e3	910.69563e-6	27.392833e-9	201.43854e-3

The characteristics below meet or exceed EIA/TIA Category 5 twisted-pair specifications.

Table 3.6   Category 5 Twisted Pair — Internal Wiring

Resistance	r_{0c}	r_{0s}	a_c	a_s
(value)	176.6 ohms/km	∞ ohms/km	0.0500079494	0.0
Inductance	l_o	l_∞	b	f_m
(value)	1090.8 μH/km	504.5 μH/km	0.705	32570 kHz
Capacitance	c_∞	c_0	c_e	
(value)	48.55 nF/km	0.0 nF/km	0.0	
Conductance	g_0	g_e		
(value)	1.47653 nS/km	0.91		

Frequency (Hz)	Resistance (Ω/km)	Inductance (H/km)	Capacitance (F/km)	Conductance (S/km)
5000	176.656720	967.308142e-6	48.55e-9	3.430086e-6
10000	176.826554	913.078780e-6	48.55e-9	6.445287e-6
20000	177.501041	847.551900e-6	48.55e-9	12.110988e-6
50000	182.020084	753.691218e-6	48.55e-9	27.880784e-6
100000	195.898798	687.417012e-6	48.55e-9	52.389261e-6
1.e6	475.172462	552.634084e-6	48.55e-9	425.835904e-6
10.e6	1.4954809e3	514.663928e-6	48.55e-9	3.461324e-3
30.e6	2.5901370e3	509.228994e-6	48.55e-9	9.406382e-3

3.5.6.3 Measurement of Transfer Function and "Insertion Loss"

Transmission engineers sometimes also directly measure the transfer characteristics of a transmission line at several frequencies. It is hard to measure the transfer function directly because of loading effects, but it is possible to measure easily the **insertion loss**, from which the transfer function can be computed if load and source impedances for the measurement are known.

The insertion loss is computed using a configuration in Figure 3.9 by first measuring the voltage V_{no}, the voltage with no transmission line and only Z_L connected, and then inserting the transmission line at the point where V_{no} was measured initially and again measuring V_L, the voltage across the load with the line inserted. Thus the insertion loss[3] is

$$T_{IL}(f) = \frac{V_L(f)}{V_{no}(f)} = \frac{Z_S + Z_L}{A \cdot Z_L + B + C \cdot Z_S \cdot Z_L + D \cdot Z_S} \qquad (3.48)$$

The desired transfer function is instead $H = V_L / V_S$, so

$$H(f) = \frac{V_{no}}{V_S} \cdot \frac{V_L}{V_{no}} = \frac{Z_L}{Z_S + Z_L} \cdot T_{IL}(f) \qquad (3.49)$$

Note that when $Z_1 = Z_L$, meaning the line is terminated in its own impedance as is often the practice, then Equation (3.54) can be rewritten in terms of the $T(f)$ in Equation (3.3) as

$$H(f) = \frac{V_1}{V_S} \cdot \frac{V_L}{V_1} = \frac{Z_1}{Z_S + Z_1} \cdot T(f) \qquad (3.50)$$

---

3.   Insertion "loss" is actually a poor choice of name in that $T_{IL}(f) \leq 1$ and so the "loss" is tacitly included. A better name would be insertion "transfer function," but such is not the practice in DSL.

which also then shows that in the matched-termination case, $T(f) = T_{IL}(f)$. In most cases of interest in DSL, the line is long and so the source impedance is matched to the characteristic impedance (which equals the input impedance of the line when the line is long) and all impedances are real over the higher frequencies used for DSL transmission. In this case, the transfer function is simply 6 dB lower than the insertion loss. The CD-ROM program that computes transfer functions of twisted pairs using two-port theory is also available on the World Wide Web at http://www-isl.stanford.edu/people/cioffi.

A crucial point of note: When the transfer function is computed for a circuit using *RLCG* parameters, then the insertion loss may be computed from the transfer function and is roughly 6 dB higher under the approximations above. The insertion point is exactly the point at which a transmit power constraint applies. Thus, for instance, input voltage levels computed from a power constraint for a DSL (in performance calculation or SNR computation in Chapter 8) traverse a channel that is characterized by the insertion loss, and not the transfer function. A common mistake is to compute data rates and performance as if the transmit power were 6 dB lower by incorrectly using the transfer function instead of the insertion loss.

3.5.7   Balance — Metallic and Longitudinal

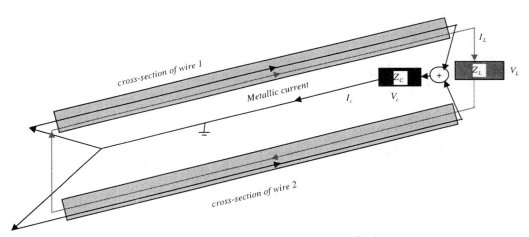

Figure 3.11   Illustration of metallic and longitudinal currents.

An excellent discussion of balance effects appears in Reeves, Ref. [3]. Here, we have a simplified tutorial. Figure 3.11 shows metallic and longitudinal currents in a twisted pair (the two wires of copper in the twisted pair are not twisted to simplify the illustration). The metallic current carries the signals intended to/from the customer. Such current encounters the load impedance Z_L as discussed previously in this section. Longitudinal currents are, with respect to earth, ground and essentially the two wires act effectively as one with the return path being through ground. Longitudinal currents can be introduced by radio waves that impinge upon the phone line or by imperfections in transmit circuits coupled to the phone line that cause differentially applied metallic voltages to leak into the longitudinal path.

The **balance** of a transmission line reflects its ability to prevent differential signals from leaking into the longitudinal path ("metallic balance") and also the corresponding reciprocal ability to prevent longitudinal signals from coupling into the metallic signals ("longitudinal balance"). The higher the balance, the better the phone line's ability to eliminate undesirable coupling effects. Balance is usually a function of frequency and reduces with higher frequencies. Tighter twisting of a twisted pair improves balance. Also, careful design of differential transmitter and receiver circuits ensures that the impedance to ground is constantly high at all points and the same for both wires. Nonetheless, practical situations see limits to balance. In the POTS band, balance is typically 50 to 60 dB, meaning that signals coupling from differential to longitudinal and vice versa are reduced by 5 to 6 orders of magnitude in power. However, at higher frequencies in ADSL/HDSL balance may reduce to 30 dB, and at even higher frequencies of VDSL still further reduction can occur.

Mathematical models for balance seem difficult to find. The authors suggest the following model for category 3 lines based on general observations that balance tends to decrease from 50 dB or better at lower frequencies, to about 35 dB at 1.5 MHz, to even less at higher frequencies (this model stops at 15 dB balance at 30 MHz), with **power** ratio:

$$B(f) = \begin{cases} 10^5 & 0 < f \leq f_b = 150\,\text{kHz} \\ 10^5 \left(f_b \middle/ f \right)^{1.5} & f_b \leq f \leq 30\,\text{MHz} \end{cases}$$

$$(3.51)$$

The balance of Category 5 twisted pair is about 20 dB better at all frequencies.

3.6 Noises

Section 3.5 discussed the computation of phone-line transmission characteristics, specifically the computation of transfer functions and impedances for metallic (differential) signals on the phone line. Noise on a phone line normally occurs because of imperfect balance (see Section 3.5.8). There are many types of noises that couple through imperfect balance (or imperfect/insufficient twisting) into the phone line, the most common of which are crosstalk noise (see Section 3.6.1), radio noise (see Section 3.6.2), and impulse noise (see Section 3.6.3).

3.6.1 Crosstalk Noise

Crosstalk noise in DSLs arises because the individual wires in a cable of twisted pairs radiate electromagnetically. The electric and magnetic fields thus created induce currents in neighboring twisted pairs, leading to an undesired crosstalk signal on those other pairs. Figure 3.12 illustrates two types of crosstalk commonly encountered in DSLs. **Near-end crosstalk (NEXT)** is the type of crosstalk that occurs from signals traveling in opposite directions on two twisted pairs (or from a transmitter into a "near-end" receiver). **Far-end crosstalk (FEXT)** results from signals traveling in the same direction on two-twisted pairs (or from a transmitter into a "far-end" receiver).

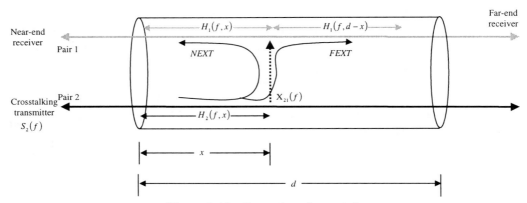

Figure 3.12   Illustration of crosstalk

Crosstalk can be the largest noise impairment in a twisted pair and often substantially reduces DSL performance when it cannot be eliminated or circumvented. In the case of a cable, the simple two-port models need generalization. Figure 3.13 illustrates the coupling between both wires in one twisted pair and both wires in another twisted pair. There is mutual inductance, M, between segments of wire, and there is also capacitance, E, between the wires themselves. In tightly controlled cables of twisted pair, one can expect the mutual inductance and capacitance to be controlled by the twisting so that adjacent segments of the twisted pair see opposite polarity and thus cancellation of induced signals. However, the twisting is not perfect, nor are the values of the mutual inductance and capacitance perfectly maintained over the length of the twisted pair. Furthermore, the variation of mutual inductance and capacitance with frequency is even greater than the variation of the $RLCG$ parameters that characterize metallic (differential) signals along particular twisted pairs. However, it is reasonable that the coupling from a metallic signal on another twisted pair to the metallic signal on the pair of interest is constant with respect to length on average (just as we assume that $RLCG$ parameters are constant per unit length). Then the coupling function (per Hz) between voltage changes on line 2 and line 1,

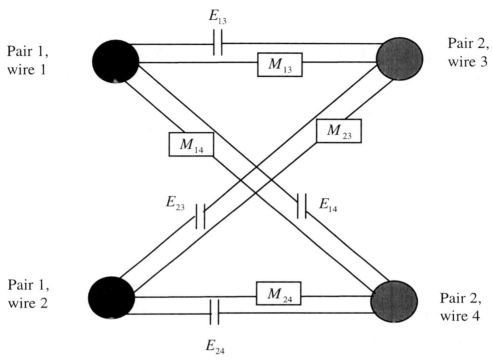

Figure 3.13   Distributed modeling of mutual coupling between twisted pairs.

$X_{21}(f)$, can be found through a generalization of the two-port theory (knowing all M and E parameters) so that

$$N_{p1}(f,x) = X_{21}(f) \cdot 2\pi j f \cdot V_{p2}(f,x)$$

(3.52)

where $N_{p1}(f, x)$ is the induced metallic voltage on pair 1 at frequency f and at position x along the transmission cable, and $V_{p2}(f, x)$ is the voltage causing the crosstalk on the second twisted pair. The factor of $2\pi j f$ represents that it is a change of voltage or current on another pair that actually leads to induced voltage/current on the first pair (this factor corresponds to differentiation). There is a similar per-unit-length crosstalk function from pair 1 into pair 2, and also for each pair in the cable into each and every other pair.

3.6.1.1 NEXT Modeling

For NEXT over a length d of cable where two pairs crosstalk, is found by summing the contributions of crosstalk over each incremental unit of length

$$N(f, d) = \int_0^d X_{21}(f) \cdot 2\pi jf \cdot V_{p2}(f) \cdot T_2(f, x) \cdot T_1(f, x) \cdot dx \tag{3.53}$$

where $V_{p2}(f)$ is the (near-end) input to pair 2, $T_2(f, x)$ is the insertion transfer function along line 2 of length x, and $T_1(f, x)$ is the corresponding transfer function back on line 1. Such an insertion transfer function tacitly presumes the line is terminated at x in its owl characteristic impedance. Most crosstalk analyses make a number of assumptions, particularly that the transmission lines are terminated in their own characteristic impedance and that the two lines have identical $RLCG$ parameters. Furthermore, as the crosstalk is viewed as noise, only the squared magnitude of the Fourier transform is of interest. In this case, the equation above becomes

$$|N(f,d)|^2 = (4\pi^2 f^2) \cdot |X_{21}(f)|^2 \cdot |V_{p2}(f)|^2 \cdot \int_0^d e^{-4\alpha x} dx$$

$$= (4\pi^2 f^2) \cdot |X_{21}(f)|^2 \cdot |V_{p2}(f)|^2 \cdot \frac{[1 - e^{-4\alpha d}]}{4\alpha} \tag{3.54}$$

Assuming further that the twisted-pair has $\alpha = \varsigma \cdot \sqrt{f}$, and that the exponential term is small for reasonable length of line d, then a common model is

$$|N(f,d)|^2 = |N(f)|^2 = \left(\frac{\pi^2 f^{1.5}}{\varsigma}\right) \cdot |X_{21}(f)|^2 \cdot |V_{p2}(f)|^2 \tag{3.55}$$

The general increase of coupling with $f^{1.5}$ is a well-known and often quoted effect. However, because of the number of assumptions about perfect line termination, uniform line characteristics, and position constancy, some more sophisticated crosstalk models fit a power of f near 1.5, say in the range 1.3 to 1.7 to measurements as well as empirically determining the constant factor.

Figure 3.14 (courtesy of C. Valenti at Bellcore [4]) shows a number of pairwise coupling transfer functions measured in a 50-pair cable. Note the general increase as $f^{1.5}$ with frequency, but significant (10 to 20 dB) variation in coupling with frequency. At any given frequency, only a few other pairs may contribute significantly to crosstalk, but over all frequencies, many lines contribute. Thus, DSL engineers average the coupling over many pairs. In this case, the sum of many coupling functions is assumed constant,

$$\sum_n |X_n(f)|^2 \approx k' \tag{3.56}$$

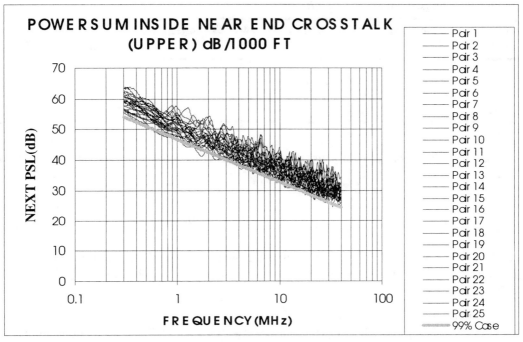

Figure 3.14 NEXT transfer loss for 25 pairs of PIC in a cable binder group.
Source: courtesy, C. Valenti, Bellcore [4].

Empirical studies then determine the value of the constant so that a 50-pair binder has coupling function

$$S_n(f) = k_{next} \cdot f^{1.5} \cdot S_2(f)$$

(3.57)

where k_{next} has been determined by ANSI studies to be

$$k_{next} = 10^{-13} \cdot \left(\frac{N}{49}\right)^6$$

(3.58)

and N is the number of pairs in the binder expected to be carrying similar DSL service. This value also appears in Figure 3.14, where it is seen to be a worst-case value (Bellcore studies have determined this value to be worse than 99% of the twisted-pair situations [4]). Table 3.7 shows the crosstalk coupling values for a few frequencies and numbers of crosstalkers.

Table 3.7   Crosstalk Attenuation in dB

number of talkers → Freq. (kHz) ↓	1	10	24	49
3	-88	-82	-79.7	-77.8
30	-73	-67	-64.7	-62.8
300	-58	-52	-49.7	-47.8
3000	-43	-37	-34.7	-32.8

Thus, for example, to find the crosstalk noise from an ISDN circuit into another twisted pair for a binder containing 24 ISDN circuits, the power spectral density on any line in the binder is modeled by

$$S_n(f) = \left(\frac{24}{49}\right)^6 \cdot 10^{-13} \cdot f^{1.5} \cdot S_{ISDN}(f)$$

(3.59)

Explicit formulas for all crosstalk types will be given in Section 3.7 on spectral compatibility. For crosstalk between binder groups, K_{next} is reduced by an additional 10dB to K_{next} (adj binders) $= 10^{-14}$.

3.6.1.2   FEXT Modeling

FEXT modeling parallels NEXT modeling. The equivalent of Equation (3.48) is now

$$F(f, d) = \int_0^d X_{21}(f) \cdot 2\pi jf \cdot V_{p2}(f) \cdot T_2(f, x) \cdot T_1(f, d-x) \cdot dx$$

(3.60)

where T_1 is now a function of the length of line from the point of coupling onward to the far-end receiver, while T_2 is from the transmitter to the point of coupling. By again assuming the two lines are terminated in the characteristic impedance and also have the same *RLCG* characterization, the squared magnitude of the FEXT signal is then

$$|F(f, d)|^2 = (4\pi^2 f^2) \cdot |X_{21}(f)|^2 \cdot |V_{p2}(f)|^2 \cdot \int_0^d e^{-2\alpha d} dx$$

$$= (4\pi^2 f^2) \cdot |X_{21}(f)|^2 \cdot |V_{p2}(f)|^2 \cdot d \cdot e^{-2\alpha d}$$

(3.61)

FEXT thus increases with the square of the frequency of the transmitted signal. Often the exponential factor at the end of Equation (3.57) is recognized as just the power transfer function of a single line and so the expression $|T(f, d)|^2$ replaces that factor [even though this assumes both lines are identical and follow the same formula — more generally, one should replace this factor with the more complicated integral formulation in Equation (3.60)]. Furthermore, the coupling function $|X_{21}(f)|^2$ again will vary significantly with frequency, possibly spanning 10 to 20 dB in magnitude (or even more at higher frequencies). Nonetheless, when several crosstalkers are summed, the approximation in (3.63) again holds. The factor ς is no longer divided, so in this case, the FEXT model accepted by ANSI is

$$S_f(f) = k_{fext} \cdot f^2 \cdot d \cdot |H(f,d)|^2 \cdot S_2(f)$$

(3.62)

where d is the length in feet, $|T(f, d)|2$ is the transfer function from the line input (insertion loss) for the length of transmission line being investigated, $S_2(f)$ is again the power spectral density input to the line (and not at the source), and finally

$$k_{fext} = \left(N/_{49} \right)^6 \cdot 9 \times 10^{-20}$$

(3.63)

Again, Bellcore has validated this value to correspond to a 1% worst-case value at frequencies to 30 MHz [4].

3.6.1.3 Distribution of Crosstalk Noise

K. Kerpez of Bellcore [5] has studied and validated the assumption for both NEXT and FEXT that the time-domain noise at the receiver is Gaussian in distribution. While this is clearly not true for a single crosstalker, because of the highly frequency-dependent nature of the crosstalk, when summed over all frequencies from different contributors on different lines, the **central limit theorem** of statistics loosely applies. This theorem states that the sum of a number of random signals tends to be Gaussian. Kerpez [5] has validated that this does hold for cases of practical interest. Such an analysis, however, may strongly depend on the size error between a Gaussian distribution and the true distribution. When thermal noise is small, this error can actually be large with respect to such noise. Ref [24] shows in this case that crosstalk effects may be mitigated in this situation, essentially meaning crosstalk is not exactly Gaussian, and the deviation is important.

While Gaussian, the noise may not be stationary.

3.6.1.4 Cyclostationarity of Crosstalk Noise

A number of studies by Pederson and Falconer [6] have noted that the distribution of crosstalk may not be stationary unless sampled at exactly the same rate as the generating crosstalk signal. So, for investigation of crosstalk between different types of DSLs, the DSL with the higher sampling rate will see a periodicity in the crosstalk from the line with the lower sampling rate. Since all DSL circuits are often timed to the same central office clock, the overall crosstalk can be cyclostationary with period equal to the least common multiple of the periods of the two sampling clocks.

This periodicity can be exploited by receiver filtering that essentially does adaptive linear prediction of the Gaussian noise samples over the period of the crosstalk. Substantial reductions in crosstalk, up to 12 dB [7], have been reported for situations where a service, for instance ADSL, is witnessing crosstalk from a lower-speed DSL like HDSL. One could expect similar effects from ADSL/T1 into VDSL. Any adaptive equalizer (see Chapter 8) will usually automatically exhibit this noise reduction. Unfortunately, this fact appears not to be well known in the DSL community, and some projects on spectral compatibility often ignore this effect because of the perception that service operators are not certain the cyclostationarity exists. The effect has been long exploited in high-speed computer networks [8].

3.6.2   Radio Noise

Radio noise is the remnant of wireless transmission signals on phone lines, particularly AM radio broadcasts and amateur (HAM) operator transmissions.

Radio-frequency (RF) signals impinge on twisted-pair phone lines, especially aerial lines. Phone lines, being made of copper, make relatively good antennae with electromagnetic waves incident on them leading to an induced charge flux with respect to earth ground. The common mode voltage for a twisted pair is for either of the two wires with respect to ground — usually these two voltages are about the same because of the similarity of the two wires in a twisted pair. Well-balanced phone lines thus should see a significant reduction in differential RF signals on the pair with respect to common-mode signals. However, balance decreases with increasing frequency, and so at frequencies of DSLs from 560 kHz to 30 MHz, DSL systems can overlap radio bands and will receive some level of RF noise along with the differential DSL signals on the same phone lines. This type of DSL noise is known as **RF ingress.**

Following Foster and Cook [9], the electromagnetic field strength for an ideal point-source antenna distributes power P_t evenly over the surface of a sphere, thus leading to a field strength at distance d of

$$F = \sqrt{\frac{P_t \cdot Z_f}{4\pi \cdot d^2}} = \frac{5.48 \cdot \sqrt{P_t}}{d} \quad \text{V/m} \tag{3.64}$$

The impedance of free space is $Z_f = 377\Omega$. The amount of voltage induced with respect to ground from a field F incident on a wire depends on a number of geometrical and conductive/magnetic properties of the cable. Through field experience [9], the induced voltage is equal to (in volts) the field strength when expressed in volts/meter in worst-case situations. Thus the common mode voltage is also given by the expression in (3.60). The differential mode voltage is the common mode voltage reduced by a factor of the balance $\sqrt{B(f)} = \sqrt{B}$, so that

$$V_d = \frac{5.48\sqrt{P_t}}{d \cdot \sqrt{B}}$$

(3.65)

This expression can be used to estimate ingress noise levels from both AM radio stations and from amateur radio operators.

3.6.2.1 Amateur Radio Interference Ingress

Amateur radio transmissions occur in the bands shown in Table 3.8.

Table 3.8 Amateur Radio Bands

HAM operator bands (MHz)	
Lowest frequency	**Highest frequency**
1.81	2.0
3,5	4.0
7.0	7.1
10.1	10.15
14.0	14.35
18.068	18.168
21	21.45
24.89	24.99
28.0	29.7

These bands overlap the transmission band of VDSL but avert the lower transmission bands of other DSLs. Thus, HAM radio interference is largely a problem only for VDSL.

A HAM operator may use as much as 1.5 KW of power, although such a large use of power is rare and may not occur in residential neighborhoods or areas with many phone lines. A 400 W transmitter at a distance of 10 meters (about 30 feet) leads to an induced common-mode (longitudinal) voltage of approximately 11 volts on a telephone line. With balance of 33 dB, the corresponding differential (metallic) voltage is about 300 mV, which is 0 dBm of power on a $Z_0 = 100\Omega$ transmission line. HAM operators use a frequency band of 2.5 kHz intermittently with either audio (voice) or digital (Morse code, FSK) signals, leading to a noise PSD of approximately −34 dBm/Hz. More typically, HAM operators transmit at lower levels or may be spaced more than 10 meters when transmitting at higher levels, so that more typical HAM RF ingress levels may be 20 to 25 dB less powerful. Nonetheless, this still leads to PSDs for noise in the range from −35 dBm/Hz to −60 dBm/Hz, well above the levels of crosstalk in Section 3.7. Furthermore, such high voltage levels may saturate analog front-end electronics.

HAM operators tend to switch carrier frequency every few minutes, and the transmitted signal is zero (SSB modulation) when there is no signal. Thus, a receiver may not be able to predict the presence of HAM ingress.

Fortunately, HAM radio signals are narrowband, and so transmission methods attempt to notch the relatively few and narrow bands occupied by this noise, essentially avoiding the noise (for some transmission methods, see Chapter 6) rather than try to transmit through it. Some degree of receiver notch filtering is also necessary to eliminate the effect.

3.6.2.2 AM Ingress

AM radio interference arises from commercial radio stations that continuously occupy bandwidths 10 kHz wide from 560 kHz to 1.6 MHz, thus affecting both downstream ADSL and VDSL. Many AM radio stations may be simultaneously active in an urban environment and evident on phone lines. AM radio stations may broadcast at power levels to 50,000 W and may transmit the largest power at night. AM radio signals thus might appear to be 20 dB (if not more) higher than HAM signals, but one needs to keep in mind that cable balance is typically better at lower frequencies (10 to 15 dB reduction). Also, the distance from an AM radio tower to a phone line is likely to be 1 km or more rather than 10 meters (40 dB reduction), and the energy is spread over four times the bandwidth, (6 dB reduction). Thus, AM radio signals typically have differential noise PSDs of –80 dBm/Hz to –120 dBm/Hz on phone lines. AM radio signals tend to be continuous because of the double side band plus carrier nature. ADSL and VDSL noise specifications, thus use a 10-frequency model, where all noises are sinusoids (the state in between utterances on the radio station) [10].

The AM ingress level is again comparable to or above crosstalk and background noise levels on a DSL and thus cannot be ignored by designers. AM radio signals, however, do not appear to be sufficiently large to saturate analog front ends of DSL receivers.

3.6.3 Impulse Noise

Impulse noise is nonstationary crosstalk from temporary electromagnetic events in the vicinity of phone lines. Examples of impulse generators are as diverse as the opening of a refrigerator door (the motor turns on/off), control voltages to elevators (phone lines in apartment buildings often run through the elevator shafts), and ringing of phones on lines sharing the same binder. Each of these effects is temporary and results in injection of noise into the phone line through the same basic mechanism as RF noise ingress, but typically at much lower frequencies.

Differential (metallic) induced voltages are typically a few millivolts, but can be as high as 100 mV. Such voltages may sound small, but the severe attenuation of high frequencies on twisted pair means that an impulse can appear enormous to a receiver in comparison to received DSL signal levels. The common-mode voltages caused by impulses can be 10s of volts in amplitude. Typical impulses last 10s to 100s of microseconds but can span time intervals as long as 3 ms.

Numerous studies of impulses have resulted in both analytical models for impulses based on statistical analysis of over 100,000 impulses by various groups. However, others insist that impulses defy analysis and prefer to just store representative worst-case waveforms. The area of impulse modeling thus remains controversial, probably because the causes of impulses are so diverse that any distillation for engineering and measurement purposes necessarily has some bias. The most widely used analytical model is the **Cook pulse**, after John Cook of BT (formerly British Telecom) [11]. Cook recorded over 100,000 impulses and via computer analyzed some 89,000 of them on many different phone lines as a basis for his model of the next subsection. The ADSL standard, however, uses two measured impulses [12] instead of the Cook pulse and another empirical formula from Bellcore to relate test results to performance (see Section 8.3).

We will concentrate here on the Cook pulse and refer the reader to Ref. [12] or the summary of testing methods in Section 8.3. Early ADSL testing used both the Cook pulse and a set of 12 worst-case impulses suggested by NYNEX [13].

3.6.3.1 The Cook Pulse

Statistically, Cook found that the amplitude of the impulse increases with the bandwidth of the DSL under test. This nominally accrues to the wider bandwidth of the DSL receiver filter, which simply means less impulse attenuation. The differential voltage induced by an impulse was found to have a peak value that increased as

$$V_{imp,peak} = \lambda \cdot f_{DSL}^{\frac{3}{4}} \cdot \tau^{\frac{1}{3}} \quad \text{mV} \tag{3.66}$$

where λ is a constant expressing degree of confidence or occurrence, f_{DSL} is the bandwidth used by the DSL, and τ is the time interval for impulse observation. A typical value for the confidence constant is $\lambda = 0.28$ with a worst-case value being $\lambda = 1.4$. Thus, wider-bandwidth systems see larger impulses (increasing as the three-fourths power of the bandwidth), and the longer one observes, the more likely a large impulse will occur. Curiously, the common mode voltage distribution is not frequency dependent and is

$$V_{common} = \mu \cdot \sqrt{\tau} \quad \text{mV} \tag{3.67}$$

with μ a common-mode confidence constant with typical value of 1100 and worst-case value of 4400. The frequency independence is apparently an indication of decaying balance at higher frequencies, apparently with $f^{1.5}$, validating again the crosstalk model that shows increase of crosstalk noise coupling with this same factor.

The continuous-time **Cook pulse** has a mathematically abstract model:

$$v(t) = V_p \cdot |t|^{-3/4} \cdot \text{sgn}(t) \tag{3.68}$$

which is discontinuous at $t=0$, making it seem unrealistic. However, the impulse is sampled and stored for test purposes at some discrete-time rate $1/T$, which leads to a reasonable sampled-time pulse

$$v(kT) = \frac{V_p}{T^{.75}} \cdot k^{-.75} \cdot \text{sgn}(k)$$

(3.69)

The absolute magnitude of the pulse at its peak value

$$V_{imp,peak} = \left| \frac{V_p}{T^{.75}} \right|$$

(3.70)

at $t = T$ measures the size of the Cook pulse. As T decreases with wider-bandwidth systems, the peak pulse magnitude can increase. However, it appears that this amplitude should be bounded at a value at about a sampling rate of 1 to 2 MHz.

Thus, a stored Cook pulse can be injected into a phone line at random time instants for testing purposes to represent the type of disturbance that might randomly occur on phone lines in service. The occurrence of impulses varies with the phone line. However, peak impulse activity usually occurs at midday on Monday through Friday in phone lines in business environments. In residential environments, impulses tend to occur in the early morning and at night. The modeling of interarrival time uses large sets of data. The probability that the interarrival time, T_i, between impulses exceeds t seconds can be estimated by

$$\text{Pr}\{T_i > t\} = P_r \cdot u(T_r - t) + (1 - P_r) \cdot \left(1 - e^{-T_c/t}\right)$$

(3.71)

where P_r is the probability that the first impulse was caused by a ring signal (about 0.7, or 70% of the impulses), T_r is the ring "cadence" (time between ring bursts when ringing a phone, typically about 3 seconds), and T_c (typical value of 70 seconds) is a time constant associated with the exponential distribution modeling the remaining types of impulses other than ring. This model tells us that for small interarrival times of less than 3 seconds, the dominant effect is ringing. However, this distribution is superimposed on one that generally models interarrival times as an exponential distribution. This distribution is plotted in Figure 3.15 for the nominal values given in this paragraph.

The distributions for peak voltage, interarrival time probability, and the shape of the Cook pulse (or other impulse shape) allow measurement or computation of the probability that a certain transmission technique will have errors in a given interval of time. One must first determine the peak voltage of the Cook pulse that will cause one or more bit errors with the transmission technique, which may be a function of the time between impulses if the system uses interleaved error correction. The probability that the peak is exceeded then must be multiplied by the corresponding probability of interarrival time at that peak amplitude (one may assume the second impulse is just as high as the first, perhaps a worst-case assumption) to get the probability that an error occurs over a time period equal to the interarrival time. The consequent probability can be

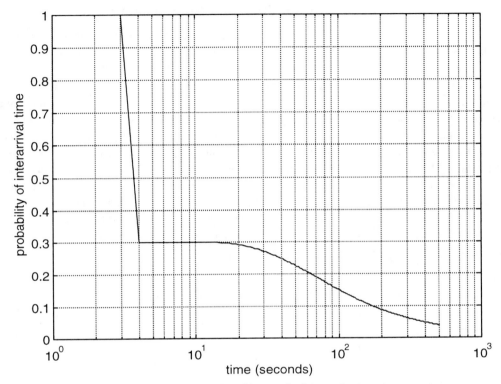

Figure 3.15   Probability of interarrival times for impulses.

Source: From Cook, Ref. [11].

divided by the interarrival time to determine the probability of an errored second. Section 8.3 illustrates some standardized similar calculations for non-Cook impulses.

3.7 Spectral Compatibility

Spectral compatibility is an often used (and perhaps over- and misused) term referring to the degree of mutual crosstalk between various DSL services. It also often refers to RF emissions. As increasing numbers of DSL services are deployed, the concern is that assumptions made in the design of modem equipment for one type of service will lead to errors in another type of modem equipment also sharing the cable. With ISDN, HDSL, ADSL, VDSL, and perhaps non-standard equipment also deployed, the possibility increases that a given situation of unacceptable crosstalk might be overlooked. The ANSI group has a standards project just to study this potential. This section only defines the issues and enumerates known power spectral densities.

3.7.1   Interference Between DSLs and Multiplexing

Chapter 5 covers various multiplexing methods for DSLs. Spectrum compatibility refers to the possible overlap of transmission bands on different DSLs sharing the same cable or, even worse, the same binder. Section 3.6 studied crosstalk coupling functions. The level of crosstalk can be sufficiently large at high frequencies to disrupt another service. Particularly noteworthy examples are T1 circuits. T1 circuits have been deployed for years by phone companies and were designed and standardized at a time when transmission techniques were not cognizant of spectral compatibility issues. T1 circuits essentially blast energy into a channel from DC to 3 to 4 MHz to transmit 1.544 Mbps digital service using a dated transmission technique that is wasteful of bandwidth and energy. The crosstalk from this particular service is larger than any other. Fortunately, T1 lines are becoming less common because of their gradual replacement using a newer, more efficient method in HDSL. (T1 lines are replaced only when removed from service.) The HDSL crosstalk is far less invasive to other services. A 6 Mbps ADSL transmission is four times the rate of T1, but not quite as invasive as a crosstalker. It still is quite invasive, especially to even newer services like VDSL.

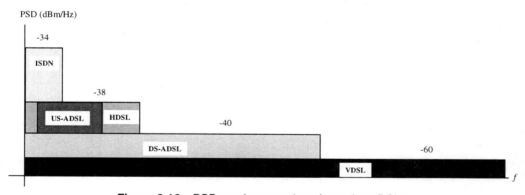

Figure 3.16   PSD mask comparison for various DSLs.

Thus, the problem is to allocate used bandwidth to different services in a way that will be least invasive to other services expected in the same cable. A rule often used is that a new service should be no more invasive than any existing service would be into itself. Figure 3.16 shows the various bandwidths of xDSL signals and approximate power levels. As can be seen, while newer services tend to use more bandwidth, their power spectrum is less than existing services in the band of those older services. Figure 3.17 shows the crosstalk spectra of ISDN, HDSL, and upstream ADSL. Figure 3.18 shows the crosstalk spectrum of a downstream ADSL signal.

Nominal background noise on a twisted pair is no larger than −140 dBm/Hz, so that these noises are clearly significant. Wider bandwidth DSLs typically exploit transmission at higher

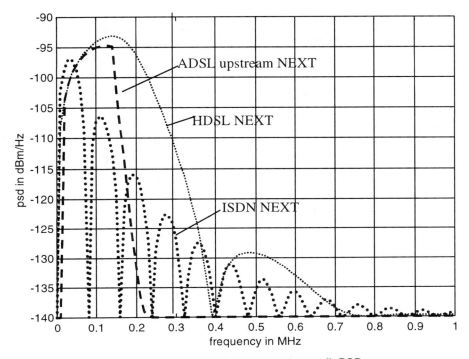

Figure 3.17 ADSL downstream crosstalk PSD.
(Assumes full 1.104 mHz bandwidth of ADSL is –40 dBm/Hz.)

frequencies on shorter lines where crosstalk from existing services appears relatively smaller because the line signals are less attenuated.

3.7.2 Self-Interference

Self-crosstalk is crosstalk induced into a service by an identical DSL. This type of spectral compatibility is the most important when, as expected, a service provider elects to provide a certain DSL service on a wide scale. Then DSLs of the same type will crosstalk into one another.

Asymmetry was first introduced by Bellcore's Joe Lechleider in ADSL. ADSL transmits over a much wider downstream bandwidth than upstream, and thus most of the downstream signal is free of self-crosstalk. This allows the downstream signal to run at a much higher rate than is possible with symmetric transmission (with all other aspects being equal). Since both entertainment (live TV and movies on demand) and Internet applications' asymmetric bandwidths match the asymmetry of ADSL, the use of asymmetric transmission for technical reasons also matches the market for services. Newer DSLs like g.lite and VDSL also have at least some modes of operation that are asymmetric.

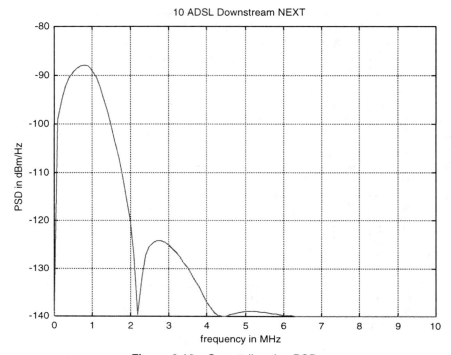

Figure 3.18 Crosstalk noise PSDs.

Self near-end crosstalk can be mitigated in either the frequency domain (by using non-overlapping transmission spectra) or in the time domain (by synchronizing all DSLs to the network clock so that they transmit downstream and upstream in different time slots as a group). FEXT will be present, however, in either time-domain or frequency-domain separation methods.

3.7.3   Crosstalk FEXT and NEXT Power Spectral Density Models

NEXT and FEXT for various DSLs is determined by applying the NEXT and FEXT power transfer functions:

$$PSD_{NEXT} = PSD_{disturber} \cdot (N/49)^6 \cdot 10^{-13} \cdot f^{1.5} \tag{3.72}$$

$$PSD_{FEXT} = PSD_{disturber} \cdot | H(f) |^2 \cdot (N/49)^6 \cdot (9 \times 10^{-20}) \cdot d \cdot f^2 \tag{3.73}$$

(where N is the number of crosstalkers and d is the length of the loop in feet) of Section 3.6 to the following PSDs for various DSLs.[4]

3.7.3.1 ISDN NEXT and FEXT

For 2B1Q ISDN NEXT or FEXT,

$$PSD_{ISDN}(f) = K_{ISDN} \cdot \frac{2}{f_0} \cdot \frac{\left[\sin\left(\frac{\pi f}{f_0}\right)\right]^2}{\left(\frac{\pi f}{f_0}\right)^2} \cdot \frac{1}{\left[1 + \left(\frac{f}{f_0}\right)^4\right]}$$

(3.74)

where $f_0 = 80 kHz$, $K_{ISDN} = \frac{5}{9} \cdot \frac{V_p^2}{R}$, $V_p = 2.5$ Volts, and $R = 135\Omega$.

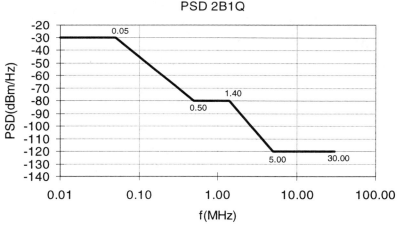

Figure 3.19 Latest ISDN PSD mask taken from ETSI document at
http://docbox.etsi.org/tech-org/tm/Document/tm6/983lulea/Hilv/960p04r8.doc.

Recently, ETSI has revised the PSD to be used for ISDN crosstalk studies to a lower PSD level than that implied by the above equation. This new PSD mask mandates the ISDN PSD maximum be below a curve beginning at –30 dBm/Hz below 50 kHz, with dBm/Hz linearly decreasing with logarithm of frequency to –80 dBm/Hz at 500 kHz, remaining flat to 1.4 MHz,

---

4.   For Category 5 twisted pair, which is listed in Section 8.5 of the appendix for information purposes, the coefficient $K_{next} = 10^{-13}$ changes to $K_{next-cat5} = 3.30 \times 10^{-16}$ and the coefficient $K_{fext} = 9 \times 10^{-20}$ changes to $K_{fext-cat5} = 2.44 \times 10^{-22}$. The Category 5 numbers may be more appropriate for new buildings or intracampus/corporate symmetric transmission.

decreasing linearly again with log of frequency to –120 dBm/Hz at 5 MHz and flat at –120 dBm/Hz thereafter. The mask is shown in Figure 3.19. The measurement window is 10 kHz wide below 300 kHz and 1 MHz wide above 300 kHz. The 1 MHz window measurements starting at lowest bottom-frequency 300 kHz, must also always be below –60 dBm total power, with 10 kHz bottom-frequency offsets to 29 MHz. See Ref. [23] for more details.

3.7.3.2 HDSL NEXT and FEXT

For 2B1Q HDSL NEXT or FEXT,

$$PSD_{HDSL}(f) = K_{HDSL} \cdot \frac{2}{f_0} \cdot \frac{\left[\sin\left(\pi f / f_0\right)\right]^2}{\left(\pi f / f_0\right)^2} \cdot \frac{1}{\left[1 + \left(f / f_{3db}\right)^8\right]}$$

(3.75)

where $f_0 = 392$ kHz, $f_{3dB} = 196$ kHz, $K_{HDSL} = \frac{5}{9} \cdot \frac{V_p^2}{R}$, $V_p = 2.7$ Volts, and $R = 135\Omega$.

3.7.3.3 ADSL NEXT and FEXT

Upstream ADSL: For upstream ADSL NEXT or FEXT,

$$PSD_{ADSL,US}(f) = K_{ADSL,US}(f) \cdot \frac{\left[\sin\left(\pi f / f_0\right)\right]^2}{\left(\pi f / f_0\right)^2} \quad , \quad 0 \le f \le \infty$$

(3.76)

where

$$f_0 = 270 \; kHz$$

$$K_{ADSL,US}(f) = \begin{cases} -38 \; dBm / Hz & 28kHz \le f \le 138kHz \\ -38 - 24\left(\dfrac{f - 138000}{43125}\right) dBm / Hz & f \ge 138 \; kHz \end{cases}$$

Issue 2 of the ANSI T1.413 ADSL Standard has just respecified the upstream PSD mask as in Figure 3.20:

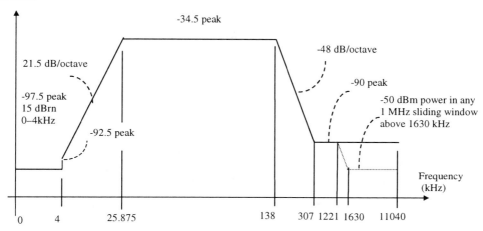

FREQUENCY BAND (kHz)	EQUATION FOR LINE (dBm/H)
0–4	-97.5, +15 dBrn, in 0–4 kHz
4–25.875	$-92.5 + 21.5 \times \log_2(f/4)$
25.875–138	-34.5
138–307	$-34.5 - 48 \times \log_2(f/138)$
307–1221	-90
1221–1630	$-90 - 48 \times \log_2(f/1221)$
1630–11040	-90 peak, with max -50 dBm power in any 1 MHz sliding window above 1630 kHz

NOTES:

1. All PSD measurements are in 100 ohms; the voice band total power measurement is in 600 ohms.

2. All PSD and power measurements shall be made at the U-R interface; the signals delivered to the POTS are specified in ANNEX I.

3. The breakpoint frequencies and PSD values are exact; the indicated slopes are approximate.

4. Above 25.875 kHz, the peak PSD shall be measured with a 10 kHz resolution bandwidth.

5. The power in a 1 MHz sliding window is measured in 1 MHz bandwidth, starting at the measurement frequency.

Figure 3.20 ADSL Issue 2 — upstream PSD mask.

Downstream ADSL: For downstream ADSL NEXT or FEXT,

$$PSD_{ADSL, DS}(f) = K_{ADSL, DS} \cdot \frac{\left[\sin\left(\pi f / f_0 \right) \right]^2}{\left(\pi f / f_0 \right)^2} \cdot |LPF(f)|^2 \cdot |HPF(f)|^2 \quad , \quad 0 \le f \le \infty$$

(3.77)

where

$$f_0 = 2.208 \, MHz \quad , \quad K_{ADSL, DS} = 110.4 \, mW$$

$$|LPF(f)|^2 = \frac{1}{1 + \left(f / f_{3db} \right)^8} \quad f_{3db} = 1.104 \, MHz$$

$$|HPF(f)|^2 = \frac{f^8}{f^8 + f_{3db}^8} \quad f_{3db} = 20 \, kHz$$

Issue 2 of the ANSI T1.413 ADSL Standard has recently more precisely specified the spectral mask as in Figure 3.21.

3.7.4   Emissions from DSLs

Emissions from twisted-pair phone lines carrying DSL signals have become of increased importance. As the bandwidths increase in DSLs, increasing numbers of radio bands are overlapped. ADSL transmission overlaps the bands used by AM radio. VDSL and the old T1 circuits overlap both AM radio bands and amateur radio bands.

The FCC is in the process of specifying the spectral emission limits for ADSL's ATU-R, customer-premises equipment [14]. These specifications correspond to the same spectral densities used in ADSL.

Numerous studies have shown that VDSL signals can cause a significant hazard to amateur radio service. However, these same studies (including several with measurements, see Ref. [15] and the references therein) also show that by lowering the power spectral density to −80 dBm/Hz in the known radio bands eliminates the effect and, fortunately, leads to little VDSL performance loss in well-designed systems. The nominal PSDs for VDSL otherwise are typically −60 dBm/Hz and can be as high as −50 dBm/Hz at low frequencies, leading to significant transmitter amateur-band notching of spectrum by various means.

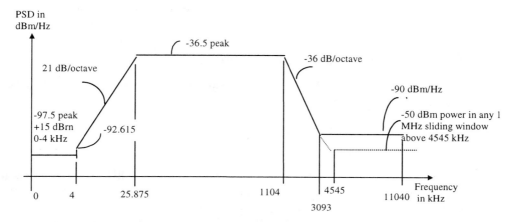

FREQUENCY BAND (kHz)	EQUATION FOR LINE (dBm/Hz)
0–4	-97.5, +15 dBrn in 0–4 kHz
4–25.875	$-92.5 +21 \times \log_2 (f/4)$
25.875–1104	-36.5
1104–3093	$-36.5 - 36 \times \log_2(f/1104)$
3093–4545	–90
4545–11040	-90 peak, with max -50 dBm power in any 1 MHz sliding window above 4545 kHz

NOTES:

1. All PSD measurements are in 100 ohms; the POTS band total power measurement is in 600 ohms.

2. All PSD and power measurements shall be made at the U-C interface; the signals delivered to the PSTN are specified in ANNEX I.

3. The breakpoint frequencies and PSD values are exact; the indicated slopes are approximate.

4. Above 25.875 kHz, the peak PSD shall be measured with a 10 kHz resolution bandwidth.

5. The power in a 1 MHz sliding window is measured in 1 MHz bandwidth, starting at the measurement frequency.

Figure 3.21 ADSL Issue 2 — downstream revised mask.

3.8 More Two-Port Networks

For proper function of DSL physical-layer transceivers, the analog interfaces to the digital signal processing must be well designed. Fortunately, a number of simple-to-use computer tools have been developed in recent years. Ultimately, component tolerances, exact circuit uses, and layout become important and are not addressed here, so the assistance of an experienced analog-circuit

designer (who may otherwise know little of DSL) may be of help to DSL system designers who
have specified filters.

3.8.1   Reciprocal and Lossless Two-Port Circuits

Reciprocal circuits are two-port circuits for which the transfer matrix Φ_r in the reverse
direction is the same as in the forward direction,

$$\Phi_f = \Phi_r \tag{3.78}$$

if the load and source impedances and the source voltage are switched. The DSL transmission
lines of Section 3.5.2 are examples of reciprocal circuits. Since the determinant of the ABDC
matrix is unity,[5] its inverse is trivially found to be

$$\Phi_f^{-1} = \begin{bmatrix} \cosh(\gamma \cdot d) & -Z_0 \cdot \sinh(\varphi \cdot d) \\ \frac{-1}{Z_0} \cdot \sinh(\gamma \cdot d) & \cosh(\gamma \cdot d) \end{bmatrix} \tag{3.79}$$

which, if one recalls the reversal in the definition of the direction of the currents if the circuit is
reversed, leads to the desired reciprocal behavior, agreeing with the engineer's intuition that the
twisted pair should look the same in both directions as long as the source and load impedances
are the same. Clearly, a cascade of reciprocal networks will also be reciprocal.

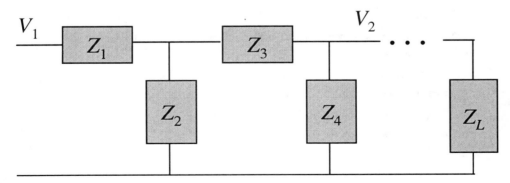

Figure 3.22   Lossless ladder circuit.

---

5.   This is simply proved by taking the determinant of the matrix in Equation (20) and noting the hyperbolic identity
     $\cosh^2(x) - \sinh^2(x) = 1$.

A specific type of circuit often encountered in DSL designs, particularly in splitter filters, is the **lossless two-port**. A lossless two-port is constructed entirely of passive elements (capacitors and inductors/transformers only are used in each of the impedances within the two-port network shown) and so therefore cannot dissipate any energy internal to the two-port. Lossless networks are also reciprocal. The most common type of lossless two-port in DSLs is the ladder circuit in Figure 3.22. Each impedance is constructed of a series or shunt-reactive (real part is zero) impedance. An ABCD matrix for each successive section is trivially monic (ones on diagonal) and triangular (zero entry either above or below diagonal) and clearly reciprocal. Thus, a cascade of such reactive elements is also reciprocal — it can be shown that any passive two-port can be realized as an equivalent ladder circuit.

The reciprocity of the lossless two-port, especially the ladder, is applicable to bidirectional circuits. Since DSLs often carry signals bidirectionally, the reciprocity of the lossless two-port makes it a good candidate for analog filter realization. First, to construct two-ports, one needs to have a desired transfer characteristic (perhaps bidirectionally), which is the subject of the following section.

3.8.2 Analog Filter Design and T(s)

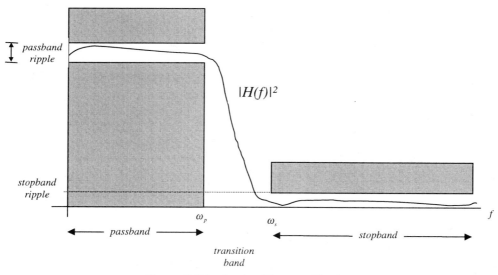

Figure 3.23 Analog filter specification.

Analog filter specification typically has ranges of frequency known as the **passband** and the **stopband,** as illustrated in Figure 3.23. Signals in the passband are to be passed with minimal distortion, while signals in the stopband are to be suppressed. Signals in the passband cannot deviate from some nominal gain (say unity) by more than a tolerance often referred to as a **passband ripple.** Output signals in the stopband may not rise above a level often known as the **stopband ripple.** Sometimes stopband filter requirements more stringently require the output signal's energy to continue to decrease as the frequency deviates further from those in the passband. The frequency ω_p is often called the **cut-off** frequency.

The transition band is not specified and allows for the transfer function to change from the passband level to the stopband level. Throughout this section, as with the text, any load impedance is presumed to be included, whether explicit or tacit, in all transfer functions $H(f)$.

Section 3.9 lists several filters of specific interest in DSL where analog filter design is necessary. Classical filter design essentially reduces the types of analog filters to a set of four that are largely used by all designers today, at least as a first step. The four types of design methods used in DSLs have different strengths and weaknesses that are discussed here in terms of the types of analog filter specifications they can accommodate. These classical design methods are known as

1. Butterworth
2. Chebychev I
3. Chebychev II
4. Elliptical/Cauer

A fifth method, known as the Bessel design, finds little use (at least yet) in DSLs, so will not be addressed here.

Some concepts used in the next few subsections may require reference to [20] or [21].

3.8.2.1 Butterworth Designs

Butterworth analog filters have a magnitude characteristic:

$$\left| H(f) \right|^2 = \frac{1}{1 + \left(\omega / \omega_c \right)^{2n}}$$

(3.80)

where n is the order of the filter (number of poles) and ω_c is the 3 dB cut-off frequency (this power transfer function is easily seen to have value 0.5 when $\omega = \omega_c$). An example of a 8th-order Butterworth filter magnitude is plotted in Figure 3.24. The transfer function in both passband and stopband is smooth, and the stopband magnitude gradually decreases as $20n$ dB/decade with frequency.[6] The higher the order of the filter, the faster the reduction with frequency and the more complex the realization. The Butterworth filter has no zeros and n equally spaced

---

6.   A decade is a factor of 10 increase in frequency.

poles around a semicircle of radius ω_c in the left half-plane.

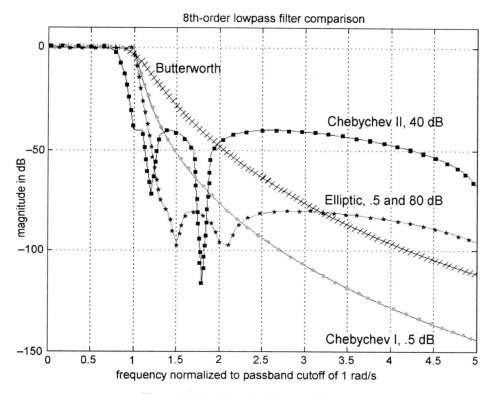

Figure 3.24 8th-order lowpass filters.

Matlab contains a function called "buttap" that determines the poles and zeros of a Butterworth filter with $\omega_c = 1$. This function is typically called as

$$[z,p,k] = \text{buttap}(n) \tag{3.81}$$

where n is the order of the Butterworth filter desired and z, p, and k are vectors of the respective zeros (z will contain nothing since there are zeros), n poles, and gain. The filter is given by

$$H(s) = \frac{k}{\displaystyle\prod_{i=1}^{n}(s - p_i)} \tag{3.82}$$

(Matlab also has a function "zp2tf" that will convert this transfer function into a numerator/denominator polynomial in s and also another function "buttord" that translates a filter specification of passband, stopband, and ripples into the minimum necessary order n for a Butterworth filter to meet these specifications.) Correction of the cut-off frequency and absolute magnitude of the passband is accomplished through frequency and magnitude scaling as in the next section. The filter can also be converted to a highpass filter with the same number of poles through a frequency transformation, or to a bandpass or bandstop filter with twice as many poles, as in the next subsection on frequency transformations. Thus, it is always sufficient to start with a lowpass design and then scale it as necessary. Then the resulting transfer function can be converted into an appropriate circuit according to the methods in Section 3.8.3.

3.8.2.2 Chebychev I Designs

Chebychev I lowpass analog filters have a magnitude characteristic:

$$|H(f)|^2 = \frac{1}{1 + \varepsilon^2 T_n^2\left(\omega \Big/ \omega_p\right)}$$

(3.83)

where n is the order of the filter (number of poles), and ω_p is the edge frequency of the passband after which all frequencies exhibit magnitude less than $1/(1 + \varepsilon^2)$ (which means the passband ripple is $\varepsilon^2/(1 + \varepsilon^2)$). The Chebychev function is given by

$$T_n(\omega) = \begin{cases} \cos\left(n \cdot \cos^{-1}[\omega]\right) & |\omega| \leq 1 \\ \cosh\left(n \cdot \cosh^{-1}[\omega]\right) & |\omega| > 1 \end{cases}$$

(3.84)

more simply determined by the recursion $T_n(\omega) = 2\omega T_{n-1}(\omega) - T_{n-2}(\omega)$ with $T_0 = 1$ and $T_1(\omega) = \omega$. Note that its value at $\omega = 1$ is always unity.

An example of a 8th-order Chebychev I filter's magnitude is also plotted in Figure 3.24. The minimum necessary order to meet a filter specification with a Chebychev I filter design can be determined with the use of the function "cheb1ord" in Matlab. The transfer function ripples in the passband but is smooth in the stopband, which gradually decreases as $20n$ dB/decade with a frequency like the Butterworth, except that the transition at lower frequencies is quicker. Thus, passband distortion (ripple) is traded for sharper transition with Chebychev I with respect to Butterworth. The higher the order of the filter, the faster the reduction with frequency and the more complex the realization. The Chebychev filter has no zeros and n poles spaced around a semi-ellipse in the left half-plane. Matlab contains a function called "cheb1ap" that determines the poles and zeros of an nth-order Chebychev I filter with $\omega_p = 1$. This function is typically called as

$$[z,p,k] = \text{cheb1ap}(n,\ pr) \tag{3.85}$$

where n is the order of the filter desired; z, p, and k are again vectors of the respective zeros (nothing since there are none), n poles, and gain; and pr is the passband ripple that can be tolerated in dB. The filter is given by

$$H(s) = \frac{k}{\displaystyle\prod_{i=1}^{n}(s - p_i)} \tag{3.86}$$

Again, frequency scaling, magnitude scaling, and frequency transformation can be used to convert this function into the appropriate filter for each increasing value of n until the filter specification is met.

3.8.2.3 Chebychev II Designs

Chebychev II lowpass analog filters have a magnitude characteristic:

$$|H(f)|^2 = \frac{1}{1 + \varepsilon^2 \left[\dfrac{T_n\left(\dfrac{\omega_s}{\omega_p}\right)}{T_n\left(\dfrac{\omega_s}{\omega}\right)} \right]^2} \tag{3.87}$$

where n is the order of the filter (number of poles and number of zeros in this case), ω_p is the passband edge, and ω_s is the stopband edge frequency. The zeros of this type of filter cause a ripple in the stopband, but not in the passband, allowing smooth passband and sharp transition at the expense of a passband that does not gradually decay with frequency. An example of a 8th-order Chebychev II filter's magnitude is also plotted in Figure 3.24. This time, note that the low-pass cut-off edge is not specified and that the frequency of the passband edge is reduced in exchange for rapid roll-off. This trade-off often makes Chebychev II filters undesirable for DSL. Matlab also contains a function called "cheb2ap" that determines the poles and zeros of a Chebychev II filter with $\omega_p = 1$. This function is typically given as

$$[z,p,k] = \text{cheb2ap}(n,\ sr) \tag{3.88}$$

where n is the order of the filter desired; z, p, and k are again vectors of the respective zeros (will contain nothing since there are none), n poles, and gain; and sr is the stopband ripple that can be tolerated in dB. The filter is given by

$$H(s) = \frac{k \cdot \prod_{i=1}^{n}(s - z_i)}{\prod_{i=1}^{n}(s - p_i)}$$

(3.89)

Again, frequency scaling, magnitude scaling, and frequency transformation can be used to convert this function into the appropriate filter for each increasing value of n until the filter specification is met. Again, the function "cheb2ord" can be used to determine a minimum value of n from a given filter specification.

3.8.2.4 Elliptic (Cauer) Designs

Elliptic lowpass analog filters have a magnitude characteristic:

$$|H(f)|^2 = \frac{1}{1 + \varepsilon^2 R_n^2\left(\omega/\omega_p\right)}$$

(3.90)

where n is the order of the filter (number of poles and number of zeros in this case) and ω_p is the passband edge and R_n is an elliptic function [20]. Elliptic filters have ripple in both pass and stop bands. They provide the sharpest transition in frequency of the four types but admit passband ripple distortion and do not further decay with increasing frequency. Elliptic filters are probably the most popular in use for DSL POTS splitters. An example of an 8th-order Elliptic filter's magnitude is plotted in Figure 3.24. The transition band is much sharper than in the previous filters, but the stopband attenuation is limited at the highest frequencies. Elliptic filters are used heavily in POTS splitter designs. The sharpest transition band also leads to severe phase distortion near the passband edge, which may or may not be a problem for the transmission line code selected — with most methods internal digital signal processing is used to offset the effect of the phase distortion for the DSL signals themselves, while phase distortion often has inaudible effect on POTS audio signals. Matlab also contains a function called "ellipap" that determines the poles and zeros of a Elliptic filter with $\omega_p = 1$. This function is typically called as

$$[z,p,k] = \text{ellipap}(n, pr, sr)$$

(3.91)

where n is the order of the filter desired; z, p, and k are again vectors of the respective zeros (will contain nothing since there are none), n poles, and gain; pr is the passband ripple and sr is the stopband ripple that can be tolerated in dB. The function "ellipord" can be used to determine the minimum value of n necessary to meet a given filter specification. The filter is given by

$$H(s) = \frac{k \cdot \prod\limits_{i=1}^{n}(s - z_i)}{\prod\limits_{i=1}^{n}(s - p_i)}$$

(3.92)

Again, frequency scaling, magnitude scaling, and frequency transformation can be used to convert this function into the appropriate filter for each increasing value of n until the filter specification is met.

3.8.3   Lossless Realization of H(s)

Once a desired transfer function has been obtained using the design methods of Section 3.8.2 or any other method (rational and stable), a designer can somewhat automatically convert this transfer function into a ladder circuit. This subsection again works only with lowpass designs because the frequency transformations of Section 3.8.4 allow conversion to highpass, bandpass, or bandstop designs. This subsection further assumes that purely resistive source and loads are used. Any realizable transfer function for a two-port network, $H(j\omega)$, can be converted into the return loss of a lossless circuit via

$$|A(j\omega)|^2 = \left| \frac{R_S - Z_1}{R_S + Z_1} \right|^2 = 1 - 4 \cdot \frac{R_S}{R_L} \cdot |H(j\omega)|^2$$

(3.93)

which can be easily proven using algebra — however, one might also note that Equation (3.93) is also a "conservation of energy" relation, observing that the sum of reflected energy and energy transferred to the load must be equal to the total energy. The function in Equation (3.93) should be factored ("analytic continuation") to generate a function $A(s)$ so that

$$|A(j\omega)|^2 = A(s) \cdot A(-s)\big|_{s=j\omega}$$

(3.94)

which is easily executed using, for instance, the "roots" command in matlab to separate the poles into left and right half-plane sets and the zeros into any symmetric sets. The actual transfer function $H(j\omega)$ may need reduction by a constant scaling factor in order for the function $|A|^2 \geq 0$ at all frequencies, but this scaling has no effect on the filter's basic effect.

Once the return loss is computed via Equations (3.93) and (3.94), the input impedance of the two-port network readily follows according to

$$Z_1(s) = \left(\frac{1 - A(s)}{1 + A(s)} \right)^{\pm 1}$$

(3.95)

A ladder circuit with the desired impedance is easily then determined by continued fraction expansion of $Z_1(s)$. Continued fraction expansion follows by dividing the higher-degree polynomial in the numerator or denominator of $Z_1(s)$ by the lower-degree polynomial, then continuing the polynomial division process for the remainder polynomial ratio. This continued process will produce a ladder circuit that can be synthesized directly from the continued fraction by reading lead coefficients. Either of the polarity exponents in Equation (3.95) can be used to generate dual circuits with the same transfer function realization. The decision as to which circuit to use is then up to the designer. For detailed examples of the continued-fraction computations, the reader is referred to the text by Van Valkenburg [20], pp. 405–410.

3.8.4   Frequency/Magnitude Scaling and Frequency Transformations

Frequency and magnitude scaling are often used to convert the resistance, capacitance, and inductance values for realization of a desired transfer function as in Section 3.8.3 into more practical values. For more details, see Ref. [20], Appendix A, and Chapter 11.

Magnitude Scaling Rule: The resistance, capacitance, and inductance values in any circuit realization of a transfer function may be scaled by the same constant value k_m according to

$$
\begin{aligned}
R_{new} &\leftarrow k_m \cdot R_{old} \\
L_{new} &\leftarrow k_m \cdot L_{old} \\
C_{new} &\leftarrow \tfrac{1}{k_m} \cdot C_{old}
\end{aligned}
$$

(3.96)

without change.

Frequency Scaling Rule: Given a certain filter transfer function and realization, the frequency access of the transfer function may be stretched/contracted by a factor k_f by scaling the component values according to

$$
\begin{aligned}
R_{new} &\leftarrow R_{old} \\
L_{new} &\leftarrow \frac{L_{old}}{k_f} \\
C_{new} &\leftarrow \frac{C_{old}}{k_f}
\end{aligned}
$$

(3.97)

Frequency scaling can be used to scale cut-off or passband/stop edge frequencies from some nominal value like 1 for lowpass designs to a desired actual frequency, thus sometimes simplifying tabular listing of well-known filter types and (frequency-normalized) component values.

Frequency Transformations: These allow the direct conversion of a lowpass filter into a highpass filter, into a symmetrical bandpass filter, or into a symmetrical bandstop filter. Thus, a designer only need design lowpass filters with passband cut-off frequency $\Omega_p = 1$ and relative stopband frequency of $\Omega_s = \omega_s/\omega_p > 1$. For a lowpass filter with alternative cut-off frequency $\omega_p \neq 1$, the designer can frequency and magnitude scale as described above. For a highpass or symmetrical bandpass/stop filter, the designer frequency transforms and magnitude scales.

highpass Transformation: To obtain a highpass filter with cut-off frequency $\omega_{h,p}$ and stop frequency $\omega_{h,s}$, a designer designs the lowpass filter with $\Omega_p = 1$ and $\Omega_s = \omega_{h,p}/\omega_{h,x}$ and then replaces the variable s in the resultant transfer function $H(s)$ by

$$S = \frac{\omega_{h,p}}{s}$$

(3.98)

Bandpass Transformation: To obtain a bandpass filter with passband cut-off frequencies ω_1 and $\omega_2 > \omega_1$ and stop frequencies ω_3 and $\omega_4 > \omega_3$, the design specification must first be tightened to a symmetrical specification where $\omega_1 \cdot \omega_2 = \omega_3 \cdot \omega_4$ by movement of one or both of the stopband frequencies into the transition band. One defines the quantities

$$\omega_0^2 = \omega_1 \cdot \omega_2 \quad \text{and} \quad B = \omega_2 - \omega_1$$

(3.99)

Then, one designs the lowpass filter with $\Omega_p = 1$ and $\Omega_s = \frac{\omega_4 - \omega_3}{\omega_2 - \omega_1}$ and then replaces the variable s in the resultant transfer function $H(s)$ by

$$S = \frac{1}{B} \cdot \frac{s^2 + \omega_0^2}{s}$$

(3.100)

Bandstop Transformation: To obtain a bandstop filter with passband cut-off frequencies ω_1 and $\omega_2 > \omega_1$ and stop frequencies ω_3 and $\omega_4 > \omega_3$, the design specification must first be tightened to a symmetrical specification where $\omega_1 \cdot \omega_2 = \omega_3 \cdot \omega_4$ by movement of one or both of the stopband frequencies into the transition band. One defines the quantities

$$\omega_0^2 = \omega_2 \cdot \omega_1 \quad \text{and} \quad B = \omega_2 - \omega_1$$

(3.101)

Then, one designs the lowpass filter with $\Omega_p = 1$ and $\Omega_s = \dfrac{\omega_2 - \omega_1}{\omega_4 - \omega_3}$ and then replaces the variable s in the resultant transfer function $H(s)$ by

$$S = B \cdot \frac{s}{s^2 + \omega_0^2} \tag{3.102}$$

Table 3.9 summarizes frequency transformations and also what happens to each inductor or capacitor in a circuit realization (resistors remain the same under frequency transformations or frequency scaling).

Table 3.9   Frequency Transformation Summary[*]

Lowpass	Highpass	Bandpass	Bandstop
L	$1/\omega_0 L$	$L/B \quad B/\omega_0^2 \cdot L$	$B \cdot L/\omega_0^2 \quad 1/B \cdot L$
C	$1/\omega_0 C$	$B/\omega_0^2 \cdot C \quad C/B$	$1/B \cdot C \quad B/\omega_0^2 \cdot C$

[*]Source: Ref. [20]

3.8.5   Active Filters

Active filters consume power, unlike the passive filters of Section 3.8.4. Nevertheless, active filters are also used in DSLs. While all filters in Section 3.8.4 can be realized passively, most manufacturers desire highly integrated DSL transceiver realizations in VLSI. To date, the inductors necessary in the passive designs cannot be directly synthesized in VLSI. Active filters, while consuming power, avoid the use of inductors in realizing any filter design, often leading to lower-cost implementation of the transceiver. Section 3.9 discusses splitter circuits, which sometimes must be passive and therefore must use inductors. However, most other analog filters in a DSL system can be synthesized actively with lower cost than a passive realization, despite the additional power consumption of the filter.

Furthermore, designs so far have assumed that the two-port filter network has control of the load impedance Z_L, when in fact this impedance can vary with each telephone line, rendering the design less effective when the wrong load impedance was estimated. Active filters permit designs that can be much less sensitive to knowledge of the load impedance. When the terminating load is not well known, a bidirectional filter can also exhibit undesirable return-loss variations that lead to larger echoes (see Chapter 5). Active circuits can force the return loss to appear much more constant.

In addition to power consumption, the drawbacks of active filters include limitations on the bandwidth of signals (which fortunately has become of less concern for DSLs with recent advances in semiconductor technology) and less linearity than passive circuits.

Usually, considerable analog filtering occurs in a DSL transmitter after the DAC but before the analog hybrid/driver circuit (see Section 3.9) and correspondingly after the hybrid/ driver but before the ADC in the receiver. The DSL-nearest stages of this filtering are usually implemented in VLSI on an analog-integrated circuit (some manufacturers combine this analog circuit with digital circuits). These circuits will use active filters to avoid inductors. These circuits most often use bipolar technology used to handle the large voltages on the twisted pair in DSL designs. Some designs attempt partial CMOS realization instead so that analog and digital circuits may be more easily combined since most digital signal-processing circuits use CMOS — in that latter case, switched-capacitor active filters are the usual choice.

There are essentially two approaches to active filter synthesis for DSLs. The first is based on factoring the transfer function into quadratic sections with direct realization of each section with an active circuit. The second is based on the ladder designs of Sections 3.8.3 and 3.8.4, with active elements replacing the reactive elements — this is known as "leap-frog" design.

3.8.5.1 Biquadratic Designs

The transfer function derived from a filter design $H(s)$ can be factored into quadratic sections

$$H(s) = \prod_i H_i(s)$$

where each factor has the general form

$$H_i(s) = \frac{as^2 + bs + c}{s^2 + \frac{\omega_0}{Q} \cdot s + \omega_0^2}$$

where $a, b, c, Q,$ and ω_0 are all real constants. Each factor can be realized through the use of resistors, capacitors, and operational amplifiers. The determination of such factors today is easy using root-solving software like the roots command in Matlab. Usually, considerable trial and error is used in design to associate the best combinations of two-poles and two-zeros for each section $H_i(s)$, but clearly the worst such subsection should have its degree of frequency variation minimized by such choice. For a good discussion of operational amplifier usage in analog filters, see Ref. [20]. Fortunately, a cascade of such individual realizations in series do not significantly load each other and thus realize the filter transfer function.

The study of active filtering often concentrates in part on the question of best circuits to implement a given quadratic transfer function. The answer in general is that it depends on the values of the constants a, b, c, Q, and ω_0. Sensitivity, accuracy, and the number of operational amplifiers (translating to cost and power consumption) are often traded for different types of filter designs.

For quadratic transfer functions like those arising in lowpass Butterworth and Chebychev I designs where $a = b = 0$, the so-called lowpass Sallen-Key circuit uses only one operational amplifier and has good sensitivity (loss in filter accuracy for given inaccuracy in component values). There are several variations (see Ref. [20]). This circuit can also be used for Butterworth and Chebychev-I highpass designs *(b = c = 0)* by switching the role of resistors and capacitors.

For quadratic transfer functions arising from Butterworth and Chebychev I bandpass designs *(a = c = 0)*, an alternative circuit is the Delyiannis-Friend circuit (see Ref. [20]), which again has good sensitivity and uses only one operational amplifier.

Elliptic filters require special care and are candidates for circuits employing more than one operational amplifier because of the more general nature of the resultant subtransfer functions (fewer if any zeroed constant values). Some of these more general circuits appear in Chapter 4 of Ref. [20]. Operational amplifiers can also be used to take differences or sums of DSL voltages. In general, the area of active circuit design is one that has entire books devoted to it, and so the reader is referred to Refs. [20] and [21].

3.8.5.2 Leap-Frog Designs

Leap-frog active designs often find use in DSLs. These designs typically use a larger number of operational amplifiers, but are based on passive ladder designs, allowing ease of partial active and passive design (in particular allowing passive components in stages containing the largest signal levels). Further, ladder designs are well known to have low sensitivity to component variances, and this low sensitivity is preserved in leap-frog designs, allowing a more repeatable mass production of accurate filter realization. For a good introduction to leap-frog design, see Chapter 15 of Ref. [20].

3.8.5.3 Switched Capacitor Designs

CMOS integration allows implementation of only capacitors and switches, along with operational amplifiers. Resistor values are not reliably repeated across a CMOS circuit, while capacitors are. With clock speed considerably in excess of signal bandwidths, the rapid charging and discharging of a capacitor can be made to appear resistive at lower signal bandwidths. Switched-capacitor designs allow realization of analog filters in CMOS. With today's CMOS clock speeds of 250 to 500 MHz and beyond in the future, switched-capacitor technologies can be used to implement a growing fraction of active analog filters. The basic cascaded-quadratic and leap-frog design technologies are again used, but with switched-capacitor realization of resistors. The drawback of switched-capacitor technologies are the voltage levels of DSL signals — sometimes as high as 10 to 30 volts. CMOS supply voltages have decreased from 5 volts to 3.3 volts, now 2.5 volts, and continue to decrease. Thus, DSLs using larger voltages will at least have to have some bipolar filter realization with greater supply voltages. At present, most DSLs use at least a small amount of bipolar realization. For more on switched-capacitor filtering, see Ref. [21].

3.8.5.4 Generalized Impedance Conversion

Of particular concern is the input impedance of a two-port network used in bidirectional filters. If the load impedance cannot be accurately predicted, as happens often with the lower frequencies on a twisted pair (higher-frequency line impedance tends to a relatively constant resistive value for most gauge wires, regardless of bridge-taps or remote load), the input impedance may behave unpredictably. This is especially true at frequencies below 20 kHz. The upredictable impedance causes poorer return loss, leading to undesirable echoes. This effect is most pronounced in the situation concerning lowpass filters in POTS splitters as described in Ref. [16]. In this case, a generalized impedance converter, which is an active circuit, can be employed to force the reflected impedance to look much more constant (also described in Ref. [16]).

3.9 Three-Port Networks for DSLs

Figure 3.25 shows two three-port networks of interest in DSLs — POTS splitters and hybrid circuits. POTS splitters separate DSL signals from telephony signals on twisted pairs where the three ports are the telephone line (LINE), the connection to the phone (TELE), and the connection to the DSL modem (DSL). Hybrid circuits separate transmit (XMIT) and receive (RCVR) signals from the bidirectional line (LINE). There are two types of hybrids for POTS and for DSL. Typically, POTS (or ISDN) hybrid circuits have their LINE output fed into the TELE input of a splitter, while DSL hybrid circuits have their LINE output fed into the DSL port of a splitter. Splitters are typically used with ADSL and VDSL, although it is possible to use them for HDSL or ISDN with some consequent complexity increase for the associated receivers.

Instead of the two insertion-loss, balance, and return-loss specifications of the two-port (one of each for the two directions), there are up to six connections of interest for specification of insertion loss, return loss, and balance in the three-port. This section discusses each of their specifications for the two types of three-port networks.

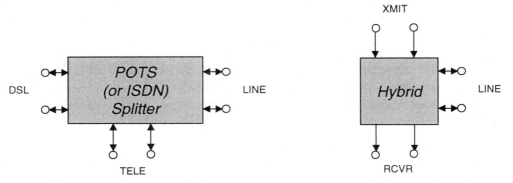

Figure 3.25   Three-port networks for DSL.

3.9.1   POTS Splitters

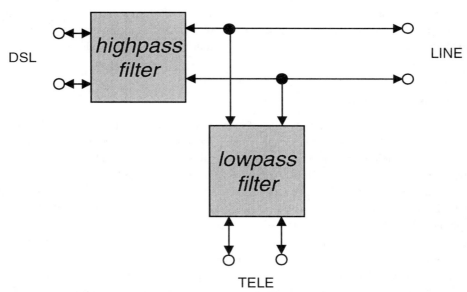

Figure 3.26   Splitter circuit (both filters are bidirection LPF or HPF as shown).

This section separately addresses each of the three possible two-port connections in a splitter. Figure 3.26 further divides the POTS splitter functionality into lowpass and highpass filters. Typically, the highpass filter is located within a DSL transceiver, while the lowpass filter may be separate from the DSL transceiver and may be within the DSL transceiver, or may be within/co-located with the telephone handset.

A more complete discussion of ADSL splitters can be found in the excellent article by Cook and Sheppard, Ref. [16].

3.9.1.1 TELE to LINE

The TELE to LINE connection bidirectionally passes only the low-frequency POTS (or ISDN) signals between the telephone line and the TELE port. DSL signals should not appear at the TELE port, nor should TELE signals appear at the DSL port. The lowpass filtering should attenuate all but the lowest DSL frequencies as DSL signals attempt to pass to the TELE port. The highpass filter should thus appear as a high impedance (meaning it starts with series capacitance) in the low-frequency band of POTS/ISDN and as a low impedance in the high-frequency band of DSL, so that the three-port appears as a two-port between TELE and LINE for frequencies used by POTS/ISDN. For testing and design purposes, the DSL port of a splitter is typically terminated in an approximately 100 to 135 ohm resistive load, which is more specifically modeled by the circuit of Figure 3.27. At high frequencies, the circuit looks like 100 ohms with U.S. and Asian cable. In Europe, cable with the 135 ohm impedance is more commonly used.

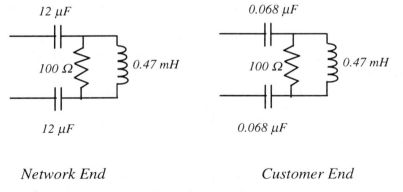

Network End   *Customer End*

Figure 3.27   DSL port termination impedances for TELE-to-LINE evaluation.

The lowpass filter should also attenuate the higher-frequency components of on-ring, trip-ring, and any other signaling voltages that originate in POTS or PSTN equipment. It is usually these latter type of ring-transient signals that determine the amount of stopband rejection necessary by the lowpass filter of splitter, rather than the requirements of rejecting the high-frequency components of DSL. This is because the ring signals can have temporary levels of 10s to over 100 volts, and even at higher frequencies the resultant transients can otherwise overwhelm a DSL transceiver. (Such transients are one source of the impulse noise discussed in Section 3.6.3.)

DC Characteristics   As an initial constraint, the TELE-to-LINE connection must carry the maximum of 100 mA of DC current that is used to power telephones. DC voltages of up to 105 volts must be able to pass, as should ringing signals (frequencies typically 20 to 30 Hz) superimposed on the DC voltage with rms voltages of 103 VAC rms. The DC resistance of the DSL port external termination should exceed 5 Megaohms when testing or designing, which means the input DC resistance of the highpass filter in a splitter should exceed 5 Megaohms when terminated in the DSL transceiver impedances shown in Figure 3.27. When the highpass filter is separated from the lowpass, series capacitors (12 µF in each lead in the United States [12]) may be inserted into the leads on the LPF that go to the HPF and need to be remembered in the design of any remote HPF filter, essentially ensuring that the DC resistance requirement is met no matter what is attached to the leads.

Insertion Loss   The insertion loss of the TELE-to-LINE connection should be small across the voice-band from 300 Hz to 3300 Hz, with attenuation increasing to 80 dB or more at some frequency just above the POTS or ISDN bands but below the DSL band, i.e., a lowpass filter. Typically stop-band edge frequencies are from 30 kHz for lower-speed DSLs with POTS splitters to 150 to 300 kHz for higher-speed DSLs with ISDN splitters. A passive version of such a filter will likely be implemented with series inductance and parallel capacitance, a lossless network. Lossless passive networks of this type have symmetrical two-port models, which means that the insertion loss and corresponding transfer function is the same in both directions as long as the terminating impedances at each end are the same, but often the actual insertion loss characteristic is not strongly sensitive to the terminating impedance values. Table 3.9 shows some choices for termination impedances for testing. Typically, the purely resistive values are used for lowpass filter design.

Table 3.10   POTS-Band Splitter Termination Impedances for Lowpass Filter Insertion and Return Loss Calculation and Design[*]

TELE PORT – CO splitter — USA	900 Ω
TELE PORT – RT splitter — USA	600 Ω
LINE PORT – CO splitter	800 Ω resistor in parallel with series connection of 100 Ω resistor and 50 nF capacitor
LINE PORT – RT splitter	1330 Ω resistor in parallel with series connection of 348 resistor and 100 nF capacitor
LINE PORT – ITU Blue Book	320 Ω resistor in series with the parallel connection of a 1050 Ω resistor and 230 nF capacitor

[*]The handset or central office impedance may not be purely resistive by design (as a better match to line characteristics) in many countries of the world. The USA network is older and has traditionally used the purely resistive values of 900 ohms at the customer end and 600 ohms at the central-office end.

The delay distortion of a lowpass filter is typically of second-order concern for voiceband audio/telephony signals but may be of more importance if voiceband modems are used. Increase in delay caused by insertion of the POTS splitter should be limited to less than 250 µs to ensure that voice-band modems are not unduly compromised.

Return Loss   Return loss is the ratio of reflected signal power to incident signal power as described in Equation (3.31) in Section 3.5. While insertion loss may be relatively insensitive to termination impedance of the splitter, return loss is not. This is because the impedance reflected through the splitter can be considerably more reactive than the line itself, meaning the original POTS or PSTN equipment was designed for a different impedance (see Section 3.9.2 on hybrid circuits). Capacitive loads can lead to significant return-loss problems if the designed cut-off frequency of the lowpass filter is tight (i.e., the DSL uses bandwidth well below 100 kHz). The reflection of voice signals from the splitter can have an audibly noticeable, if not annoying, effect on a user of a telephone. The magnitude of the problem depends largely on the cut-off frequency and the actual line to which the LINE port of the splitter is connected. This text will not pursue this problem in earnest here, but the reader is referred to Ref. [16] for a sampling of the magnitude of the problem. As Ref. [16] also shows, the problem can be virtually eliminated with active splitter designs or with passive designs where the cut-off frequency is set sufficiently high (thus imposing some performance loss, acceptably or not, on the DSL system).

Balance   As discussed in earlier sections, balance in the telephony band of a phone line is typically 50 to 60 dB or more. As frequency increases, this balance decreases on the phone line (see Section 3.5.8). Telephones on customer premises can be very poorly balanced, with the amount of balance being particularly low for ringing and signaling signals. High-frequency common-mode signals from the LINE to the TELE are unlikely to have significant effect on the phone, especially if the DSL transmitter is reasonably well balanced itself. However, if ring signals and associated transients were to pass through a common-mode (longitudinal) path to the DSL transceiver, then impulse noise has been created through a path that may bypass the well-designed differential lowpass filter in the splitter.

Figure 3.28(a) shows a single-ended passive lowpass filter section, while Figure 3.28(b) shows its associated balanced implementation. Any common-mode (longitudinal) signals on each of the wires in a twisted pair input to the filter (or on the two wires input to the TELE port from the phone) thus undergo a common impedance through the filter. The longitudinal components in each of the paths through the filter thus remain the same. Then, balanced differential processing of signals should retain the level of balance to the common-mode signal.

Typically, to ensure that the inductors are identical in upper and lower paths, they are often implemented with transformers, as shown in Figure 3.28(c). In the configuration shown, the total inductance in both leads is $2(L'+M')$ when the transformer "dots" are as indicated and the transformer is 1:1. L' is the self-inductance of either of the coils, while M' is the mutual inductance between coils. Since the coils in the transformer share a common magnetic core, then it is more likely that the upper and lower paths are identical than in a realization with separate inductors as in Figure 3.28(b). Furthermore, the transformer should provide more inductance per size and per cost than the two independent inductors. This configuration will give maximum inductive impedance, but will more easily let common-mode components pass (especially when $M \gg L$). Thus, the first inductor closest to the TELE port is usually implemented with a transformer as shown in the **common-mode-choke (CMC)** of Figure 3.28(d). In this case, common-

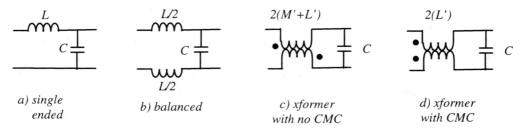

Figure 3.28   Single-ended versus balanced filter realization.
(a) single ended; (b) balanced; (c) xformer with no CMC; (d) xformer with CMC

mode components are those that are primarily reduced and see the largest inductance (especially at high frequencies). The ensuing inductors are then realized as in Figure 3.28(c) to minimize size for a given amount of impedance.

Sample Lowpass Splitter Designs   Two sample lowpass filter designs, one passive and one active, for ADSL (see Ref. [16]) and another for POTS/ISDN-splitting in VDSL (see Ref. [17]) appear in Figure 3.29 and Figure 3.30. The active ADSL design still is balanced and bidirectional, but uses an active circuit known as a generalized immitance converter (GIC) to cause the input impedance to better match the TELE port impedance and thus reduce return loss. The active designs are typically more effective in ADSL deployments in countries where telephone equipment (central office PSTN and/or telephones POTS) uses nonresistive termination impedances by design, for instance the UK and Australia.

The VDSL splitter [17] passes either POTS or ISDN signals, in this case up to 600 kHz, and meets requirements of insertion and return loss with passive components because of the relatively high cut-off frequency. This design appears in Figure 3.31.

Note that all designs have first component equal to a large inductance when viewed by looking into the LINE port. This large inductance value makes the lowpass filter appear as a very high impedance within the passband of the highpass filters to be discussed next. Thus, the lowpass filter's high impedance at high frequency makes the splitter appear as a two-port for the design of the highpass filter in the next section.

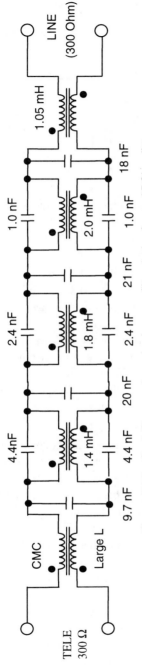

Figure 3.29 Cook-Sheppard passive lowpass filter design for an ADSL splitter.

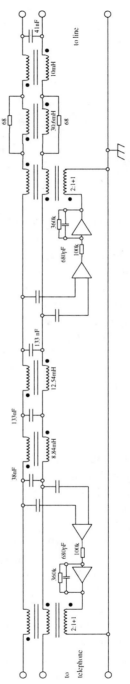

Figure 3.30 Cook-Sheppard active lowpass filter design for an ADSL splitter.

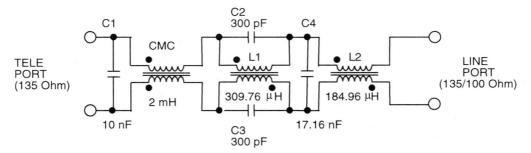

Figure 3.31 Cook-Foster passive lowpass filter design for a VDSL splitter.

3.9.1.2 DSL to LINE

The DSL to LINE interface is the highpass filter shown in Figure 3.26. The purpose of this filter is to prevent low-frequency DSL signal energy from entering the passband of the POTS (or ISDN) signals, while simultaneously preventing the low-frequency POTS and ringing signals from entering the DSL receiver. DSL systems commonly implement this highpass filter within the DSL transceiver, although not always. Splitters with separated lowpass filter sometimes implement the first (series capacitance) stage of highpass filtering within the lowpass filter. The capacitors ensure DC blocking in case the DSL port "short-circuits" accidentally (and thus protects against accidental disruption of POTS service powering). However, this practice seems to have become less common due to the increased interest in splitterless operation, in which case the highpass design would need to be robust to either case, complicating its design.

Insertion Loss The highpass filter should appear as a high impedance at low frequencies to prevent loading of the lowpass TELE-to-LINE filtering. A typical POTS specification might be that [12], the POTS-band noise from DSL must be less than 18 dBrnC. The unit, dBrnC is referenced to –90 dBm/Hz with a filter that has 3dB cut-off frequencies at 300 Hz and 3000 Hz. It roughly corresponds to a noise PSD of about –110 dBm/Hz, so that DSL signals typically at –34 dBm/Hz to –50 dBm/Hz (VDSL), depending on the DSL type, must be reduced by 60 to 76 dB. Thus, stopband rejection of 60 to 80 dB is typically specified for the HPF, with the larger attenuation being common with lower-speed DSLs. These numbers are also applicable to ISDN roughly. A desire for DSL use of lower-frequency spectrum imposes a more narrow transition band of this highpass filter (as well as of the associated lowpass filter). For frequency-division multiplexed DSLs, one direction of transmission may have a very tight constraint with respect to POTS/ISDN splitters, typically upstream. 6th- to 8th-order or even higher-order highpass filters are common (some FDM systems use 12th- to 13th-order elliptical filters).

The insertion-loss of the filter for DSL transmission depends on the transmission line code and the DSL type, but typically the passband has an edge frequency of 30 to 200 kHz, with the upper frequencies being for ISDN-splitters or VDSL and the lower frequency being for ADSL.

The filter is roughly flat to the edge of the DSL passband (552 kHz for ADSL-LITE, 1.104 MHz for ADSL, 20 MHz for VDSL) and then must reduce again. So, the highpass filter is really a bandpass filter when implemented at the DSL location. Insertion loss should look flat across the band, with typical ripple of 0.5 dB to 2 dB. Delay distortion in the band edges can be a problem for most transmission methods, so sometimes Chebychev filters are used instead of Elliptic filters. The level of signal reduction should also prevent any POTS-band low-frequency ring-transient signal energy from saturating the front-end electronics of the DSL receiver.

The reference impedance attached to the TELE port for design of the highpass filter should be one of those from Figure 3.32. It may be desirable simply to assume a high upper-frequency impedance.

Return Loss   The return loss is only of major importance for echo-canceled DSLs (as long as saturation of receiver front-ends will not occur). Insertion of a passive circuit can cause the appearance of a complex impedance to the hybrid circuit where a resistive impedance of about 100 ohms may have otherwise been assumed. Typically, since the highpass filter is located with the DSL, the impedance match can be better controlled because both DSL and filter have been designed by the same engineer.

Design Example   A passive separate highpass filter for a POTS + ISDN splitter with cut-off frequency of 500 kHz for VDSL has been designed by Cook and Foster [17] and appears in Figure 3.32. It is a 4th-order Chebychev design. Note that the transformer/inductor and capacitance elements have been interchanged in position with respect to the lowpass filters discussed earlier.

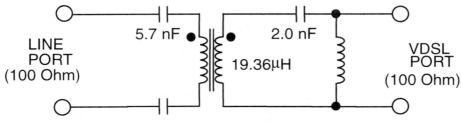

Figure 3.32   Cook/Foster separate passive highpass filter for VDSL splitters.

3.9.1.3   TELE to DSL
The TELE to DSL path through the splitter ideally passes no signals. The stopband of the highpass filter should overlap the passband of the lowpass filter and vice versa. The level of stop-band insertion loss of 60 to 80 dB is typically sufficient to prevent harmful energy transfer. Each filter should appear as a high impedance (to prevent loading) in the other's passband. Good design of the TELE-to-LINE lowpass and DSL-to-LINE highpass will essentially ensure that the TELE-to-DSL path is opaque.

3.9.2 Hybrid Circuits

The function of a hybrid circuit is to separate single-ended transmit and receive circuits from the line (the line may be the TELE port of a splitter when splitters are used). Figure 3.33 illustrates basic hybrid circuit operation using the so-called "**bridge**" circuit. The three ports are XMIT, RCVR, and LINE. Signals nominally flow from the XMIT to LINE port, but ideally "echo" signals should not flow from the XMIT to the RCVR. Signals also flow from the LINE port to the RCVR port. The hybrid is thus a four-wire to two-wire interface.

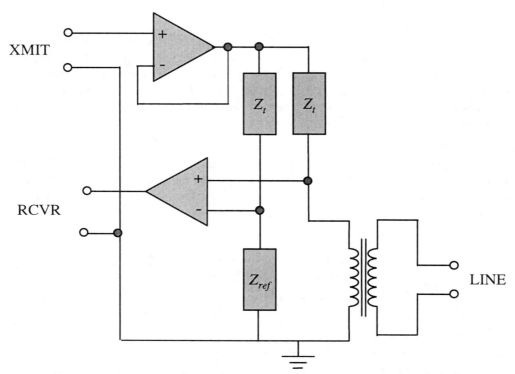

Figure 3.33 Basic hybrid circuit, "bridge."

There are two hybrid circuits of interest in DSL: one is for the telephone when DSL and POTS occupy the same line, and the other is the hybrid for the DSL transceiver. The hybrid circuits in telephones are addressed in many basic books on telephony (see Ref. [3] for example).

Of crucial importance to proper function of the hybrid is the reference impedance, which ideally should equal the line impedance. In this case, simple tracing of currents and voltages in the hybrid will find that no signal leaks from the XMIT port to the RCVR port. Further, the transmit impedance should be the conjugate impedance of the line impedance for maximum energy transfer to the line. The RCVR port–terminating impedance is usually chosen to eliminate bouncing (and so is equal to the line impedance, not its conjugate). Thus, some energy is lost in the LINE-to-RCVR transition (the gain on the insertion loss is less than unity while the return loss is ideally infinite) to reduce echoes.

DSL hybrid circuits need to be balanced in their realization to prevent leakage of longitudinal signal components to the RCVR. Also, the balancing further helps eliminate any common-mode noises the transmit path. Some of the highpass and also bandpass filtering of the transmit and receive path may be absorbed into the hybrid circuit or may actually precede it in implementation if implemented actively. The hybrid usually involves line-driving transformers and the final-stage driver amplifiers so that any filters that follow would need passive implementation. A sample hybrid/driver circuit has been published by Analog Devices [18] and appears in Figure 3.34. Some other examples appear in Ref. [19].

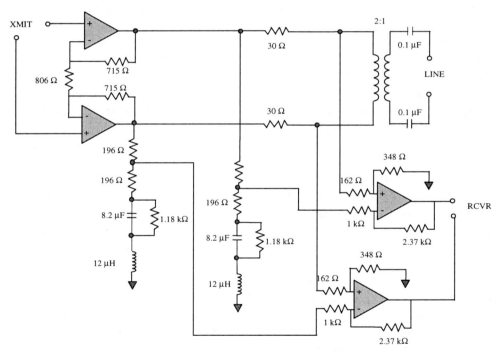

Figure 3.34 Sample ADSL hybrid/driver circuit (From Ref. [18]).

References

[1]   D. K. Cheng, *Field and Wave Electromagnetics*, 2nd Ed. Addison Wesley: Menlo Park, CA, 1992.

[2]   J. J. Werner, "The HDSL Environment," *IEEE Journal on Selected Areas in Communications,* Vol 9, No. 6, August 1991, pp. 785–800.

[3]   W. D. Reeve, *Subscriber Loop Signaling and Transmission Handbook – Analog.* IEEE Press: New York, 1992; and *Subscriber Loop Signaling and Transmission Handbook – Digital.* IEEE Press: New York, 1995.

[4]   C. Valenti, "Cable Crosstalk Parameters and Models," *ANSI Contribution T1E1.4/97-302,* Bellcore, Minneapolis, MN, September 1997.

[5]   K. J. Kerpez, "Near End Crosstalk is Almost Gaussian," *IEEE Transactions on Communications,* Vol. 41, No. 5, May 1993, pp. 670–672.

[6]   B. Pederson and D. Falconer, "Minimum Mean Square Equalization in Cyclostationary and Stationary Interference Analysis and Subscriber Line Calculations," *IEEE Journal on Selected Areas in Communications,* Vol. 9, No. 6, August 1991, pp. 931–40.

[7]   M. Abdulrahman and D. Falconer, "Cyclostationary Crosstalk Suppression by Decision Feedback Equalization on Digital Subscriber Loops," *IEEE Journal on Selected Areas in Communications,* Vol. 10, No. 3, April 1992, pp. 640–649.

[8]   G. H. Im and J. J. Werner, "Bandwidth Efficient Digital Transmission over Unshielded Twisted-Pair Wiring," *IEEE Journal on Selected Areas in Communication,* Vol. 13, No. 9, December 1995, pp. 1643–1655.

[9]   K. T. Foster and J. W. Cook, "The Radio Frequency Interference (RFI) Environment for Very High-Rate Transmission Over Metallic Access Wire-Pairs," *ANSI Contribution T1E1.4/95-020,* Dallas, TX, February 27, 1995.

[10]   J.M. Cioffi, "VDSL System Requirements Document," *ANSI Contribution T1E1.4/98-043R3,* June 1998, available at Web site: ftp://ftp.t1.org/pub/t1e1/e1.4/dir98/8e14043x.pdf (x is revision number, use highest value).

[11]   J. Cook, "Wideband Impulsive Noise Survey of the Access Network," Vol. 11, No. 3, July 1993, *BT Technical Journal,* pp. 155–162.

[12]   American National Standard T1.413-1999, *Asymmetric Digital Subscriber Line (ADSL) Metallic Interface,* for draft see web site: ftp://ftp.t1.org/pub/t1e1/e1.4/dir98/8e14007x.pdf (x is revision number, use highest value).

[13]   R. W. Lawrence, "Impulse Noise Test Results for Three Line Code Technologies Proposed for ADSL," *ANSI Contribution T1E1.4/93-078,* March 10, 1993, Miami, FL.

[14]   P. Walsh, "Proposed FCC Part 68 Requirements for Customer Premises ADSL Equipment," *Telecommunications Industries Association (TIA) Contribution TR41.9/98-02-010,* November 19, 1997, San Antonio, TX.

[15]   R. Roberts, "VDSL Out-of-Band Spectral Roll-Off: FCC Part 15 Compliance Based Upon Flat-Pair Distribution-Cable Radiation," *ANSI T1E1.4/97-224,* Minneapolis, MN, September 1997.

[16]   J. Cook and P. Sheppard, "ADSL and VADSL Splitter Design and Telephony Performance," *IEEE Journal on Selected Areas in Communication,* Vol. 13, No. 9, December 1995, pp. 1634–42.

[17]   J. Cook and K. Foster, "Splitter for VDSL," *ANSI Contribution T1E1.4/96-014,* Irvine, CA, January 1996.

[18]   AD816 Application Note "AD816 as an ADSL Transceiver," Analog Devices, Rev 0, 1997.

[19]   D. Macq, "DMT ADSL Circuits and Systems," Alcatel, IEEE ISSCC '98 Conference, San Francisco, CA, April 1998, xDSL Broadband Interactive Communications (short course).

[20]   M. E. Van Valkenburg, *Analog Filter Design.* Holt, Reinhart, & Winston: New York, 1982.

[21]   M. S. Ghausi and K. R. Laker, *Modern Filter Design.* Prentice-Hall, Inc: Englewood Cliffs, NJ, 1981.

[22]   J. A. C. Bingham, *Multicarrier Modulation and xDSL.* Wiley: New York. In press.

[23]   ETST TM6 Permanent Document No. 4, 960p04 revision 8, Living List (revision 8) for amendment of TS 080 (ETR 080), June 2, 1998.

[24]   J. Cioffi et. Al., "Mitigation of DSL Crosstalk via Multiuser Detection and CDMA," ANSI Contribution T1E1.4/98-253, August 31, 1998, San Antonio, TX.

Comparison with Other Media

DSLs allow broad-bandwidth signals to be delivered to a customer over existing twisted pairs. Competing alternatives to deliver the same signals to the same customers are manifold, but usually require installation of new media or reservation of scarce wireless bandwidth. This chapter briefly reviews the relative strengths and weaknesses of the principle alternatives to DSLs: fiber-to-the-home, coax, terrestrial radio, and satellite.

4.1 Fiber-to-the-Home

Fiber-to-the-home (FTTH) (see Refs. [1], [6]) was an alternative of great excitement in the late 1980s and early 1990s when phone companies had definite plans to upgrade their existing twisted pairs to fiber. Support for FTTH was fueled by the notion that twisted pairs did not have sufficient bandwidth to deliver wideband services. The advent of HDSL and ADSL did much to show that indeed the fundamental bandwidth constraint was in the public switched telephone network and not in the access network of twisted pairs. Asynchronous transfer mode (ATM) switches will allow this constraint to be averted in the future, thus enabling technologies like DSL access to wider switching bandwidth on an individual call (and even packet) basis. Nonetheless, there is little doubt that fiber will be laid to many customers' premises in the future and the issue is only when. Realistically, estimates by phone companies show that at least 15 years would be necessary to convert half their networks from copper to fiber, and these estimates are aggressive in their use of human resources and financial capital. Indeed, even today in new homes, copper twisted pairs are laid for phone service to new homes in residential areas. However, a growing trend is to feed the copper loops from a multiplexer (i.e., ONU) that connects to the CO via optical fiber. New ONUs are being placed ever closer to the homes.

Nonetheless, research efforts into optical transmission and networks continue worldwide, and fiber is rapidly deploying in the network where many users share a common physical transmission link and where the cost of fiber can be averaged over the corresponding large number of users. Fiber differs from twisted pair in that a very thin strand of glass/silicon replaces a twisted pair, and light is injected at one end to represent a 1, and no light a 0. This type of modulation is called OOK (on/off keying). The waveguide created by the fiber carries the light pulses to a receiver very efficiently and with relatively tiny dispersion with respect to twisted pair. Speeds of fiber to an individual customer range from 600 Mbps to 2.4 Gbps, data rates easily carried over short fiber optic links. Theoretically, hundreds of times higher data rates could be obtained on fiber.

Optical transmission has the advantages of extremely high bandwidth with simple electronics, immunity to crosstalk, absence of emissions, small size per user, and low power. Duplex is easily achieved because the light pulses travel in different directions noninvasively to one another. The main disadvantage is that fiber needs to be deployed to replace twisted pairs and the endeavor is of gargantuan proportions. Other disadvantages include the relative difficulty in splicing fibers with respect to twisted pairs and the inability to deliver power to the customer (leading to a reliability problem). The powering of customer-premises equipment (like a phone) requires a twisted pair to be used anyway for carrying the power.

Passive optical networks (PONs) [2] attempt to reduce the complexity of a fiber by using a single (possibly group of) fiber(s) to carry service to a large group of customers distributed over a wide area. The large bandwidth of the fiber is shared over all the users. Upstream transmissions must be timed so that they do not interfere with other upstream transmissions from other points on the fiber. In such networks, smaller Central Offices or distribution points can be eliminated, allowing network costs to be reduced. While PONS were originally proposed for FTTH, they are now candidates for fiber-to-the-curb (FTTC) with VDSL on twisted pairs for the last few hundred meters.

The deployment of fiber for customer access is growing rapidly, but not for fiber-to-the-home. Fiber is deployed to most major business sites today, often connecting to a multiplexer placed on the business customer's premises. Fiber is also being used to an increasing extent from the Central Office to an optical network unit (ONU) in the vicinity of residential customers. The placement of the ONU is changing from a 1000-line unit with copper lines up to 3 km long serving each customer to "deep fiber positioning" with a 16-line ONU with copper line up to a few hundred meters long serving each customer.

4.2 Coax and Hybrid Fiber Coax

Service providers can potentially provide DSL-like service on cable TV networks [3]. However, on average, 90% of cable installations are currently unidirectional. Cable companies are upgrading to bidirectional coax in selected areas. Since this upgrade is expensive, the proportion of bidirectional coax is increasing slowly. Cable companies in some areas are also upgrading the reliability of the infrastructure, which, in some cases, has been less than the standards expected by telephone customers. Data service via unidirectional coax is possible via the use of a telephone line and a modem for upstream transmission for rates up to 28.8 kb/s. This coax-telephone-modem hybrid is unattractive for the following reasons: (1) the cost of a coax modem plus the telephone modem, (2) the inability to make phone calls during data sessions, and (3) the limit of downstream throughput to about 300 kb/s due to the slow upstream transport of acknowledgment packets.

Hybrid fiber coax (HFC) networks followed FTTH as the firmly intended technology for broadband delivery of digital service to phone company customers, perhaps reaching peak interest in late 1994. Again, the perception was that HFC could be more economically deployed than DSL and had greater bandwidth. HFC is an active version of a PON where fiber carries broad-bandwidth to a distribution node, and then coaxial cable is deployed to loop through several customers. A coaxial cable typically has 500 MHz to 1000 MHz of bandwidth with theoretical limits of 10 Gbps or more shared over a few hundred to a few thousand customers in a geographical area served by the cable. Powering and splicing issues no longer are problems, and coaxial electronics are typically economical. HFC also offered phone companies an opportunity to compete with cable companies for entertainment services as well as offering bidirectional service. Interest by phone companies in HFC decreased in 1996 and 1997, and simultaneously satellite TV has increasingly decreased the revenue of cable companies.

A loss of broadcast television revenue has increased cable companies' interest in what are known as *cable modems*. A recent effort (MCNS) and standardization activities in the IEEE802.14 group have produced two (different) standards/methods for cable modem implementation. The intended service is Internet access through cable lines, which is again frustrated by the unidirectional nature of existing cable networks. Again, the cost of infrastructure rebuilding to make cable networks bidirectional is being weighed with respect to the potential revenue increase from Internet service. Many argue that cable modems must be implemented or cable companies will see the consequences of their service revenue becoming very small as satellite TV increasingly attracts their customers. When converted to bidirectional service, cable modems operate by sending 384 kbps to 2 Mbps downstream in a cable system in a few frequency-indexed slots above 350 MHz to 500 MHz, depending on the operator. The modulation type is some form of QPSK in both standards. The coax must be amplified at these frequencies for successful transmission, which is part of the rebuilding operation for the coaxial system. Upstream bandwidth is limited to transmission between 5 and 40 MHz after cable loops have been reengineered to allow signals to pass in this direction. Bandwidths of 384 kbps to 2 Mbps are in this case shared, often over hundreds of users, with the expectation that customers will not all try to send IP packets for internet access activity at exactly the same time. Some cable modem systems transmit 40 Mb/s downstream in each 6 MHz channel; this is statistically shared among up to several hundred users. Privacy, reliability, and performance concerns abound for coax modems. A medium-access control protocol allows contention resolution between the various upstream users. Cable modems are beginning trials and at a similar stage to DSLs in terms of modem technology and integration, both being largely complete. Cable modems will compete and provide some level of motivation for phone companies to also provide these same services over DSL. The phone company's technical advantage is the ability to use existing copper, thus avoiding additional deployment restructuring costs and much higher upstream bandwidth from customers. Each customer also has individual and more secure transmission, rather than the shared coaxial medium. Nonetheless, complicated financial and political issues remain to be resolved before the true relative use of cable modems versus DSL will be better understood and projected.

4.3 Wireless Alternatives

Wireless digital telephony service [4] has benefited from a worldwide explosion of interest and revenue. While digital wireless service today is narrowband (usually about 8 kbps allocated per customer), emerging third-generation wireless networks — perhaps best exemplified by the IMT 2000 project of the ITU — are encompassing capabilities of distribution of up to 1 Mbps to a customer willing to pay for the delivery. These digital wireless services will compete with some of the same services as DSL. Additionally, *wireless local loop (WLL)* initiatives have been announced by several major telecommunications service suppliers at speeds from 144 kbps in early deployment to 26 Mbps in eventual deployment.

The authors believe that wireless access will at some time capture a significant position in the data services marketplace. Nonetheless, behind all the excitement, are some fundamental technical issues that will be resolved as wireless alternatives progress:

- Antenna deployment
- Bandwidth crowding
- Reliability

These issues are somewhat related. Antenna deployment can be used to solve bandwidth crowding and reliability problems, but requires expense to do so. Today, the issue of scarce bandwidth has been mitigated through the use of cellular division of geographical areas. The more, and consequently the smaller, the cells, the greater the bandwidth reuse — essentially exploiting a fundamental tradeoff between bandwidth and space. Early wireless networks operate in the 800 MHz to 1000 MHz carrier band, where attenuation of signals through nominal wireless environments is much less than at higher frequencies. Even at user data rates of less than 10 kbps per user, such networks can rapidly saturate. More sophisticated code-division access networks exploit the statistical nature of calls and speech itself to effect greater reuse factors. Indeed, CDMA (code-division multiple access) is beginning to live to its early fundamental promises and conception (see history in [5]), and the third-generation networks worldwide standard has recently been selected as CDMA. The spreading of a wider-bandwidth signal by the factor of 100 or more typically in CDMA is just not practical when signals have data rates of 1 Mbps or more, thus requiring alteration of the CDMA concept for wideband access. Nonetheless, even this method has limitations on bandwidth. Furthermore, severe macroscopic fading effects (basically signal blockage by objects) frustrate reliable connection in many places, of which any user of mobile telephony service is well aware. Wireless connection today is highly unreliable and insecure, but nevertheless is used heavily for voice telephony.

Early WLL efforts resolve this issue by careful antenna placement. At higher carrier frequencies like the LMDS band of 28 GHz and beyond, signal blockage effects can be enormous. Tree leaves are a major problem, and the present solution is to increase power and reduce the maximum distance. At lower bandwidths of 2 to 5 GHz, these effects are less, but some bandwidth has been allocated for data communications. Nonetheless, successful demonstration of 100 to 200 kbps networks, and standardization in ETSI of hyperlan transmission (using COFDM, a wireless variant of ADSL's DMT), progress. Ultimately, adaptive antenna solutions will increasingly resolve the spatial transmission problems, along with placement of many antennas in many places will lead to viable WLL. Some efforts suggest that as much as 26 Mbps per user is possible within residential neighborhoods with sufficient adaptive antenna capability, thus competing with VDSL.

Suffice it to say, numerous competent efforts proceed worldwide and the problems will be resolved, perhaps not within the scale of quarterly earnings reports and consequent press/stock-market exposure, but certainly within reasonable time spans. Such progress has been the fulfilled expectation of technology overall over the past five decades and will continue for wireless. Nonetheless, in the author's opinion, viable competitive wireless alternatives to DSL presently lag behind DSL efforts by three to five years.

4.4 Satellite Services

Digital satellite television broadcast has seen faster growth than any communications service, with millions of customers subscribing over just the first few years of introduction and enormous growth ongoing. Broadcast television entertainment is likely to be dominated by satellite delivery by the beginning of the next century. There are three major networks — Direct TV (the largest), Primestar, and Echostar.

For DSL, cable, or WLL to exist, their services need distinction from that of satellite. There is no upstream capability on a satellite, and providing one is enormously expensive and difficult. Thus, upstream capability is provided through a terrestrial link, which is today a voiceband modem link in the TV receiver that is connected to a phone line (perhaps a DSL in the future).

The upstream facility of DSL fundamentally will distinguish it from satellite service and thus preserve its competitive opportunity with respect to digital satellites.

References

[1]   B. C. Lindberg, *Digital Broadband Networks and Services.* McGraw-Hill: New York, 1995, Chapter 5.

[2]   D. Mestdagh, *Fundamentals of Multiaccess Optical Fiber Networks.* Artech House: Norwood, MA, 1995.

[3]   R. K. Jurgen, *Digital Consumer Electronics Handbook.* McGraw-Hill: New York, 1997, Chapters 16–18.

[4]   T. S. Rappaport, *Wireless Communications – Principals and Practice.* Prentice-Hall: Upper Saddle River, NJ, 1996.

[5]   S. G. Glisic and P. A. Leppanen, *Code Division Mutiple Access.* Kluwer: Boston, 1995, Chapter 1.

[6]   P. W. Schumate and R. K. Snelling, "Evolution of Fiber in the Residential Loop Plant," *IEEE Communications Magazine,* March 1991, pp. 68–74.

[7]   S. Perkins and A. Gatherer, "Two-Way Broadband CATV-HFC Networks: State-of-the-Art and Future Trends," *to appear, special issue of "Computer Networks and ISDN Systems,"* Elsevier, December 1998, see also *http://www.elsevier.nl/inca/publications/store/5/0/5/6/0/6/ menu=gen.aimsandscope.*

CHAPTER 5

Transmission Duplexing Methods

Almost all DSL services require bidirectional, or "duplex," transmission of data, even if the bit rates in opposite directions are asymmetrical. DSL modems separate the signals in opposite directions using *duplexing methods*. There are four distinct duplexing methods: four-wire duplexing, echo cancellation, time-division duplexing, and frequency-divsion duplexing. The last three methods all use the same twisted-pair for both directions of transmission.

5.1 Four-Wire Duplexing

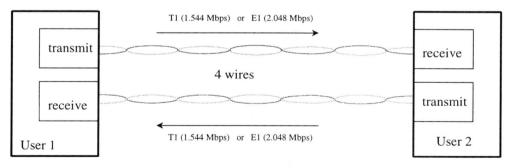

Figure 5.1 Four-wire duplexing.

Figure 5.1 illustrates *four-wire duplexing*, which uses two twisted-pair channels, one for each direction of transmission. Four-wire duplexing is also known as "dual-simplex" transmission, because there are two simplex (unidirectional) transmission channels. The obvious disadvantage of four-wire duplexing is the need for two twisted pairs instead of one as in the other duplexing methods. Four-wire duplexing is the least costly duplexing method if extra copper is freely available. However, since extra copper is often expensive, the overall system cost is often highest with four-wire duplexing, even though the electronics cost may be less. This tradeoff of electronics cost to save copper is a general theme that often occurs in DSL engineering.

The earliest DSLs, T1 (DS1, ANSI-T1.403 [1]) and E1 (G.703, ITU-T [2]) circuits, carry symmetric 1.544 Mbps and symmetric 2.048 Mbps, respectively, with four-wire duplexing. Typical maximum distance for T1 and E1 circuits (without repeaters) is approximately 6,000 ft (2,000 meters).

Four-wire duplexing is also often used in some modern 2.048 Mbps HDSL modems with approximate range of 12,000 ft [3] when copper twisted pairs are abundant and low-cost transceivers are desirable. Such DSLs are known as dual-simplex HDSL. The directions of dual simplex HDSL transmission typically exist in different binder groups to avoid near-end crosstalk (see Chapter 3) between the signals in opposite directions.

5.2 Echo Cancellation

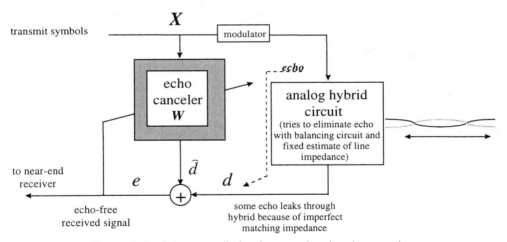

Figure 5.2   Echo cancellation for two-wire signal separation.

The echo-cancellation duplexing, shown in Figure 5.2, achieves the data rates of four-wire duplexing on a single twisted pair, thus saving copper or equivalently increasing the bandwidth of existing twisted-pair wires. Echo cancellation is the most prevalent form of multiplexing in modern DSLs, being standardized for use in ISDN, HDSL, and ADSL. An "echo" is a reflection of the transmit signal into the near end received. Echoes are of concern because the signals that correspond to both directions of digital transmission simultaneously coexist on the twisted-pair transmission line, so that the echo is unwanted noise. The echo is a filtered version of the transmitted signal. An echo canceler adaptively constructs a replica of the filtered echo transmit signal and subtracts it from the received signal. The arrow drawn through the echo canceler box in Figure 5.2 is used universally in digital signal processing to denote a filter that is trained adaptively using the signal on the line as a control signal.

The analog hybrid circuit in Figure 5.2 decouples transmit signals from receive signals when the line impedance is exactly matched by a corresponding hybrid impedance, as discussed in Chapter 3. When these impedances cannot be exactly matched, as is almost always true in practice, a remnant echo of the transmit signal appears at the receive-signal output of the hybrid. Good hybrid designs achieve 20 dB of attenuation of the transmit signal before it emanates from the hybrid receive-signal output. Digital line signals may be attenuated by as much as 40 dB, so that the echo may still be 20 dB larger than the desired far-end signal, resulting in unacceptable signal-to-noise ratio for detection of the far-end desired signal. The echo canceler input is the sampled digital transmission signal. This input and its past samples are stored in a digital delay line that often must span 100 to 200 microseconds in time. The echo canceler multiplies these stored samples by echo canceler coefficients and sums the resulting products to form an estimate of the echo, which is then subtracted from the hybrid output. The coefficients for best cancellation depend on the transmission line and are therefore determined adaptively.

Echo cancelers must be able to reject echo by 50 dB or more for ISDN, by 60 dB or more for HDSL, and by more than 70 dB for ADSL. The levels of cancellation are different because increasingly higher frequencies with greater attenuation are used by HDSL and then ADSL, meaning that the receiver must reduce high-frequency echo to a lower level to make it less than the smallest of received frequencies. To achieve this high level of echo rejection, the echo canceler must determine the adaptive echo coefficients with high precision and accuracy. Some echo cancelers also use nonlinear construction of echo components that cannot be represented as a sum of products of the transmit signal samples. Echo cancellation is more complex than four-wire mutiplexing. Nonetheless, with the advent of high levels of digital-signal processing in VLSI, echo cancellation even for the most strenuous cases (ADSL) often have negligible cost and thus often appear in practice.

The echo estimate produced by the echo canceler can be succinctly written as

$$\hat{d} \; = \; W^* X \tag{5.1}$$

where $\hat{d}$ is the estimate of the echo component, d, of the received signal at the hybrid output. W is a column vector of the echo canceler coefficients, and X is a corresponding vector (a superscript of * denotes (conjugate) transposition) of the samples of the transmit signal. The inner product of the two vectors mathematically describes the weighted sum formed by the echo canceler. The values in the vector of inputs, X, depends on the line code selected (see Chapter 7) as does the performance and achievable accuracy of the echo canceler. To illustrate complexity, the sampling rate of an ISDN transmitter is at least 80 kHz, corresponding to a sampling interval of

12.5 µs. Thus, an echo response length (corresponding to the size of **W** and **X**) of 400 µs would require at least 32 coefficients, and thus 32(80,000) = 2.56 million operations per second. Low-pass filters used for noise rejection typically increase the length of echoes. HDSL and ADSL use wider bandwidths, and so their lowpass filters have less lengthening effect, but their sampling rates are also higher, usually leading to even greater computation. Echo cancelers for HDSL may need 64 to 128 coefficients and as much as 128(800,000) = 100 million operations per second. Echo cancelers for ADSL may need as many as 300 coefficients, but exploit the asymmetry and the "circular" aspects of the DMT line code, typically requiring 20 to 30 million operations per second.

5.2.1   Adaptive Echo Cancellation

The values of the coefficients can be determined by the so-called least-mean square (LMS) updating algorithm [8], a stochastic-gradient method that iteratively determines and tracks the correct coefficient values. The echo canceler error signal is

$$e = d - \hat{d} \tag{5.2}$$

and is equal to the received signal if the echo canceler perfectly reconstructs the echo and removes it. This error signal also is used in the LMS update:

$$W_{k+1} = W_k + \mu e_k X_k^* \tag{5.3}$$

where the subscript of k is a time index for updates. The constant μ determines the tradeoff between estimate accuracy and tracking speed.

Nonlinear echo cancellation augments the above echo cancellation by adding a look-up table (RAM addressed by the vector X_k) output to the echo estimate, d. The address for the look-up table is determined by the most significant bits of the transmit signal samples for only a few samples that correspond to the receive time periods when most echo energy is reflected. The updating of the addressed look-up table location is simple averaging (new value = old value + λ (error)). λ is a constant that determines the tradeoff between accuracy and tracking for the non-linear echo canceler.

5.3 Time-Division Duplexing

Time-division duplexing (TDD) silences one direction of transmission while the other is active. Control of the link alternates at regular intervals between the directions of transmission. TDD avoids the need for echo cancellation on a two-wire loop, as illustrated in Figure 5.3. The "switches" alternate between the two directions of transmission to allow the transmitter/receiver pairs to both use the same two-wire line. However, for symmetric transmission, the data rate must be at least halved because only half the connection time is available for transmission. Typically, the turnaround of the loop imposes some additional small rate loss because it cannot be instantaneous in practice. TDD is sometimes affectionately called "ping-pong," an accurate simile between table tennis and the control of transmission.

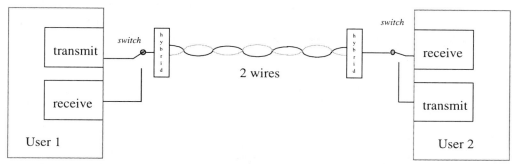

Figure 5.3   llustration of TDD (switches alternate to allow time slots for transmission in both directions).

Example 1 — ISDN in Japan   The Japanese telecommunications suppliers sometimes use a TDM alternative ([4] JISDN) to the ANSI T1.601 ISDN transmission standard ([5] T1.601) (the standard is echo canceled). In this system, the directions of transmission each take one turn at transmission during a 400 μs interval.

Example 2 — VDSL   Nominally, TDM causes unacceptable bandwidth loss for services like ADSL or VDSL that attempt to achieve high transmission performance on a twisted pair. However, in a DSL, crosstalk becomes increasingly important at high frequencies. Near-end crosstalk (see Chapter 3) would dominate all noises at frequencies above 300 kHz in a DSL. VDSL signals occupy a much wider bandwidth and would crosstalk unacceptably into one another if echo cancellation were used. Thus, while TDM does inevitably lose transmission time, this loss is more than offset by the zeroing of NEXT if all lines in a binder can be coordinated (that is they "ping" and "pong" at the same times). Some recent VDSL proposals ([6], VDSL-SDMT) use TDM to both reduce cost and improve performance with respect to an echo-canceled implementation. VDSL is not yet standardized, and other proposals for multiplexing (FDM) are also being studied.

5.4 Frequency-Division Multiplexing

Frequency-division multiplexing (FDM) alternatively transmits different directions in nonoverlapping frequency bands, as in Figure 5.4. FDM has not been heavily used in practice mainly because the variation in transmission-line attenuation causes an uncertainty in the amounts of bandwidth that need to be allocated to the two directions. FDM, like TDM, avoids NEXT if all lines use the same bandwidth assignments. An FDM option (compatible with echo cancellation) for ADSL permits reservation of up to the first 138 kHz for upstream transmission and complies with the standard ([7] T1.413). This option is most often used in the United States. However, performance is compromised in this configuration and upstream bandwidth limited to data rates below those that may be desired for some services (i.e., Internet access). Some new ADSL standards from the ITU known as "g.dmt," g.922.1, and "g.lite," g.922.1, actually have 3 appendices, each allowing for FDM, echo cancellation, or TDM in these ADSL modems (see Chapter 7).

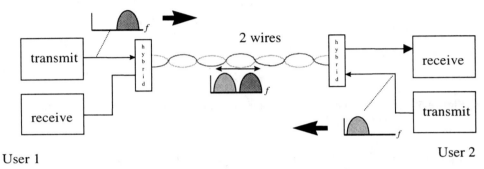

Figure 5.4   Frequency-division multiplexing (FDM).

References

[1]   American National Standard T1.403-1995, *Network-to-Customer Installation — DS1 Metallic Interface.* ANSI: New York.

[2]   International Telecommunications Union —Telecommunications Standard G.703 *Physical/Electrical Characteristics of Hierarchical Interfaces.* ITU: Geneva.

[3]   R. Segev, "Line-Code for a 2048 kbps HDSL," ETSI TM3 Contribution TD1-26, March 13, 1992, Stockholm; see also, J. Cioffi, J. A. C.Bingham, and P. Chow, "Discrete Multi-tone Transceiver for E1-HDSL," *ANSI Contribution T1E1.4/92-075*, Williamsburg, VA, May 1992.

[4]   *Technical Reference for High-Speed Digital Leased Circuit Services (transmission on metallic subscriber lines),* Nippon Telegraph and Telephone Corporation, Japan.

[5]   American National Standard T1.601-1992, *Integrated Services Digital Network (ISDN) – Basic Access Interface for Use on Metallic Loops for Application on the Network Side of the NT (Layer 1 Specification).* ANSI: New York.

[6]   J. M. Cioffi, "VDSL System Requirements Document," *ANSI Contribution T1E1.4/98-043R3,* June 1998, available at web site: ftp://ftp.t1.org/pub/t1e1/e1.4/dir98/8e14043x.pdf (x is revision number, use highest value).

[7]   American National Standard T1.413-1999, *Asymmetric Digital Subscriber Line (ADSL) Metallic Interface.* For draft see web site: ftp://ftp.t1.org/pub/t1e1/e1.4/dir98/8e14007x.pdf (x is revision number, use highest value).

[8]   B. Widrow and S. D. Stearns, *Adaptive Signal Processing,* Prentice-Hall: Englewood Cliffs, NJ, 1985.

Basic Digital Transmission Methods

6.1 Basic Modulation and Demodulation

All transmission channels are fundamentally analog and thus may exhibit a variety of transmission effects. In particular, telephone lines are analog, and so DSLs use some form of **modulation**. The basic purpose of modulation is to convert a stream of DSL input bits into equivalent analog signals that are suitable for the transmission line.

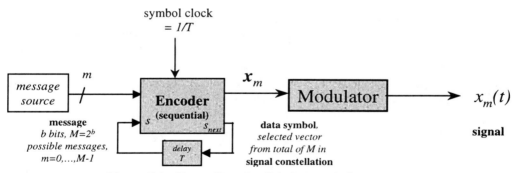

Figure 6.1 Transmitter of a digital transmission system.

Figure 6.1 depicts the transmitter of a digital transmission system. The transmitter converts each successive group of b bits from a digital bit stream into one of 2^b **data symbols, x_m** , via a one-to-one mapping known as an **encoder.** Each group of b bits constitutes a message m, with $M = 2^b$ possible values $m = 0,...,M − 1$. The data symbols are N-dimensional (possibly complex) vectors, x_m and the set of M vectors forms a **signal constellation**. **Modulation** is the process of converting each successive data symbol vector into a continuous-time analog **signal** $\{x_m(t)\}_{m = 0, \ldots, M-1}$ that represents the message corresponding to each successive group of b bits. The message may vary from use to use of the digital transmission system, and thus the message index m and the corresponding symbol x_m are considered to be random, taking one of M possible values each time a message is transmitted. This chapter assumes that each message is equally likely to occur with probability $1/M$. The encoder may be **sequential**, in which case the mapping from messages to data symbols can vary with time as indexed by an **encoder state**, corresponding to v bits of past state information (function of previous input bit groups). There are 2^v possible states when the encoder is sequential. When $v = 0$, there is only one state and the encoder is **memoryless.**

Linear modulation also uses a set of N, orthogonal unit-energy basis functions, $\{\varphi_n(t)\}_{n = 1:N}$, which are independent of the transmitted message, m. The basis functions thus satisfy the orthonormality condition:

$$\int_{-\infty}^{\infty} \varphi_n(t)\varphi_l^*(t)dt = \begin{cases} 1 & n = l \\ 0 & n \neq l \end{cases} \tag{6.1}$$

The nth basis function corresponds to the signal waveform component produced by the nth element of the symbol x_m.[1] Different **line codes** are determined by the choice of basis functions and by the choice of signal constellation symbol vectors, x_m, $m = 0,\ldots,M-1$. Figure 6.2 depicts the function of linear modulation: for each **symbol period** of T seconds, the modulator accepts the corresponding data-symbol-vector elements, x_{mn}, and multiplies each by its corresponding basis function, $\varphi_1(t),\ldots,\varphi_N(t)$, respectively, before summing all to form the modulated waveform $x_m(t)$. This waveform is then input to the channel.

The **average energy**, $\mathcal{E}_x$, of the transmitted signal can be computed as the average integrated squared value of $x(t)$ over all the possible signals,

$$\mathcal{E}_x = \frac{1}{M} \cdot \sum_{m=0}^{M-1} \left\{ \int_{-\infty}^{\infty} |x_m(t)|^2 \, dt \right\} \tag{6.2}$$

or more easily by finding the average squared length of the data symbol vectors,

$$\mathcal{E}_x = \frac{1}{M} \cdot \sum_{m=0}^{M-1} \|\mathbf{x}_m\|^2 \tag{6.3}$$

---

1.   Complex basis functions occur only in the mathematically abstract case of baseband-equivalent channels, as in Section 6.3.5, and a superscript of * means complex conjugate (and also transpose when a vector).

The **digital power** of the transmitted signals is then $S_x = \mathcal{E}_x/T$. The **analog power,** P_x, is the digital power at the source driver output divided by the input impedance of the channel when the line and source impedances are real and matched (see Section 3.5.2.1). Generally, analog power is more difficult to calculate than digital power, and Section 3.5.2.1 describes the correct analog-power calculations for twisted-pair transmission lines. Transmission analysts usually absorb the gain constants for a specific analog driver circuit into the definitions of the signal constellation points or symbol vector values x_m and the normalization of the basis functions. The digital power is then exactly equal to the analog power, effectively allowing the line and analog effects to be viewed as a 1 ohm resistor.

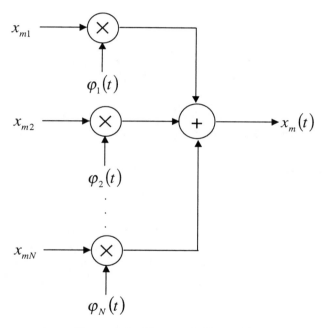

Figure 6.2 Linear modulator.

The channel in Figure 6.3 consists of two potential distortion sources: bandlimited filtering of the transmitted signals through the filter with transform function $H(f)$ and additive Gaussian noise[2] (unless specifically discussed otherwise) with zero mean and power spectral density

2. Gaussian noise has probability distribution $Pn(x) = \dfrac{1}{\sqrt{2\pi\sigma^2}}e^{-x^2/(2\sigma^2)}$ where σ^2 is the variance of this zero mean noise. When the noise is also "white," it is uncorrelated with itself when shifted in time by any nonzero time offset (i.e., the power spectral density is constant at $S_n(f) = \sigma^2$ for all frequencies).

$S_n(f)$. The designer should analyze the transmission system with an appropriately altered $H(f)$

$$\left(H(f) \rightarrow \frac{H(f) \cdot \sigma}{S_n^{.5}(f)} \right)$$

to include the effects of spectrally shaped noise, and then it is sufficient to investigate only the case of equivalent **white noise** where the noise power spectral density is a constant, σ^2.

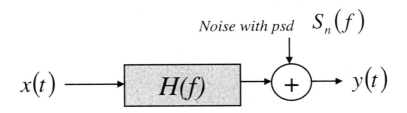

bandlimited channel

Figure 6.3   Bandlimited channel with Gaussian noise.

6.1.1   The Additive White Gaussian Noise Channel

The additive white Gaussian noise (AWGN) channel is the most heavily studied in digital transmission. This channel simply models the transmitted signal as being disturbed by some additive noise. This channel[3] has $|H(f)| = 1$, which means there is no bandlimited filtering in the channel (clearly an idealization). If the channel is distortionless, then $|H(f)| = 1$ and $\sigma^2 = 0$. On a distortionless channel, the receiver can recover the original data symbol by filtering the channel output $y(t) = x(t)$ with a bank (set) of N parallel matched filters with impulse responses $\varphi_n(-t)$ and by sampling these filters' outputs at time $t = T$, as shown in Figure 6.4. This recovery of the data symbol vector is called **demodulation.** A bidirectional digital transmission apparatus that implements the functions of "modulation" and "demodulation" is often more succinctly called a **modem.** The reversal of the one-to-one encoder mapping on the demodulator output vector is called **decoding.** With nonzero channel noise, the demodulator output vector y is not necessarily equal to the modulator input x. The process of deciding which data symbol is closest to y is known as **detection.** When the noise is white Gaussian, the demodulator shown in Figure 6.4 is optimum. The optimum detector selects $\hat{x}$ as the symbol vector value x_m closest to y in terms of the vector distance/length,

---

3.   Any issues of load impedance are assumed contained within this transfer function.

$$\hat{m} = i \text{ if } \left\| \boldsymbol{y} - \boldsymbol{x}_i \right\| \leq \left\| \boldsymbol{y} - \boldsymbol{x}_j \right\| \text{ for all } j \neq i \quad , i,j = 0,\ldots, M-1 \tag{6.4}$$

Such a detector is known as a **maximum likelihood detector** and the probability of erroneous decision about $\boldsymbol{x}$ (and thus the corresponding group of b bits) is minimum. This type of detector is only optimum when the noise is white, Gaussian, and the channel has very little band-limiting (essentially infinite bandwidth) and is known as a **symbol-by-symbol** detector. Each matched-filter output has independent noise samples (of the other matched-filter output samples), and all have mean-square noise sample value σ^2. Thus, the signal-to-noise ratio (SNR) is

$$SNR = \frac{\mathcal{E}_x/N}{\sigma^2} \tag{6.5}$$

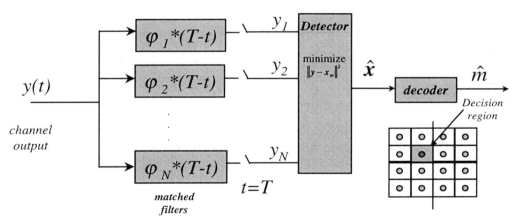

Figure 6.4   Demodulation, detection, and decoding.

Detector implementation usually defines regions of values for $\boldsymbol{y}$ that map through the maximum likelihood (ML) detector into specific symbol values or the associated b bits. These regions are often called **decision regions**.

An error occurs when $\hat{m} \neq m$, that is, $\boldsymbol{y}$ is closer to a different symbol vector than to the correct symbol vector. An error is thus caused by noise being so large that $\boldsymbol{y}$ lies in a decision region for a point $\boldsymbol{x}_j$, $j \neq m$ that is not equal to the transmitted symbol. The probability of such an error on the AWGN channel is less than or equal to the probability that the noise is greater than half the distance between the closest two signal constellation points. This **minimum distance** between two constellation points, d_{min}, is easily computed as

$$d_{min} = \min_{i \neq j} \left\| x_i - x_j \right\|$$

$$(6.6)$$

Symbol vectors in a constellation will each have a certain number of **nearest neighbors** at, or exceeding, this minimum distance, N_m. The **average number of nearest neighbors** is

$$N_e = \frac{1}{M} \cdot \sum_{m=0}^{M-1} N_m$$

$$(6.7)$$

which essentially counts the number of most likely ways that an error can occur. So, the **probability of error** is often accurately approximated by

$$P_e \cong N_e Q \left[\frac{d_{min}}{2\sigma} \right]$$

$$(6.8)$$

where the **Q-function** is often used by DSL engineers. The quantity $Q(x)$ is the probability that a unit-variance ($\sigma^2 = 1$) zero-mean Gaussian random variable exceeds the value in the argument, x,

$$Q(x) = \int_x^\infty \frac{1}{\sqrt{2\pi}} e^{-u^2/2} du$$

$$(6.9)$$

The Q-function must be evaluated by numerical integration methods, but Figure 6.5 plots the value of the Q-function versus its argument ($\log(x)$) in decibels. For the physicist at heart,

$$Q(x) = .5 \cdot erfc\left(x / \sqrt{2}\right)$$

$$(6.10)$$

To compare performance for transmission formats with different dimensionality, performance measures are often normalized, leading to the **normalized probability of symbol error**

$$\overline{P}_e = \frac{P_e}{N}$$

$$(6.11)$$

and the normalized energy per symbol

$$\overline{\mathcal{E}}_x = \frac{\mathcal{E}_x}{N}$$

$$(6.12)$$

leaving $SNR = \dfrac{\overline{\mathcal{E}}_x}{\sigma^2}$. A related measure to P_e is the **probability of bit error**, which is given by

$$\overline{P}_b = \frac{N_b}{b} \cdot Q\left(\frac{d_{min}}{2\sigma}\right)$$

(6.13)

where N_b is the **average number of bit errors per symbol error** and given by

$$N_b = \sum_{m=0}^{M-1} \frac{1}{M} \cdot \sum_{j \neq m} n_b(m, j)$$

(6.14)

and $n_b(m, j)$ is the number of encoder-mapping bit errors if message m is incorrectly decoded as message j.

DSL service providers sometimes prefer the switching measure of **errored seconds** to the transmission engineer's preferred probability of error. An errored second is a second in which one or more errors occur. Often, errors in digital transmission occur in bursts because a single symbol error can cause several bit errors in high-performance designs. To investigate any interval, say $\tilde{T} = 1$ second for the occurrence of one or more bit errors, can be complicated for sophisticated designs, because a symbol error may not lead to bit errors (when forward error correction is used, see Section 7.4) or may lead to several bit errors. A simple method is to assume that bit errors are isolated and occur on a binary channel with a probability of $\overline{P}_b = p$. The number of bits in $\tilde{T}$ is $\tilde{N} = R \cdot \tilde{T}$. Then the probability that one or more errors occurs in this block of bits is the **probability of an errored second,**

$$P_{es} = \sum_{k=1}^{\tilde{N}} \binom{\tilde{N}}{k} p^k (1 - p)^{\tilde{N}-k} \approx \tilde{N}p$$

(6.15)

with the approximation improving for the typical case of small p. The probability of an errored second is usually expressed as a percentage. Similarly, there is a measure of reliable operation called the **percentage of error-free seconds,**

$$P_{EFS} = (1 - P_{es}) \times 100 \quad \%$$

(6.16)

With forward error correction, the sum is essentially over the number of symbol errors that cannot be corrected by the capability of the error-correcting code. The reader is deferred to Section 7.4 for more on error correction with DSLs.

Later sections of this chapter study the modifications of the ML detector necessary to accommodate the imperfections of the twisted-pair transmission line that were discussed in Chapter 3. These imperfections include spectrally shaped noise, non-Gaussian crosstalk noise distributions, and the severe frequency dependence of attenuation on a twisted pair. Before proceeding with that discussion, the choice of basis functions and signal constellations for several DSLs will be reviewed in Sections 6.2 and 6.3.

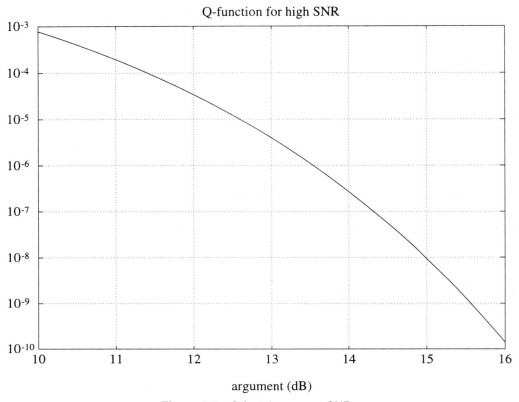

Figure 6.5 Q function versus SNR.

6.1.2   Margin, Gap, and Capacity

It is desirable to characterize a transmission method and an associated transmission channel simply. Margin, gap, and capacity are related concepts that allow such a simple characterization. Many commonly used line codes are characterized by a **signal-to-noise Ratio Gap** or just **gap**. The gap, $\Gamma = \Gamma(P_e, C)$, is a function of a chosen probability of symbol error, P_e, and the line code, C. This gap measures efficiency of the transmission method with respect to best possible performance on an additive white Gaussian noise channel and is often constant over a wide range of b (bits/symbol) that may be transmitted by the particular type of line code. Indeed, most line codes are quantified in terms of the achievable bit rate (at a given P_e) according to the following formula:

$$\bar{b} = \tfrac{1}{2} log_2 \left(1 + \frac{SNR}{\Gamma} \right)$$

(6.17)

Thus, to compute data rate with a line code characterized by gap Γ, the designer need only know the gap and the SNR on the AWGN channel. An experienced DSL engineer usually knows the gaps for various line codes and can rapidly compute achievable data rates in his or her head.

An optimum line code with a gap of $\Gamma = 1$ (0 dB) achieves a maximum data rate known as the **channel capacity** (after Shannon, 1948 [1]). Such an optimum code necessarily requires infinite complexity and infinite decoding/encoding delay. However, it has become practical at DSL speeds to design coding methods for which the gap is as low as 1 to 2 dB, thus allowing recent DSL designs to fulfill the promises of Shannon, some 50 years after his projection, on twisted-pair telephone lines.

Often, transmission systems are designed conservatively to ensure that a prescribed probability of error occurs. The **margin** of a design at a given performance level is the amount of additional signal-to-noise ratio in excess of the minimum required for a given code with gap Γ. The margin can be computed according to

$$\gamma_m = \frac{SNR}{\Gamma \cdot \left(2^{2\bar{b}} - 1 \right)}$$

(6.18)

6.2 Baseband Codes

Baseband codes are distinguished from **passband codes** (see Section 6.3) in that baseband codes can transmit energy at DC ($f = 0$), while passband codes transmit at a frequency spectrum translated away from DC. Baseband line codes in DSLs have $N = 1$ and appear in the earliest DSLs: examples include T1, ISDN, and HDSL.

6.2.1   The 2B1Q Line Code (ISDN and HDSL)

The 2B1Q baseband line code was heavily used in early DSLs. The 2B1Q line code of Figure 6.6 *ideally* uses a basis function[4]

$$\varphi(t) = \frac{1}{\sqrt{T}} \text{sinc}\ (t/\ T\ )$$

(6.19)

which leads to the independence of successively transmitted data symbols when sampled at multiples of the symbol period. Thus, in the modulation of Section 6.1, $N = 1$, $b = 2$, $M = 4$, and the

---

4.   The sinc function is $\text{sinc}(x) = \sin(\pi x)/(\pi x)$.

encoder is memoryless. On an ideal channel with no distortion, a corresponding matched filter in the receiver also maintains the independence of successive symbol transmissions. In practice, the sinc function cannot be exactly implemented, and transmission through the bandlimited channel distorts this basis function anyway. Thus, compromise pulse masks that attempt to predistort the transmit basis function to incur minimum distortion have been incorporated into the ISDN and HDSL standards documents [2, 3]. These masks appear in Figures 6.7 and 6.8. An update to the ISDN PSD mask has recently been suggested where the high frequencies are more constrained in Ref. [4] and the corresponding mask appears in Figure 6.9. These pulses represent the combined effect of approximating the ideal basis function, lowpass filtering in the transmitter, and attempted compensation of anticipated channel bandlimiting effects. The transmit ISDN spectrum must decrease 50 dB/decade above 50 kHz, and the HDSL spectrum must also decrease 50 dB/decade above 192 kHz.

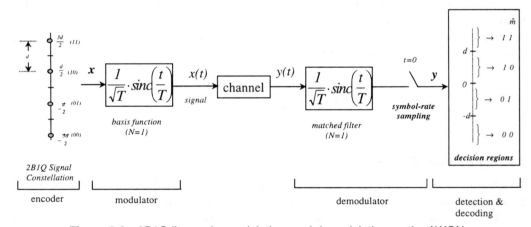

Figure 6.6   2B1Q line code, modulation, and demodulation on the AWGN.

The signal constellation for 2B1Q appears in Figure 6.6. The number of bits in a group is $b = 2$, and each group of bits maps into one of the four data symbol values illustrated in Figure 6.6, where a one-dimensional value for x could be computed as $x = 2m - 3$, $m = 0, 1, 2, 3$. The name "2B1Q" derives from the mnemonic "2 bits per 1 quarternary" symbol. Since the transformer/hybrid coupling to the transmission line does not pass DC (see Chapter 3), baseband line codes like 2B1Q undergo severe distortion, and a straightforward implementation of a symbol-by-symbol detector will likely not perform acceptably. The receiver must compensate for the transformer and line distortion, as described in Chapter 7. HDSL and ISDN 2B1Q encoding actually use the mapping of the bits in Table 6.1 to ensure that decoding error to a nearest neighbor results in only one corresponding bit error, whence the unusual bit mapping from input bits to m, is often known as **gray coding**.

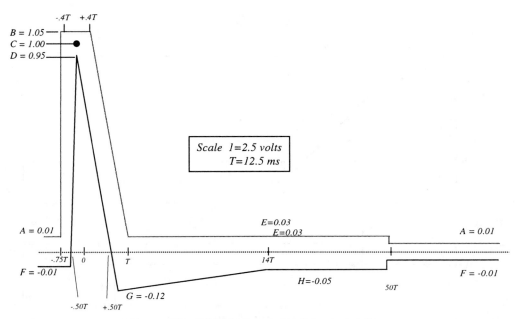

Figure 6.7 ISDN (unnormalized) basis function.

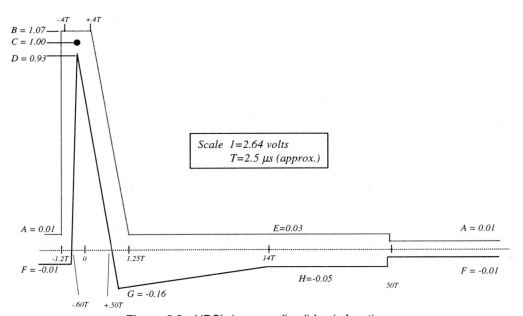

Figure 6.8 HDSL (unnormalized) basis function.

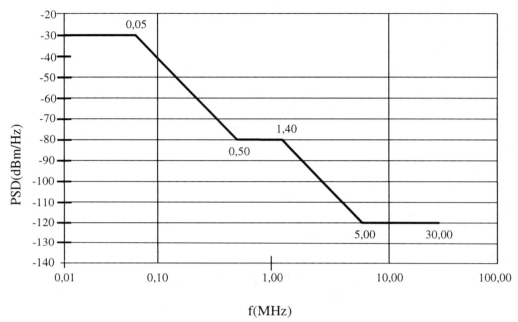

Figure 6.9 NEW ISDN PSD mask for 2000 and beyond

Source: Extracted from Ref. [4].

Table 6.1 Gray Coding ISDN BR and HDSL Encoder

m	Bit Combination	ISDN (HDSL) Signal Level (Volts)
3	10	2.5 (2.7)
2	11	0.833 (0.9)
1	01	−0.833 (−0.9)
0	00	−2.5 (−2.7)

For purposes of analysis, the following function characterizes the one-sided output power-spectral density of the transmit signal where the use of a rectangular basis function (with additional gain $\sqrt{T} = 1/\sqrt{f_0}$ was assumed, along with a unity DC gain nth-order Butterworth filter (see Section 3.8),

$$PSD_{2B1Q}(f) = K_{2B1Q} \cdot \frac{2}{f_0} \cdot \frac{\left[sin\left(\frac{\pi f}{f_0} \right) \right]^2}{\left(\frac{\pi f}{f_0} \right)^2} \cdot \frac{1}{\left[1 + \left(\frac{f}{f_{3dB}} \right)^{2n} \right]} \qquad (6.20)$$

where $K_{2B1Q} = \frac{5}{9} \cdot \frac{V_p^2}{R}$ and $R = 135 \ \Omega$. For ISDN: $n = 2$, $V_p = 2.5$ V, $f_0 = f_{3dB} = 80$ kHz, so the data rate is 160 kbps; and for HDSL: $n = 4$ $V_p = 2.7$ V, $f_0 = 392$ kHz, $f_{3dB} = 196$ kHz, so the data rate is 784 kbps (per line).

The probability of error, given an AWGN channel, is

$$P_e = P_b = 1.5Q\left(\sqrt{\frac{SNR}{5}} \right) \qquad (6.21)$$

with the encoder mapping in Table 6.1. The gap for 2B1Q is 9.8 dB at $P_e = 10^{-7}$. ISDN and HDSL also mandate an additional 6 dB of margin, leaving ISDN and HDSL at least 16 dB below the best theoretical performance levels. Additional loss of several dB occurs with ISI (see Section 7.1) on most channels. Thus, uncoded 2B1Q is not a high-performance line code, but clearly the transmitter is fairly simple to implement.

6.2.2 Pulse Amplitude Modulation

2B1Q generalizes into what is known as pulse amplitude modulation (PAM) but has the same basis function as 2B1Q modulation, and simply has $M = 2^b$ equally spaced levels symmetrically placed about zero, $b = 1,..., \infty$. The number of dimensions remains $N = 1$ and PAM also uses a memoryless encoder, $x = 2m - (M - 1)$, $m = 0,...,M - 1$ or a code like gray code in 2B1Q and Table 6.1. Approximately a 6 dB increase in transmit power is necessary for the extra constellation points associated with each additional bit in a PAM constellation if the performance is to remain the same (i.e., P_e is constant). Eight-level PAM is sometimes called 3B1O and is likely to be used in single-line HDSL or HDSL-2 standards.[5] The symbol rate for HDSL-2 3B1O is 517.3 kHz, so that the data rate is 1.552 Mbps (there is thus 8 kbps overhead with respect to DS1 service at 1.544 Mbps). HDSL-2 also may use the one-dimensional 512-state trellis code discussed in Section 7.3, so the 9.8 dB gap is reduced by about 5.5 dB of coding gain to 4.3 dB. Furthermore, margin requirements are likely to be relaxed to only 3 dB in HDSL-2, meaning that the loss from theoretical levels will be a minimum of 7.3 dB, nearly an order of magnitude better than HDSL (ignoring ISI) (see Section 7.1 for how to compute ISI loss for HDSL's 2B1Q). The general transmit analog power across a resistive load R is

---

5. The original HDSL makes use of two twisted pairs (or three in Europe) while HDSL-2 uses only one twisted pair.

$$P_x = \frac{\varepsilon_x}{T \cdot R} = \frac{M+1}{3(M-1)} \cdot \frac{V_p^2}{T \cdot R}$$

(6.22)

where $T = 1/f_0$ is the symbol period, R is the line impedance (typically 100 to 135 ohms), and V_p $= (M-1)d_{min}/2$ is the peak voltage. The power spectral density of the transmitted signal similarly generalizes to

$$PSD_{PAM}(f) = K_{PAM} \cdot \frac{2}{f_0} \cdot \frac{\left[\sin\left(\pi f/f_0\right)\right]^2}{\left(\pi f/f_0\right)^2} \cdot \frac{1}{\left[1 + \left(f/f_{3dB}\right)^{2n}\right]}$$

(6.23)

with $K_{PAM} = \frac{M+1}{3(M-1)} \cdot \frac{V_p^2}{R}$ and f_{3dB} is equal to the cut-off frequency of an nth-order (Butterworth) transmit filter. The data rate is $\log_2(M)/T$ bps. The probability of error is

$$P_e = P_b = 2(1 - 1/M) \cdot Q\left(\sqrt{\frac{3 \cdot SNR}{M^2 - 1}}\right)$$

(6.24)

assuming again an input bit encoding that leads to only 1 bit error for nearest neighbor errors.

6.2.2.1 Differential Encoding

PAM systems sometimes use **differential encoding** when the channel may introduce a polarity reversal. This type of encoding does not materially affect the performance of the PAM system, but does allow correct decoding in the presence of the unknown phase reversal. With differential encoding, one of the PAM bits specifies a sign change: a zero implies no sign change between successive PAM symbols, while a one implies a sign change. Thus, for differential encoding, the encoder becomes sequential (see left part of Figure 6.10). The box with a D in it is a unit-delay flip-flop.[6]

The remaining $b - 1$ bits specify magnitude in the usual way. When decoding, the decoding device compares the sign of successively decided PAM signal levels to decode this sign bit, while the rest of the bits are decoded according to the magnitude of the detected PAM signal. Such decoding performs the same in terms of symbol error as memoryless encoding, but because one symbol error causes two successive sign-bit errors, the probability of bit error can be double that of memoryless encoding.

---

6.   Sometimes later the same box D means an entire register instead of a flip-flop.

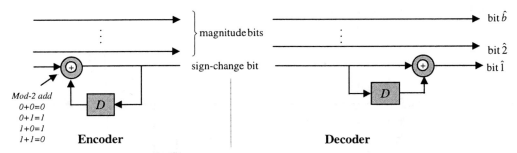

Figure 6.10 Differential encoding for PAM.

6.2.3 Binary Transmission with DC Notches

One is tempted to design a binary ($b = 1$) PAM constellation using a simple bipolar signal constellation and the sinc basis function. However, in the simplest transmission systems, significant receiver processing to avert the DC-notch characteristic of telephone lines was not implementable at the initial time of these methods' use. Thus, alternative basis functions with no DC content were used. Some of these are described in the following subsections.

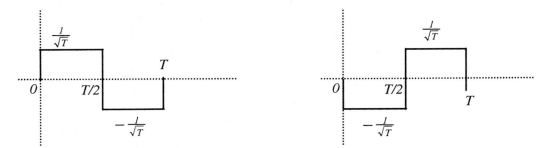

Figure 6.11 Basis function for Manchester modulation.

6.2.3.1 Manchester Modulation

Manchester modulation, which is often referred to as **Manchester encoding**, uses the transmit signals of Figure 6.11 and has $N = 1$, $b = 1$, and $M = 2$. The encoder is memoryless, $x = 2m - 1$. The function is DC-free and provides a transition in the middle of each symbol period, which is useful for robust timing recovery. Manchester modulation, however, requires approximately twice the distortion-free signal bandwidth as the sinc function, so variants were derived in early DSL line-code developments. Manchester encoding is used in **10BaseT Ethernet transmission** on twisted pairs, which are usually not considered to be DSLs, but nevertheless worthy of note since there are tens of millions in existence.

6.2.3.2 Alternate Mark Inversion

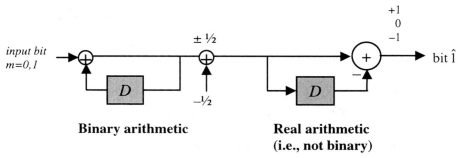

Figure 6.12   Encoder/modulator for AMI.

Alternate mark inversion (AMI) uses somewhat less bandwidth than Manchester encoding, but is not as efficient as binary PAM. The modulation basis function is dependent on past data symbols in that a 0 bit is always transmitted as a 0 level, but the polarity of successive 1s alternate, whence the name when one recalls that early data communications engineers called "1" a "mark." Thus, AMI is a second example of sequential encoding. The basis function can contain DC because the alternating-polarity symbol sequence has no DC component, thus the sinc basis function can again be used. However, the distance between the signal level representing a 0, 0 volts, and 1 (either plus or minus a nonzero level) is 3 dB less for the same transmit power as binary PAM, so

$$P_e = 1.5Q\ (SNR/2) \tag{6.25}$$

This 3 dB performance loss simplifies receiver processing. An AMI transmitter is shown in Figure 6.12. Differential encoding of binary PAM signals ("1" causes change, "0" causes no change) provides signal output levels of ½ and –½. The difference between successive outputs (+1, 0, or –1) is then input to the basis function shown to complete modulation. The receiver (not shown) maps

$$\begin{aligned} \pm 1 &\to 1 \\ 0 &\to 0 \end{aligned} \tag{6.26}$$

The transmit power spectral density of AMI is shaped as

$$S_{ami}(f) = \frac{V_p^2}{R} \cdot \frac{2}{f_0} \cdot \left[\frac{\sin\left(\frac{\pi f}{2 f_0}\right)}{\left(\frac{\pi f}{f_0}\right)}\right]^2 \cdot \sin^2\left(\frac{\pi f}{2 f_0}\right) \cdot \frac{1}{1 + \left(\frac{f}{f_{3dB}}\right)^{12}} \tag{6.27}$$

assuming 6th-order Butterworth transmit filtering and no transformer highpass. A highpass filter that represents a transformer is sometimes modeled by an additional factor of $[1 + (f_1/f)^2]^{-1}$ where f_t is the 3 dB cut-off frequency of the highpass.

ANSI T1 Transmission (T1.403) AMI code is essentially that used for both T1 transmission at symbol rate 1.544 MHz and for E1 transmission at symbol rate 2.048 MHz. The exact transmit pulse shape for T1 is shown in Figure 6.13 [5].

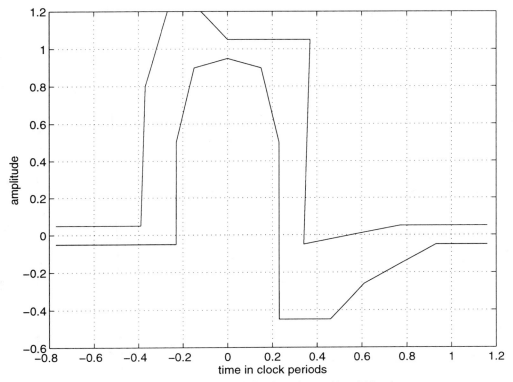

Figure 6.13 T1.403 "T1" pulse shape. (1 = 648 ns)

Japanese ISDN Japan uses AMI for ISDN transmission [6] with time-division multiplexing (TDM) as described in Chapter 5. When transmitting, the symbol rate is 320 kHz. The pulse shape shown in Figure 6.14 is also half the width of a symbol period (1.5625 μs) in an attempt to reduce interference between successive symbols when transmitted over a dispersive channel. The maximum transmitted amplitude is 6 volts. The transmit power spectral density (when on) is then

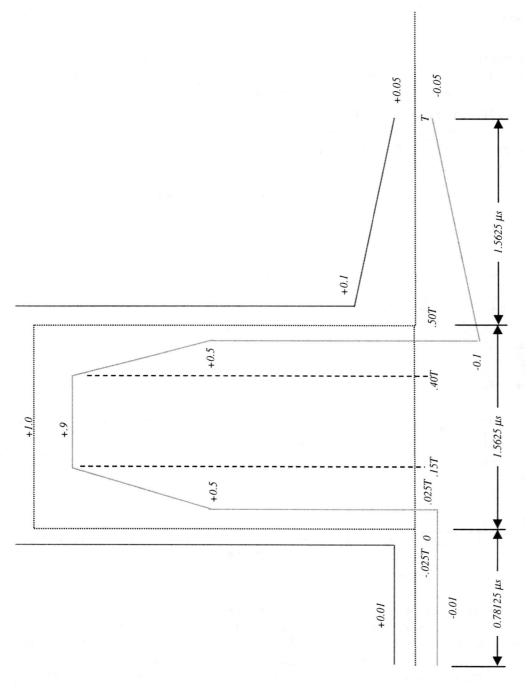

Figure 6.14 Japanese ISDN basis function (unnormalized).

$$S_{J-ISDN}(f) = K_{J-ISDN} \cdot \left| \frac{sin(\omega T / 4)}{\omega T / 4} \cdot sin(\omega T / 2) \right|^2$$

(6.28)

where $1/T = 320$ kHz and $K_{J\text{-}ISDN} = (0.25)36/110$.

Binary Zero Substitution (BnZS)   AMI is not quite robust as a line code because a long string of zeros would result in no energy transmitted through the line. Timing recovery loops could slip phase lock when a sufficiently long string of zeros occurred. Furthermore, gain control circuits might be "fooled" by a long quiet period, as might certain network protocols.[7]

Since AMI has three-level encoding with nonzero signals alternating in polarity, it is possible to exploit these three levels with "code violations," which are successive nonzero signals of the same polarity. Essentially, the encoder looks for strings of n zeros and replaces them by a string with code violations. The substitution occurs at the interface of the encoder and modulator shown in Figure 6.12. This type of modified AMI is also an example of a sequential encoder and usually is called a BnZS line code, where n is the number of successive zeros in a string that is to be replaced by a code violation. The most common line codes of this type are B8ZS and B3ZS, which are used in DS1 (T1) [5] and DS3 (T3) [7] transmission, respectively. In B8ZS, a string of eight successive zeros is replaced by alternate nonzero strings with two violations. If the last nonzero sample preceding the string was positive, then the transmitter substitutes (+ − + − + − +), which ends in a positive sample and results in two code violations in the first and seventh positions. A receiver would look for this pattern and replace it by eight zeros instead of the eight ones to which it would otherwise decode. If the last nonzero sample was negative, then the opposite polarity sequence is sent in B8ZS. The first violation must occur in the first bit of the inserted sequence, allowing the receiver to know where the 8-bit sequence started. The inserted pattern also has zero-sum (is DC-free).

As an example, the following sequence encodes as shown

$$+ 0\ 0\ 0\ 0\ 0\ 0\ 0\ 0\ 0 - \rightarrow + + - + - + - - + 0\ 0 -$$

(6.29)

with B8ZS. For B3ZS the substitution patterns are:

Last Mark Polarity	Last Substitution Pattern		
	0 0 + or + 0 +	0 0 − or − 0 −	
+	− 0 −	0 0 +	
−	0 0 −	+ 0 +	

---

7.   Simple scrambling of the input bit stream is not sufficient to ensure that the potential problems are of sufficiently low probability with AMI.

Note that a single violation signifies the end of a 3-bit string that should be reset to three zeros by the B3ZS decoder.

The code violations do not change the probability of a symbol error, however, an erroneous detection of a violation can lead to as many as n-bit error propagation, causing a slight increase in the probability of bit error with respect to AMI without BnZS.

HDB (High-Density Bipolar) Coding   High-density bipolar (HDB) coding is a sequential encoding used by the European equivalents of T1 and T3, often called E1 [8] and E3. HDBm codes correspond to $B(m + 1)$ ZS codes, only a single violation occurs in the $m + 1$ position for the substitution pattern. The rest of the symbols are zero except possibly the first bit in the substitution string. The first bit is set so that two successive violations cannot occur and the number of polarity changes (ones) between successive violations is odd.

6.2.4   4B3T Line Code

The 4B3T block code is a three-dimensional sequential encoding and baseband modulation method that selects successive three-dimensional vectors for both minimum distance improvement and elimination of any DC component. The basis functions can contain DC and thus can be in general a baseband sinc function like 2B1Q or PAM. There are 16 possible three-dimensional symbol vectors corresponding to $b = 4$ and $N = 3$ and $M = 16$. The "3T" stands for three tertiary symbols (the three levels are plus/minus a constant and zero), corresponding to a maximum of $27 = 3^3$ values of three successive three-level symbols. Of these 27, 16 are chosen for zero DC content and maximum separation. A table summarizing these values appears below, and the encoder/modulator has two states (running sum[8] negative and running sum positive):

Input bits	Ternary level sent when running sum is	
	Negative	Positive
0000	$0 + +$	$0 - -$
0001	$+ 0 +$	$- 0 -$
0010	$+ + 0$	$- - 0$
0011	$+ + +$	$- - -$
0100	$+ + -$	$- - +$
0101	$+ - +$	$- + -$
0110	$- + +$	$+ - -$
0111	$+ 0 0$	$- 0 0$
1000	$0 + 0$	$0 - 0$
1001	$0 0 +$	$0 0 -$

(continued)

---

8.   The running sum is defined as the sum of all components of the symbol vector over all time. A condition for DC-free transmit signals with linear modulation is that the running sum be bounded.

Input bits	Ternary level sent when running sum is	
	Negative	Positive
1010	+ 0 −	+ 0 −
1011	− + 0	− + 0
1100	0 − +	0 − +
1101	+ − 0	+ − 0
1110	0 + −	0 + −
1111	− 0 +	− 0 +

Some of the remaining 11 values may be used for signaling. The three dimensions of (bandwidth used by) 4B3T are greater than the two dimensions necessary in 2B1Q for transmission. However, 4B3T zeroes DC, allowing a simpler receiver. The probability of symbol error is approximated by the nearest neighbor union bound to be

$$P_e \approx 3Q\left(\sqrt{2 \cdot SNR/3} \right)$$

(6.30)

Deutsche Telekom uses a special version of the 4B3T code called MMS43 for what are known as E3 (34.368 Mbps) and E4 (139.264 Mbps) transmission, but its use is probably most famous for the German ISDN [9]. The MMS43 code uses the values in Table 6.2, where the running sum is maintained between −1 and +2 to ensure DC-free transmission:

Table 6.2 German MMS43 Code

Input bits	Ternary level sent when running sum is			
	−1	0	+1	+2
0000	+ + +	− + −	− + −	− + −
0001	+ + 0	0 0 −	0 0 −	0 0 −
0010	+ 0 +	0 − 0	0 − 0	0 − 0
0011		0 − +		
0100	0 + +	− 0 0	− 0 0	− 0 0
0101		− 0 +		
0110		− + 0		
0111	− + +	− + +	− − +	− − +
1000	+ − +	+ − +	+ − +	− − −
1001	0 0 +	0 0 +	0 0 +	− − 0
1010	0 + 0	0 + 0	0 + 0	− 0 −
1011		0 + −		
1100	+ 0 0	+ 0 0	0 + +	− − 0
1101		+ 0 −		
1110		+ − 0		
1111	+ + −	+ + −	+ − −	+ − −

The performance is the same as 4B3T. The encoder is slightly more complex and has four states with this code. However, 4B3T modulation consequently has less low-frequency energy. The power spectral density is the product of the transmit pulse power spectral density and the factor[9] (letting $w = 2\pi f T / 3$ with $T = 3T'$ and T is the symbol period that is divided into three sampling periods of equal duration T'.)

$$\frac{PSD(w) \cdot T'}{|P(w)|^2} = \frac{1}{624}\left[479 - 48 \cdot cos(w) + 16\,cos(2w) - \frac{\sum\limits_{n=0}^{5} v_s \cdot cos(nw)}{u_0 + u_3 \cdot cos(3w) + u_6 \cdot cos(6w)}\right]$$

(6.31)

with constants:

$v_0 = 301{,}339$ $u_0 = 4{,}505$

$v_1 = 426{,}304$ · $u_3 = -2{,}440$

$v_2 = 171{,}668$ $u_6 = -384$

$v3 = -62{,}432$

$v4 = -62{,}252$

$v5 = -23{,}120$

Recently, ETSI has mandated a newer PSD mask, which is shown in Figure 6.16 (see Ref. [4]), to take effect after the year 2000 for 4B3T ISDN systems.

6.2.5   4B5B Modulation

4B5B modulation uses a memoryless encoder and selects 16 of 32 possible N = five-dimensional binary vectors for transmission/modulation, with $b = 4$ and $M = 16$. This method is used in ATM25 transmission [10] at 25.6 =(4/5)32 Mbps. Table 6.3 describes the modulation used in ATM25, for which the basis function can be constant at any lowpass filter that satisfies the masks specified in Ref. [10] or alternately can be implemented with data-dependent basis functions for each of the situations.

The ML decoder has the same performance on a channel with AWGN as if data were transmitted at 32 Mbps in that d_{min} has not been increased by the selection of five-dimensional vectors. A variety of other factors dominated the selection of patterns, including timing recovery and simple implementation of the demodulator/decoder. The ESC pattern is used to qualify control information that may follow.

A better 4B5B code simply selects the even (or odd) code words and achieves a 2 dB improvement in d_{min} on the AWGN. (The distance doubles, but there is 1 dB less energy per

---

9.   The authors would like to thank Dr. Werner Henkel of Deutsche Telekom for the power spectral density plot for the MMS43 code in Figure 6.15. This plot includes the effect of a rectangular pulse shape on each dimension within a symbol.

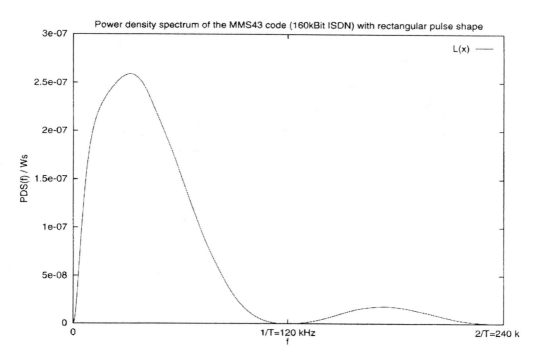

Figure 6.15 MMS43 Power spectral density.

dimension, for a total of 2 dB gain.) This ATM code selects the 5-bit patterns of Table 8.3 to match filtering of an anticipated channel.

6.2.6 Successive Transmission

Data transmission consists of successive transmission of independent messages over a transmission channel. The subscript n denoting dimensionality is usually dropped for one-dimensional transmission and a time-index subscript k is instead used to denote the message transmitted at time $t = kT$, or equivalently the kth symbol. Thus, x_k corresponds to one of M possible symbols at time k. The corresponding modulated signal then becomes

$$x(t) = \sum_k x_k \, \varphi(t - kT)$$

(6.32)

a sum of translated single-message waveforms. The symbols x_k may be independent or may be generated by a sequential encoder. A receiver design could sample the output of matched filter $\varphi(-t)$ on an AWGN channel at time instants $t = kT$. The design of the modulation filter $\varphi(t)$ may

PD 4B3T

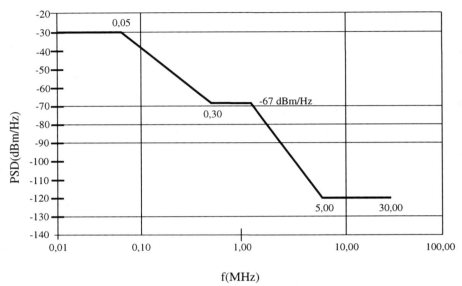

f(MHz)

Figure 6.16   New ISDN 4B3T PSD mask.

Source: Extracted from Ref. [4].

Table 6.3   4B5B Code for ATM25

Input Bits	Five-Dimensional Vector
0000	+ − + − +
0001	− + − − +
0010	− + − + −
0011	− + − + +
0100	− − + + +
0101	− + + − +
0110	− + + + −
0111	− + + + +
1000	+ − − + −
1001	+ + − − +
1010	+ + − + −
1011	+ + − + +
1100	+ − + + +
1101	+ + + − +
1110	+ + + + −
1111	+ + + + +
ESC(X)	− − − + −

not be such that successive translations by one symbol period are orthogonal to one another. However, this property is desirable and the function is called a **Nyquist pulse** when it satisfies the orthogonality constraint

$$\int_{-\infty}^{\infty} \varphi(t - kT)\varphi(t - lT)dt = \delta_{kl} = \begin{cases} 1 & k \neq l \\ 0 & k = l \end{cases} \tag{6.33}$$

Nyquist pulses exhibit no overlap or **intersymbol interference** between successive symbols at the matched-filter output in the receiver. The **Nyquist criterion** can be expressed in many equivalent forms, but one often encountered is that the combined transmit-filter/matched-filter shape $q(t) = \varphi(t) * \varphi(-t)$ should have samples $q(kT) = \delta_k$ or the "aliased" spectrum should be flat ($Q(w)$ is the Fourier transform of $q(t)$ in Equation (6.34) below):

$$1 = \frac{1}{T} \cdot \sum_{n=-\infty}^{\infty} Q\left(\omega + \frac{2\pi n}{T}\right) \tag{6.34}$$

The function $\varphi(t) = \frac{1}{\sqrt{T}} \cdot sin\, c(\frac{t}{T})$ is a Nyquist pulse, but also one that "rings" significantly with time, because the amplitude of the sinc function at nonsampling instants (when $t \neq kT$) decays only linearly with time, as $1/t$, from the maximum at time $t = 0$. This leaves a transmission system designed with sinc pulses highly susceptible to small **timing-phase errors** in the sampling clock of the receiver. There are many Nyquist pulses, but perhaps the best known are the **raised-cosine pulses,** which satisfy the Nyquist criterion and have their ringing decay with $1/t^3$ instead of $1/t$. These pulses are characterized by a parameter α that specifies the fraction of **excess bandwidth**, which is the bandwidth in excess of the minimum $1/T$ necessary to satisfy the Nyquist criterion with sinc pulses. The raised cosine pulses have response (with excess bandwidth parameter $0 \leq \alpha \leq 1$):

$$q(t) = sin\, c\left(\frac{t}{T}\right) \cdot \left[\frac{cos(\alpha\pi t/T)}{1 - (2\alpha t/T)^2}\right] \quad or$$

$$Q(\omega) = \begin{cases} T & |\omega| \leq \frac{\pi}{T}(1-\alpha) \\ \frac{T}{2}\left[1 - sin(\frac{T}{2\alpha}[|\omega| - \frac{\pi}{T}])\right] & \frac{\pi}{T}(1-\alpha) \leq |\omega| \leq \frac{\pi}{T}(1+\alpha) \\ 0 & \frac{\pi}{T}(1+\alpha) \leq |\omega| \end{cases} \tag{6.35}$$

The actual pulse shapes for transmit filter and receiver filter are **square-root raised-cosine pulses** and have transform $\sqrt{Q(\omega)}$ and the time-domain impulse response:

$$\varphi(t) = \frac{4\alpha}{\pi\sqrt{T}} \cdot \frac{cos([1+\alpha]\pi t / T) + T \cdot sin([1-\alpha]\pi t / T) / (4\alpha t)}{1 - [4\alpha t / T]^2} \tag{6.36}$$

This pulse convolved with itself (corresponding to the combination of a transmitter and receiver-matched filter) is a raised-cosine pulse. The time and frequency response of a square-root raised cosine transmit filter are shown in Figures 6.17 and 6.18, respectively.

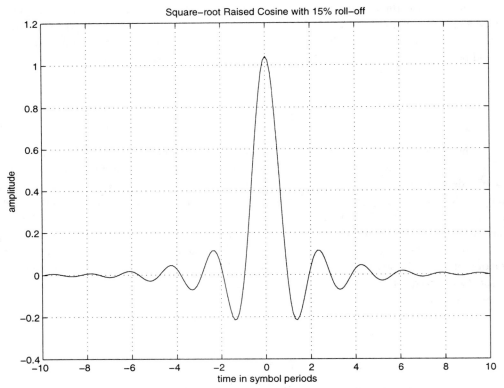

Figure 6.17   Square-root raised cosine for α=.15, time domain.

6.3 Passband Codes

Passband line codes have no energy at or near DC. The reason for their use in DSLs is because DSLs normally are transformer-coupled (for purposes of isolation) from transceiver equipment and thus the transformers pass no DC or low frequencies. This section reviews four of the most popular passband transmission line codes.

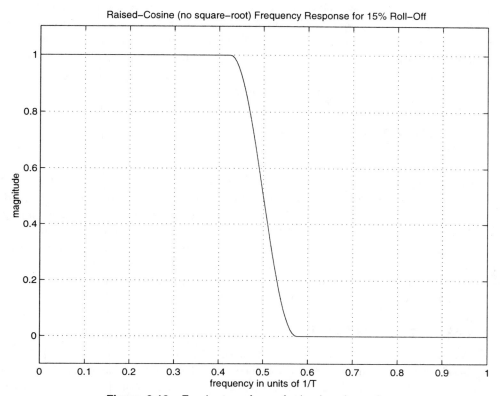

Figure 6.18 Fourier transform of raised cosine pulse.

6.3.1 Quadrature Amplitude Modulation

Quadrature amplitude modulation (QAM) is an N = two-dimensional modulation method. The two basis functions are (for transmission at time 0):

$$\varphi_1(t) = \sqrt{\frac{2}{T}} \varphi(t) \cos\left(2\pi f_c t\right)$$

$$\varphi_2(t) = -\sqrt{\frac{2}{T}} \varphi(t) \sin\left(2\pi f_c t\right)$$

(6.37)

where $\varphi(t)$ is a baseband modulation function like sinc or a square-root raised cosine pulse shape. The time and frequency domain plots of the raised cosine pulses appear in Figures 6.17 and 6.18. The multiplication of the pulse shape by sine and cosine moves energy away from baseband to avoid the DC notch of the twisted-pair transformer coupling. QAM pulses suffer severely from line attenuation in DSLs, and compensation is expensive. Some manufacturers use proprietary systems with QAM. QAM is often used in voice-band modem transmission where line characteristics have considerably less variation over the small 3 to 4 kHz bandwidth, allowing DC to be avoided and a receiver to be implemented with tolerable complexity (see Section 7.1.2 on equalization).

For successive transmission, QAM is implemented according to

$$x(t) = \sqrt{2/T} \cdot \sum_k x_{1,k}\, \varphi(t - kT) \cos\left(2 f_c \pi t\right) - x_{2,k}\, \varphi(t - kT) \sin\left(2\pi f_c t\right)$$

(6.38)

in which one notes the sinusoidal functions are not offset by kT on the kth symbol period. Because of the presence of the sinusoidal functions and the potential arbitrary choice of a carrier frequency with respect to the symbol rate, QAM functions do not appear the same within each symbol period. That is, QAM basis functions are usually not periodic at the symbol rate, $x_n(t) \neq x_n(t + kT)$, even if the same message is repeatedly transmitted. However, the baseband pulse is repeated every symbol period. This aperiodicity is not typically of concern, but the use of periodic functions can allow minor simplification in implementation in some cases with the so-called CAP methods of the next subsection.

6.3.2   Carrierless AMPM

Carrierless amplitude/phase modulation (CAP) was proposed by Werner [11], who notes that the carrier modulation in QAM is superfluous because the basic modulation is two-dimensional and a judicious choice of two DC-free basis functions can sometimes simplify the transmitter implementation. The potentially high receiver complexity of QAM is still apparent with CAP (see Section 7.1.2).

CAP basis functions appear the same within each symbol period:

$$\varphi_1(t) = \sqrt{\frac{2}{T}} \varphi(t) \cos\left(2\pi f_c t\right)$$

$$\varphi_2(t) = -\sqrt{\frac{2}{T}} \varphi(t) \sin\left(2\pi f_c t\right)$$

(6.39)

Successive transmission is implemented with

$$x(t) = \sqrt{2/T} \cdot \sum_k x_{1,k}\, \varphi(t - kT) \cos\left(2\pi f_c [t - kT]\right) - x_{2,k}\, \varphi(t - kT) \sin\left(2\pi f_c [t - kT]\right)$$

(6.40)

This form is a more natural extension of one-dimensional successive transmission and the basis-function concept. There is no carrier frequency, and f_c is simply a parameter that indicates the center of the passband used for transmission, whence the use of the term "carrierless," while nevertheless the amplitude and phase of $x(t)$ are modulated with the two-dimensional basis set. CAP transmission systems are not standardized for use in DSL, but have been used by a few vendors in proprietary implementations [12]. They are under consideration for VDSL at time of writing.

CAP and QAM are fundamentally equivalent in performance on any given channel given the same receiver complexity — only the implementations differ, and only slightly.

The Digital Audio Video Council (DAVAC) and the ATM Forum have recommended CAP implementations for transmission on Class 5 twisted pairs for short lengths of less than 50 meters, as used in local area networks. Like 10BaseT Ethernet and ATM26, these are not considered to be digital subscriber lines and so are not further studied here. (The reader is referred to Ref. [11] for an excellent description of this area.)

6.3.3   Other Quadrature Modulation Schemes

Vestigial sideband modulation (VSB) is a form of QAM where the carrier frequency is no longer centered. It is essentially one-dimensional modulation where the basis function is

$$\varphi(t) = \sqrt{\frac{1}{T}} \cdot \left[\phi(t) \cdot \cos\left(2\pi f_c t\right) - \breve{\phi}(t) \cdot \sin\left(2\pi f_c t\right) \right] \tag{6.41}$$

where the quadrature portion of the basis function uses the Hilbert transform, $\breve{\phi}(t) = \dfrac{1}{\pi t} * \phi(t)$

of the in-phase portion, $\varphi(t)$. Because such a system is single side band, the symbol rate is usually double that of QAM and yet still uses the same bandwidth — however, only one dimension is transmitted, not two, so VSB and QAM are equivalent in performance and differ only in implementation.

VSB differs in the basis function from the single side-band basis function above only in that the function $\varphi(t)$ and its Hilbert transform are approximated by functions that have vestigial symmetry with respect to the chosen carrier frequency. That is,

$$\Phi_1\left(f_c - f\right) + \Phi_2\left(f + f_c\right) = \sqrt{T} \tag{6.42}$$

and the basis function then becomes

$$\varphi(t) = \sqrt{\frac{1}{T}} \cdot \left[\varphi_1(t) \cdot \cos\left(2\pi f_c t\right) - \varphi_2(t) \cdot \sin\left(2\pi f_c t\right) \right] \tag{6.43}$$

Staggered QAM or offset QAM (OQAM) are two-dimensional modulation methods where the inphase and quadrature energy transmissions are effectively delayed by one-half a symbol period with respect to one another. The OQAM basis functions are

$$\varphi_1(t) = \sqrt{\frac{2}{T}}\varphi(t)\cos(2\pi f_c t)$$

$$\varphi_2(t) = -\sqrt{\frac{2}{T}}\varphi(t - T/2)\sin(2\pi f_c t)$$

(6.44)

A comparison of QAM, CAP, VSB, and OQAM modulation systems appears in Figure 6.21.

6.3.4   Constellations for QAM/CAP and Relation to VSB

The easiest constellations for QAM are the **QAM square constellations,** which are essentially the same as two PAM dimensions treated as a single two-dimensional quantity. Indeed, square QAM and PAM are essentially equivalent when appropriate normalization to the number of dimensions occurs, $P_e(PAM) = \overline{P}_e(SQ-QAM)$ when $M_{PAM}^2 = M_{QAM}$. Figure 6.19 shows square QAM constellations for even and odd numbers of bits per two-dimensional symbol. The expressions for probability of error are then

$$P_e(odd) < 4\left(1 - \frac{3}{2M}\right) \cdot Q\left(\sqrt{\frac{6 \cdot SNR}{2M-1}}\right)$$

$$P_e(even) \le 4\left(1 - \frac{1}{\sqrt{M}}\right) \cdot Q\left(\sqrt{\frac{3 \cdot SNR}{M-1}}\right)$$

(6.45)

VSB and SSB systems are strictly speaking one-dimensional modulation methods with time-varying basis functions. However, because they are analogous to square QAM, one often hears the term "64 VSB," which means the equivalent of a 64-SQ constellation, which is eight levels in one dimension. 16 VSB is thus four levels in one dimension. This is somewhat an abuse of terminology that arose in competitive comparisons of QAM and VSB for HDTV transmission (only in the United States — a different method is used by the rest of the world).

Another popular series of constellations for $b \ge 5$ are the cross-constellations that use the general constellation shown in Figure 6.20 and have relationship to probability of symbol error

$$P_e(cross) \le 4\left(1 - \frac{1}{\sqrt{2M}}\right) \cdot Q\left(\sqrt{\frac{3 \cdot SNR}{\frac{31}{32}M - 1}}\right)$$

(6.46)

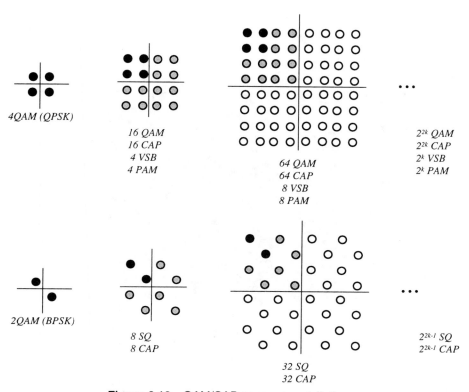

Figure 6.19   QAM/CAP square constellations.

6.3.4.1   Differential Encoding

Differential encoding for square and cross QAM/CAP systems is different than for PAM systems.[10] These constellations exhibit four-way 90-degree symmetry. Two bits are used to encode the phase change, $0°$, $90°$, $180°$, $270°$ with respect to the last transmitted symbol value. These two differential-phase bits are decoded by comparing successive quadrants of detected two-dimensional symbol constellation points. The remaining bits specify orientation within the quadrant with respect to the lowest energy point and are decoded independently for each symbol.

---

10.   For odd b square QAM, differential encoding (if used) is the same as for PAM with the sign bit specifying which side of the 180-degree symmetry line that the transmitted signal lies, and the remaining $b - 1$ bits specify an orientation with respect to that line.

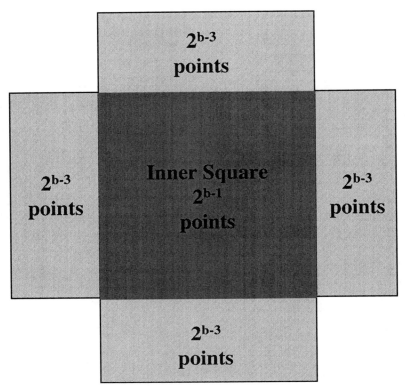

Figure 6.20 General QAM cross for *b>4* and odd.

6.3.5   Complex Baseband Equivalents

Figure 6.22 shows the PSD of the input and output signals in a passband transmission system. The designer only cares about the region of bandwidth used and not the entire transmission band.

Typically, these systems are two-dimensional, and many designers prefer to describe the transmission system with scalar complex (rather than real) functions and analysis. There are two types of complex representations in common use: **baseband equivalents** are most useful in analyzing QAM systems, and **analytic equivalents** that are most useful in analyzing CAP systems. We first define the complex data symbol $x_{b,k} = x_{1,k} + jx_{2,k}$.

A QAM waveform is determined by taking the real part of the complex waveform

$$x_a(t) = \sum_k x_{b,k}\varphi(t - kT) \cdot e^{j2\pi f_c t}$$

$$(6.47)$$

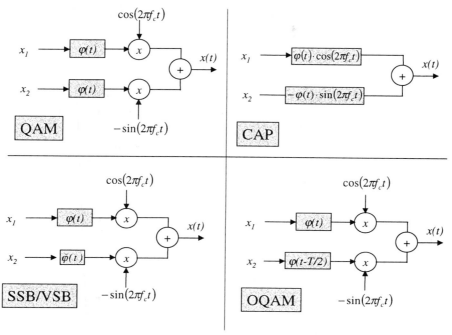

Figure 6.21   Passband modulator structures.

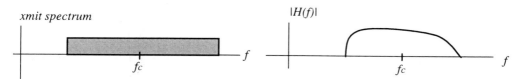

Figure 6.22   Passband channel input and output spectra.

where $x_a(t)$ is the **analytic equivalent** of the QAM signal. The **baseband equivalent** is just

$$x_b(t) = \sum_k x_{b,k}\varphi(t - kT) = x_a(t) \cdot e^{-j2\pi f_c t}$$

$$(6.48)$$

The baseband equivalent does not explicitly appear to depend on the carrier frequency and essentially amounts to shifting the passband signal down to baseband. That is

$$X_b(f) = 2 \cdot X(f - f_c) \quad \forall \, f \geq -f_c$$

$$(6.49)$$

The baseband equivalent output of a noiseless linear channel with transfer function $H(f)$ can be found as

$$Y_b(f) = H(f - f_c) \cdot X_b(f) \; \forall \; f \geq -f_c \tag{6.50}$$

Defining $P(f) = H(f - f_c) \, \Phi \, (f)$ as the pulse response of the channel, and scaling the channel output (signal and noise) down by a factor of $\sqrt{2}$ (to eliminate the extra square root of 2 factors in the normalized basis functions of QAM and the artificial doubling of noise inherent in the complex representation), produces the convenient complex baseband channel model

$$y_b(t) = \sum_k x_{b,k} \cdot p(t - kT) + n_b(t) \tag{6.51}$$

where $x_{b,k}$ are complex symbols at the channel input and $n_b(t)$ is complex baseband equivalent white noise with power spectral density $2\sigma^2$ or equivalently σ^2 per real dimension. This baseband equivalent channel has exactly the same form as with successive transmission with real baseband signals like PAM, except that all quantities are complex. Thus, PAM is a special case. This representation allows a consistent single theory of equalization/modulation as will be followed by subsequent sections of this text and generally throughout the DSL industry.

CAP systems can also be modeled by a complex equivalent system except that the channel output is still passband and is the analytic signal

$$y_a(t) = \sum_k x_{b,k} \cdot p_a(t - kT) + n_a(t) \tag{6.52}$$

where $P_a(f) = H(f) \cdot \Phi \, (f - f_c) \, f > 0$ and the noise is statistically again WGN with power spectral density σ^2 per real dimension.

In either the QAM or CAP case, a complex equivalent channel has been generated, and the effects of carrier and/or center frequencies can be subsequently ignored in the analysis, which can then concentrate on detecting $x_{b,k}$ from the complex channel output.

The generation of the complex channel outputs for QAM and CAP transmission appears in Figure 6.22.

References

[1]   Shannon, C. E., "A Mathematical Theory of Communication," *Bell Systems Technical Journal*, Vol. 27, 1948, pp. 379–423 (Part I), pp. 623–656 (Part II).

[2]   *Integrated Services Digital Network (ISDN) — Basic Access Interface for Use on Metallic Loops for Application on the Network Side of the NT (Layer 1 Specification)*, ANSI Standard T1.601-1992, New York.

[3]   *HDSL.* ANSI Technical Report TR28. See Also Bellcore Technical Advisories T!— NWT-001210 and 001211.

[4]   Zoetman, H., "Living List (Revision 8) for Amendment of TS 080 (ETR 080)," ETSI WG TM 6 Permanent Document: No. 4, 960p04, revision 8, June 2, 1998. http://docbox.etsi.org// tech-org/tm/Document/tm6/983lulea/Hilv/960p04r8.doc

[5]   *DS1 Metallic Interface.* ANSI Standard T1.403-1995, New York.

[6]   "Technical Reference for High-Speed Digital Leased Circuit Services (transmission system on metallic subscriber lines)," Nippon Telephone and Telegraph: Advanced Telecommunications Services Sector, Tokyo, 1991.

[7]   *DS3 Metallic Interface.* ANSI Standard T1.404-1994, New York.

[8]   International Telecommunication Union — Telecommunications. Recommendation G.703 (1991) — "Physical/Electrical Characteristics of Hierarchical Interfaces."

[9]   "Leitungscodes fur Digitalsignalübertragung," courtesy of Deutshe Telekom, Dr. S. Heuser and W. Henkel, pp. 22–31.

[10] European Telecommnication Standard, "Broadband ISDN TC and PM Sublayers For the S_B Reference Point at a Bit Rate of 25.6 Mbps Over Twisted-Pair Cable," ETSI STC-TM3, 1996, Sophia Antipolis, France.

[11] IM, G. H. and Werner, J. J., "Bandwidth-Efficient Digital Transmission over Unshielded Twisted-Pair Wiring," *IEEE Journal on Selected Areas in Communication (JSAC),* Vol. 12, No. 9, December 1995, pp. 1643–1655.

[12] Friedman et al., "VDSL Proposal," *ANSI Contribution T1E1.4/98-054,* March 2, 1998, Austin, Texas.

Intersymbol Interference, Equalization, and DMT

7.1 Intersymbol Interference

Intersymbol interference (ISI) is the dominant impairment in DSL transmission. The severe line attenuation and delay variation with frequency (see Chapter 3) causes successively transmitted symbols to interfere with one another. This interference from one symbol can affect hundreds of other symbols in DSLs. Such intersymbol interference renders symbol-by-symbol detection inadequate. Early transmission methods like AMI attempt to avoid ISI by using short lengths of twisted pair (AMI for T1 and E1) so that the attenuation varies less with frequency. 2B1Q ISDN uses a very low symbol rate of 80 kHz so that successive transmissions have less overlap. However, even ISDN can exhibit significant ISI. At the high speeds of HDSL and ADSL and, in the future, VDSL, ISI must be compensated adaptively and robustly. **Equalization** methods can effect significant improvement (see Section 7.1.2) but are especially expensive as data rates and symbol rates increase. Equalization methods adaptively configure the receiver to mitigate ISI. HDSL makes extensive use of equalization. Multichannel methods adapt both the transmitter and the receiver to effect additional improvement (Section 7.2). ADSL, and some forms of VDSL, uses multichannel transmission. The symbol rate and carrier/center frequency of CAP/QAM can be varied for each channel to effect additional equalization improvement. Some forms of VDSL use such transmit and receive equalization.

Figure 7.1 illustrates ISI using a simple channel response. As transmissions occur more frequently (that is, the symbol rate is increased, which corresponds to decreasing T for each successive plot in Figure 7.1), then a given channel shape exhibits increasing overlap of transmissions and thus more ISI.

Decreasing T

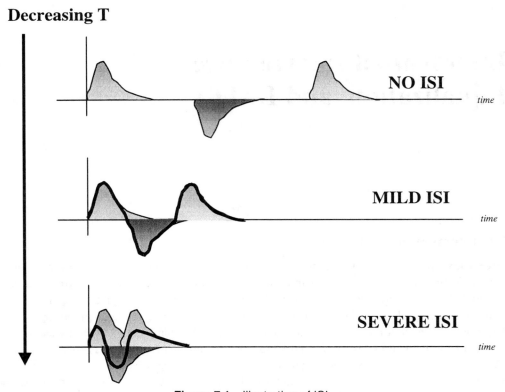

Figure 7.1   Illustration of ISI.

7.1.1   Quantifying ISI

An AWGN channel has infinite bandwidth, implying equal gain and zero phase at all frequencies, or $H(f)$, $H_b(f)$, $H_a(f)$ are constant over the used transmission bands.[1] In practice, the channel has a nonideal transfer function $H(f)$ that describes the gain and phase variation caused by the bandwidth limitations of any physical channel. Figure 7.2 modifies the ISI channel to include the impulse response of the transmission line. The channel output becomes

$$\tilde{y}(t) = \sum_{m=-\infty}^{\infty} x_m \cdot p(t - mT) + \tilde{n}(t)$$

(7.1)

where the pulse response is determined by the convolution of the channel (complex baseband)

---

1.   We will use $H(f)$ as the channel response, whether the situation is baseband (PAM), baseband equivalent (QAM), or analytic equivalent (CAP), or any other transmission method (i.e., DMT).

impulse response with the basis function

$$p(t) = h(t) * \varphi(t) \tag{7.2}$$

Successive translations of the pulse response will not generally be orthogonal to each other, leading to ISI.

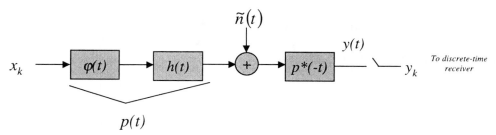

Figure 7.2   ISI channel model.

A receiver matched-filter of $p^*(-t)$ with output sampled at the symbol rate $1/T$ is sufficient to represent the channel output [1]. The output of such a matched filter has a representation

$$y(t) = \sum_{m=-\infty}^{\infty} x_m \cdot r(t - mT) + n(t) \tag{7.3}$$

The function $r(t) = p(t) * p^*(-t)$ is the **channel autocorrelation function**. The output of the matched filter is sampled at the symbol rate to obtain the sequence y_k. Any sampled sequence has a **D-transform**, for instance $Y(D) = \sum_{k=-\infty}^{\infty} y_k \cdot D^k$ [2]. The output sequence transform is related to the transform of the sequence of transmitted symbols according to

$$Y(D) = X(D) \cdot R(D) + N(D) \tag{7.4}$$

For an ISI-free channel, $r_{no-ISI}(kT) = \|p\|^2 \cdot \delta_k$. In any case, the mean-square value of the ISI is

$$\text{Mean Square ISI} \stackrel{\Delta}{=} \varepsilon_x \cdot \left\{ \sum_{k \neq 0} \|r_k\|^2 \right\} \tag{7.5}$$

which can be considered an additional noise to the filtered white noise at the matched filter output. This **mean square ISI** often exceeds Gaussian or crosstalk noise in a DSL by several orders of magnitude. Thus, it is reasonable to expect some receiver filtering to be effective in reducing this distortion, even though such filtering likely increases the noise. Receiver filtering to reduce ISI is called **equalization.**[2]

7.1.1.1 Noise Equivalent ISI Channel

Self-correlated or "colored" noise with power spectral density $\sigma^2 S_n(f)$ has an effect equivalent to ISI. To analyze a channel with correlated noise, the designer views the channel as the result of passing the channel output through a filter $1/[S_n(f)]^{0.5}$ that whitens the noise but changes the ISI. This new equivalent channel has additive white noise with flat power spectral density σ^2 and a pulse response modified to

$$P(f) \rightarrow \frac{P(f)}{\sqrt{S_n(f)}}$$

(7.6)

and thus the noise is equivalent to ISI.

Final receiver implementation should thus always contain this whitening filter even though white-noise-based analysis proceeds as if this filter were part of the channel. A channel without ISI ($P(f) = 1$), but with correlated noise, will thus have a potential improvement for symbol-by-symbol detection obtainable by using methods that mitigate ISI, like equalization, because the equivalent channel has ISI.

7.1.2 Equalization

Equalization is a term generally used to denote methods employed by DSL receivers to reduce the mean-square ISI. The most common forms of equalization for DSLs appear in this subsection.

7.1.2.1 Linear Equalization

Linear equalizers are very common and easy to understand, albeit they often can be replaced by simpler nonlinear structures that work better but are harder to understand. Nonetheless, this study begins with linear equalization and then progresses to better equalization methods in later sections. The basic idea of linear equalization is to invert the channel impulse response so that the channel exhibits no ISI. However, straightforward filtering would increase the noise substantially, so the design of an equalizer must simultaneously consider noise and ISI. Noise enhancement is depicted in Figure 7.3. The equalizer increases the gain of frequencies

---

2.   The term "equalization" has grown in meaning in recent years to encompass all transmitter and receiver filtering.

that have been relatively attenuated by the channel, but also simultaneously increases the noise energy at those same frequencies. This increase of noise is called **noise enhancement**.

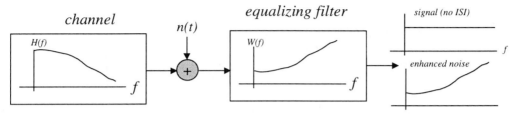

Figure 7.3 Illustration of noise enhancement with linear equalizer.

The minimum mean-square error linear equalizer (MMSE-LE) appears in Figure 7.4.

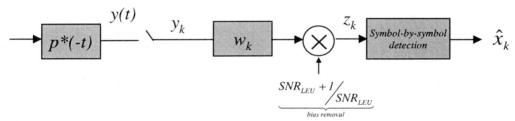

Figure 7.4 MMSE linear equalizer.

A matched filter for the channel pulse response and sampler precede a linear filter whose transfer function is selected to minimize the mean-square difference between the channel input and the equalizer output. This filter has D-transform

$$W(D) = \frac{1}{R(D) + \frac{1}{SNR}}$$

(7.7)

where SNR $= \mathcal{E}_x/\sigma^2$.[3] The equalizer need only be symbol-spaced (implemented at the symbol sampling instants in discrete time), but some implementations often combine the matched filter and linear filter into a single filter called a **fractionally spaced equalizer**. This single combined filter must be implemented at a sampling rate at least twice the highest frequency of the channel pulse response. The minimum mean-square error (MMSE) value in either implementation is

---

3. $\mathcal{E}_x$ and σ^2 must consistently be for the same number of real dimensions.

$$MMSE = \sigma_{LE}^2 = \sigma^2 \cdot w_0 = \frac{\sigma^2 \, \mathrm{T}}{2\pi} \int_{-\pi/\mathrm{T}}^{\pi/\mathrm{T}} \frac{1}{\mathrm{R}(e^{j\omega \mathrm{T}}) + 1/SNR} \, d\omega \tag{7.8}$$

where w_0 is the center tap (time zero sample) in the discrete-time symbol-spaced equalizer response w_k. A symbol-by-symbol detector processes the equalizer output. The scaling just prior to the detector removes a small bias inherent in MMSE estimation [3]. The performance of the MMSE-LE is characterized by an unbiased signal-to-noise ratio at the detector

$$SNR_{LEU} = \frac{E_x}{\sigma_{LE}^2} - 1 \tag{7.9}$$

This receiver signal-to-noise ratio is equivalent to the SNR that characterizes an AWGN channel's ML receiver, which makes independent decisions about successive transmitted symbols. Thus, a probability of error can be approximated by assuming the error signal is Gaussian and computing P_e according to well-established formulas for PAM and QAM. For instance, square QAM on a bandlimited channel using an MMSE-LE with SNR_{LEU} computed above has probability of error

$$\overline{P}_e \approx 2\left(1 - \frac{1}{\sqrt{M}}\right) Q\left[\sqrt{\frac{3 \cdot SNR_{LEU}}{M - 1}}\right] \tag{7.10}$$

Other formulas from Section 6.2 for other constellations with an MMSE-LE can similarly be used by simply substituting SNR_{LEU} for SNR in the usual AWGN channel expression of Section 6.2.

7.1.2.2 Decision Feedback Equalization

Decision feedback equalization (DFE) avoids noise enhancement by assuming that past decisions of the symbol-by-symbol detector are always correct and using these past decisions to cancel ISI. The DFE appears in Figure 7.5.

The linear filter characteristic changes so that the combination of the matched filter and linear **feedforward filter** adjust the phase of the intersymbol interference so that all ISI appears to have been caused by previously transmitted symbols. Since these previous symbols are available in the receiver, the **feedback filter** estimates the ISI, which is then subtracted from the feedforward filter output in the DFE. The DFE always performs at least as well as the LE and often performs considerably better on severe ISI channels where the noise enhancement of the LE is unacceptable. Clearly, the LE is a special case of the DFE when there is no feedback section.

The best settings of the DFE filters can be obtained through spectral ("canonical") factorization of the channel autocorrelation function, through the **key equation**:

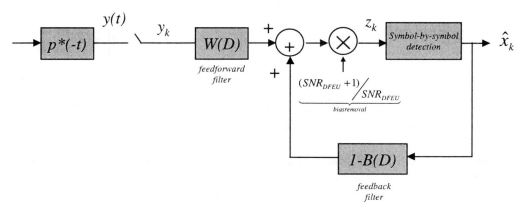

Figure 7.5   Decision-feedback equalizaton.

$$R(D) + 1 / SNR = \gamma \cdot G(D) \cdot G^* \left(D^{-*} \right) \qquad (7.11)$$

where $\gamma > 0$ is a positive constant representing an inherent gain in the channel and $G(D) = 1 + g_1 \cdot D + g_2 \cdot D^2 + g_3 \cdot D^3 + ...$ is a monic ($g_0 = 1$), causal (exists only for nonnegative time instants), and minimum-phase (all poles/zeros outside the unit circle) polynomial in D.[4] Also, $G^*(D^{-*}) = 1 + g_1^* \cdot D^{-1} + g_2^* \cdot D^{-1} + ...$ is monic, anticausal, and maximum-phase. The feedforward filter has setting

$$W(D) = \frac{1}{\gamma \cdot G^* \left(D^{-*} \right)} \qquad (7.12)$$

and the combination of the matched-filter and feedforward filter is known as a mean-square matched whitening filter (MS-WMF) [3]. It can be shown that the MS-WMF does not enhance noise [3]. The output of the MS-WMF is the sequence $X(D)G(D) + E(D)$, a causally filtered version of the channel input plus an error sequence $E(D)$ that has minimum mean square amplitude among all possible filter settings for the DFE. Since this sequence is causal, trailing intersymbol interference may be eliminated without noise enhancement by setting $B(D) = G(D)$.

The sequence $E(D)$ is "white," with variance $\sigma_{DFE}^2 = \sigma^2 / \gamma$. Thus the signal to noise ratio at the DFE detector is

---

4.   The factor $G(D)$ can be obtained by finding roots of magnitude greater than one, or it can be directly computed for finite lengths in practice through a well-conditioned matrix operation known as Cholesky factorization.

$$SNR_{DFEU} = \gamma \frac{E_x}{\sigma^2} - 1$$

$$(7.13)$$

The SNR of a DFE is necessarily no smaller than that of a linear equalizer in that the linear equalizer is a trivial special case of the DFE where the feedback filter is $B(D) = 1$. The DFE MS-WMF is approximately an all-pass filter (exactly all-pass as the SNR becomes infinite), which means the white input noise is not amplified nor spectrally shaped — thus, there is no noise enhancement. As such, the DFE is better suited to channels with severe ISI, in particular those with bridge taps or nulls. Also, recalling that the channel pulse response may be that of an equivalent white noise channel, narrowband noise appears as a notch in the equivalent channel. Thus the DFE can also mitigate narrowband noise much better than a linear equalizer. The DFE is intermediate in performance to a linear equalizer and the optimum maximum-likelihood detector.

The removal of bias is also evident again for the MMSE-DFE in Figure 7.5. This bias removal can be pushed in front of the summing junction if the feedback section is altered slightly to

$$G_U(D) = \frac{SNR_{DFEU} + 1}{SNR_{DFEU}} \cdot \left[G(D) - \frac{1}{SNR_{DFEU} + 1} \right]$$

$$(7.14)$$

This particular filter is used with Tomlinson or "flexible" precoding (see Section 7.1.2.4).

7.1.2.3 Error Propagation in the DFE

Error propagation is a major problem with QAM/CAP and baseband PAM transmission over ISI channels where DFE is used (i.e., HDSL, ISDN, some VDSL, and some nonstandard ADSL systems). The feedback filter of the DFE depends on correct previous decisions. When an incorrect decision occurs, another error becomes highly likely. Indeed, the DFE system needs to rely on probability (luck) for just the right pattern of data to occur to avoid making errors continuously. Errors thus occur in a burst until a pattern that clears the burst happens to occur. A larger number of nonnegligible coefficients in the feedback section magnifies error propagation. Error-propagation burst length thus also increases with the symbol set size M. Coded systems (see Sections 7.3 and 7.4) exacerbate the problem because the inner DFE system typically operates at a higher (worse) error rate with the expectation that the outer coding will reduce errors. However, the DFE preliminary decisions are then even more likely to be in error and thus cause bursts more often with coding, overwhelming any applied coding in most cases and often causing a catastrophic failure of the transmission channel.

Error propagation can be analyzed by recognizing that with the DFE, the feedback section of the DFE defines a partial-response channel to which the feedforward equalizer shapes the channel [4]. A feedback section with ν nonzero taps then defines a state-machine with M^ν states.

Given that the DFE is in error propagation, then each feedback input can in turn be characterized by an **error event sample** ε_k at time k. There are $(2M - 1)^v$ states in the **error propagation state machine** for PAM and roughly the same number for QAM/CAP or other passband methods. Each such state has a probability of occurring, given a first error has already occurred. This steady-state probability distribution is called a **Markov distribution** for the error-propagation state machine. One state in the state machine is the all-zeros state, meaning that the burst has ended. A tedious analysis can then enumerate all possible paths through the state machine to the all-zeros state and sum the number of errors on each path times its probability of occurrence to get an average burst length. One can also enumerate the probabilities of all bursts of all lengths in the distribution. The result is that bursts of long length (10s to 100s of bits) occur unacceptably often (1/10 or more of the bursts), overwhelming the transmission system unless the probability of a first error is very small.[5] In ISDN and HDSL transmission, the probability of bit error must be 10^{-7} with **6 dB margin**, which corresponds to a probability of first error of well below 10^{-20} so that error bursting is not a problem. However, when coding is added, the probability of a first error on the inner DFE may increase to a much higher number. For instance, a QAM system with this 6 dB margin and 4 dB of trellis coding gain will have an error rate on the DFE of only 10^{-6}, meaning error bursts could cause dramatic performance loss.

Coding should not be used unless error propagation has been mitigated by some mechanisms (a few follow in Section 7.1.2.4) in DFE-based DSLs. Some systems use external FEC with interleaving for impulse control as described in Section 7.4. Since error propagation causes bursts, which are like impulses, this FEC with interleaving may correct the problem, but of course coding gain would be reduced for this effect. This is a suspect design strategy and usually is avoided unless bursts occur with much lower probability than impulses. The occurrence of bursts is so often that the FEC only corrects for the bursts, so why use FEC since it introduces a problem for which it can only just barely correct with no other advantage? A better solution is to use Tomlinson precoding or Flexible precoding as described next.

7.1.2.4 Tomlinson-Harashima Precoding

The Tomlinson precoder [5], [6] eliminates DFE error propagation at a negligible performance loss, but at the complication that a receiver must tell the transmitter through a secure reverse channel the settings for the DFE feedback filter. A Tomlinson precoder appears in Figure 7.6.

The feedback section has essentially been moved from the receiver to the transmitter, where a modulo arithmetic device has replaced the decision element in the transmitter. The modulo arithmetic device can be considered to be adding some constant integer multiple of Md for PAM or $\sqrt{M}\,d$ for each dimension of square QAM/CAP at each symbol period such that the resultant sum falls between $[-Md, Md)$ for PAM or $[-\sqrt{M}\,d, \sqrt{M}\,d)$ for QAM. This added constant prevents the power from increasing significantly with precoding. The action of the feed-

---

5.   A classic treatment of this problem appears in Ref. [4].

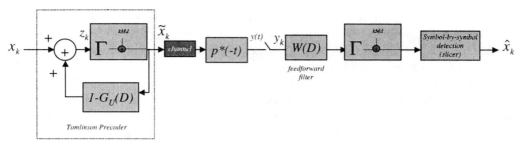

Figure 7.6   Tomlinson (Harashima) precoding.

back section in the transmitter is to add in the negative of the ISI introduced by the feedforward-equalized channel. The constant is removed by a second modulo device in the receiver, leading to the transmitted symbol plus (modulo'd) noise, and a conventional decision device (or decoder for a coded system) can thus be used.

Since no previous decisions are used, error propagation is eliminated. One drawback is that the Tomlinson precoder does increase transmit power slightly by a factor of $M^2 / (M^2 - 1)$ for PAM and $M / (M - 1)$ for square QAM. The loss is greater for shaped constellations. Another drawback is the need to transmit the feedback section values (which must be identified in the receiver for an a priori unknown channel) to the transmitter through a secure control channel during initialization, and at periodic intervals thereafter for a DSL.

7.1.2.5 Flexible Precoding

The **flexible** precoder (invented by Laroia [7], along with others, and sometimes called a Laroia precoder) has less loss than a Tomlinson precoder in most cases and allows the shape of a constellation (which could be much better than a square in high-performance coded designs) to be generally preserved after precoding. The flexible precoder is conceptually a little more difficult to understand (Figure 7.7). A slicer (symbol-by-symbol detector) is used in the transmitter to find the closest constellation point to the feedback section output, λ_k. This operation allows only the transmit symbol plus an offset, modulo the constellation, to be sent over the channel rather than a power-increasing inversion of the channel. The difference between the feedback filter output and the nearest constellation point, m_k, is added to the symbol value and transmitted. An equivalent circuit appears in Figure 7.7. Clearly the flexible precoder undoes the combined action of the feedforward filter and channel. Furthermore, since m_k by definition is a small error signal, the constellation energy is not significantly altered and any shape/gain of the constellation is largely maintained, *unlike the Tomlinson precoder*. At the receiver the symbol-by-symbol detector after the feedforward filter (not shown) has output $\hat{x}_k + \lambda_k$, which is not the transmitted symbol. But, from the equivalent circuit in Figure 7.7, the value of x_k is easily recovered by the filter $1/G_U(D)$ acting on $x_k + \lambda_k$. Any burst effect of receiver decision errors on $x_k + \lambda_k$ will decay to zero over the length of the response $\dfrac{1}{G_U(D)}$.

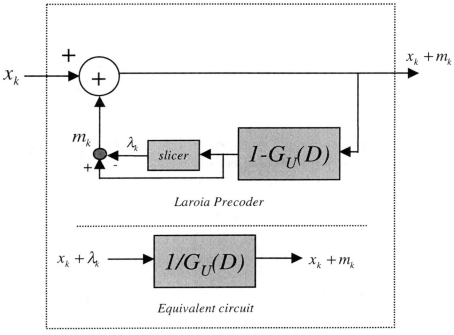

Figure 7.7 "Flexible" precoder.

7.1.2.6 Finite-Length Equalizers

In practice, the filters of either a LE or a DFE are implemented as finite-length sums-of-products (or FIR filters). Such implementations are better understood and less prone to numerical inaccuracies, while simultaneously lending themselves to easy implementation with adaptive algorithms.

The formulas for determining the settings of finite-length equalizers are of much more practical use in analyzing system performance and appear in this section. They involve the matrix operations of matrix inversion and Cholesky factorization, replacing polynomial division and spectral factorization, respectively, in the infinite-length case.

For a finite-length filter, the issue of the channel group delay becomes very important — this delay is of no consequence with infinite-length filters and was previously omitted. Thus, the equalizing receiver tries to estimate $x_{k-\Delta}$ instead of x_k where $\Delta \geq 0$ is the delay parameter used to denote the delay in symbol periods through the combination of channel and equalizer. This delay, along with the filter coefficients of the equalizer, must be optimized. Since the LE is a special case of the DFE, we will only consider the latter and the reader will understand to substitute 0 for the feedback section to get the LE results.

The error with delay Δ is written

$$e_k(\Delta) = b^* x_{k-\Delta} - w^* y_k \tag{7.15}$$

where b^* is the $1 \times N_b$ vector $b^* = \begin{bmatrix} 1 & b_1^* & \cdots & b_{N_b}^* \end{bmatrix}$ of feedback coefficients ($b = 1$ for the LE), $x_{k-\Delta} = \begin{bmatrix} x_{k-\Delta} & \cdots & x_{k-\Delta-N_b} \end{bmatrix}$ is the vector of delayed, current-and-previous channel inputs (decisions), y is the $N_f \cdot q$ vector of channel output samples (there is no matched filter, but rather a lowpass filter, whose baseband output is sampled at speed sufficiently in excess of the symbol rate [q/T in this development where q is an integer, $q = 1, 2, \ldots$] to capture useful information), and w^* is the corresponding vector of feedforward equalizer coefficients. The MMSE equalizer w^* for any choice of the feedback section b^* is

$$w^* = b^* R_{xy}(\Delta) \cdot R_{yy}^{-1} \tag{7.16}$$

For $b = 1$, the equalizer is a MMSE linear equalizer with $N_f q$ taps. The best b^* leads to the MMSE-DFE for any given number of feedback taps N_b.

For the bandlimited AWGN channel, one can always find a $qN_f \times (N_f + v)$ matrix H such that

$$y = HX + n \tag{7.17}$$

for any input and noise with given autocorrelation matrices $R_{XX} = \mathcal{E}_x I$ and $R_{nn} = \sigma^2 I$ (zero mean). Thus, in terms of known quantities H, R_{XX}, and R_{nn}, the designer can compute two new matrices:

$$R_{xy}(\Delta) = E[x_{k-\Delta} y^*] = \mathcal{E}_x J_\Delta^* H^* \tag{7.18}$$

(where J_Δ is an $(N_f + v) \times N_b$ matrix of 0s and 1s with upper $\Delta + 1$ rows zeroed and identity matrix of dimension $\min(N_b, N_f + v - \Delta - 1)$ with zeros to the right (when $N_f + v - \Delta - 1 < N_b$), zeros below (when $N_f + v - \Delta - 1 > N_b$), or no zeros to the right or below, exactly fitting in the bottom of J_Δ, (when $N_f + v - \Delta - 1 = N_b$). The autocorrelation matrix for y is

$$R_{yy} = \mathcal{E}_x HH^* + R_{nn}$$

(7.19)

The MSE is

$$MSE(\Delta) = b^* R_{le}(\Delta)b$$

(7.20)

where R_{le} is

$$R_{le}(\Delta) = q \cdot \sigma^2 \left\{ J_\Delta^* \left(H^* H + \frac{q}{SNR} I \right)^{-1} J_\Delta \right\}$$

(7.21)

Further, one computes the Cholesky factorization of the **discrete key-equation matrix**

$$R_{le}^{-1}(\Delta) = G_\Delta^* S_\Delta^{-1} G_\Delta$$

(7.22)

where G_Δ is an upper triangular, monic (ones down diagonal) matrix and $S_\Delta = diag\{S_0(\Delta), ..., S_{N_b}(\Delta)\}$ a diagonal matrix with smallest element in the upper left corner and positive, real, nondecreasing values down the diagonal, leading to the MSE

$$MSE(\Delta) = b^* G_\Delta^{-1} S_\Delta G_\Delta^{-*} b$$

(7.23)

The best $b^* = g_\Delta$ is determined by the top row of G. The corresponding MMSE value for the DFE is

$$MMSE = S_0(\Delta)$$

(7.24)

which is a function of Δ, and so the discrete-key matrix above may need to be factored for all reasonable values of Δ to find the MMSE DFE with smallest MMSE over Δ. In the LE case, the factorization trivializes to a scalar, and so the designer needs only to find the minimum value of the discrete-key equation in this scalar case to optimize over Δ. The SNR for the DFE is then

$$SNR_{DFE} = \frac{E_x}{S_0(\Delta)} - 1$$

(7.25)

The limiting value of the SNR and MMSE as the number of feedforward and feedback taps can be shown to be the same as those of the infinite-length case by Toeplitz distribution arguments. The feedforward section can be written as

$$w^* = g_\Delta J_\Delta^* \left(H^* H + \tfrac{q}{SNR} I \right)^{-1}$$

(7.26)

There are a number of matrix mathematical "tricks" that enable swift solution to the DFE problem. In practice, the coefficients are typically found adaptively without matrix inversion using gradient algorithms and a synchronization search over the best delay values. The formulas here are useful for exact analysis of how well a DFE will do. One should take care that there are a number of ad hoc approaches to computation of DFE settings that often tacitly assume infinite-length filters. These ad hoc approaches often over-estimate the true performance of a realizable DFE for use in DSLs. The formulas here are more exact and realistic (and even at that assume infinite precision and no error propagation).

7.1.3 Transmit Equalization

The transmit filter of a channel may also be optimized for best detection-point SNR. This may be done with or without a receiver equalizer. Such transmit filtering is called **transmit equalization** or **precompensation.** This section investigates optimized transmit filters for the case of no receiver equalization, a linear equalizer, and for decision feedback.

A tenet of transmitter optimization, often overlooked in virtually all previous studies (see Lucky, Salz, and Weldon [8] for a classic discussion of transmit filter optimization), is that a transmit filter must be realizable in discrete time. Such a filter must satisfy the Paley-Wiener criterion [9] for realization. In essence, this criteria does not allow a filter to have a band of nonzero measure in which all frequencies are zeroed. In lay terms this means **no dead zones** where the filter transfer function is zero except at discrete notch frequencies (which must be countable in number). Almost all transmit filter optimizations will lead to filter characteristics that violate the Paley-Wiener criterion, and thus the symbol rate (and/or carrier/center frequency) needs to be varied also until the criterion is no longer violated. It is the later symbol-rate optimization step that is often ignored in most studies. The ignorance of symbol rate optimization is often justified by a "high-SNR" assumption. However, with a high enough SNR for this ignorance to be valid, there is no need for transmit-filter optimization, resulting in a circular argument and development of limited value. The reader is cautioned against this mistake as it is prevalent in the open literature and has led many groups to false conclusions about DSL performance.

Optimization of the transmit filter always requires knowledge of the channel characteristics, either a priori or provided by a secure reverse channel from receiver to transmitter.

7.1.3.1 Optimum Transmit Filters Without an Optimized Receiver

In some system designs, a strong complexity constraint only at the receiver may cause the designer to construct an optimized transmitter instead. Figure 7.8 shows this situation where any receiver analog filtering prior to sampling is absorbed into the channel response $H(e^{-j\omega T})$. A MMSE design minimizes $E|x_k - h_k * \phi_k * x_k - n_k|^2$ over the choice of transmit filter ϕ_k subject to the constraint that the transmit filter be normalized $T\|\phi_k\|^2 = 1$. Constrained minimization using Lagrange multipliers (or calculus of variations) leads to the solution

$$\Phi\left(e^{-j\omega T}\right) = \frac{H^*\left(e^{-j\omega T}\right)}{\left|H\left(e^{-j\omega T}\right)\right|^2 + \lambda}$$

(7.27)

where λ is a frequency-independent constant determined by the filter gain constraint, $\|\phi\|^2 = 1/T$ and $H(e^{-j\omega T})$ is the (aliased) channel characteristic for the symbol rate selected.

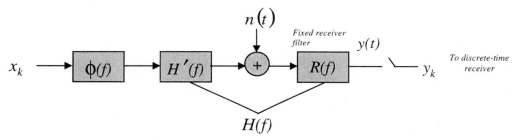

Figure 7.8 Optimizing the transmit filter.

The transmit filter must be full band, which means it must satisfy the Paley-Wiener constraint. If such a full-band filter cannot be found, then the symbol rate $1/T$ and/or carrier frequency implicit in the baseband complex equivalents must be altered until this condition is met. With DSLs, it is often necessary to decompose the channel into a set of channels, each with non-zero regions, and for each select an optimum carrier/center frequency and symbol rate.

The receiver MMSE is then

$$\sigma^2_{trans} = \frac{T}{2\pi} \cdot \int_{-\pi/T}^{\pi/T} \left(\frac{\lambda}{\left|H\left(e^{-j\omega T}\right)\right|^2 + \lambda} \right) \cdot d\omega$$

(7.28)

7.1.3.2 Optimum Transmit Filters with the MMSE-LE

The MMSE for the LE was determined in Section 7.1.2.1 and can also be optimized with Lagrange multipliers subject to the same unity gain constraint. The solution in this case is the solution of the **smile** equation:

$$\frac{1}{SNR \cdot \left|H\left(e^{-j\omega T}\right)\right|^2} + \left|\Phi\left(e^{-j\omega T}\right)\right|^2 = \frac{\lambda}{\left|H\left(e^{-j\omega T}\right)\right|}$$

(7.29)

This equation can be solved a number of ways including finding the value of λ such that the area of the "smiles" in Figure 7.9 satisfies the gain constraint. Again, the symbol rate and/or carrier/center frequency will need to be varied until the transmit filter(s) are (each) full band.

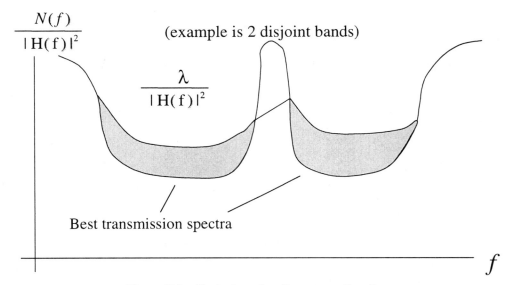

Figure 7.9   Illustration of smile energy allocation.

The MMSE is computed according to the MMSE equation in Section 7.1.2.1 with the understanding that the optimized transmit filter is now used in all dependent quantities. With multiple bands, the MMSE is individually computed for each band — the sum of the MMSEs in this case has no significant meaning.

7.1.3.3   Optimum Transmit Filters with the MMSE-DFE

The MMSE for the DFE was determined in Section 7.1.2.2 and can also be optimized with LaGrange (or really Calculus of variations) subject to the same unity gain constraint. The solution in this case is the solution of the **water-filling** equation:

$$\frac{1}{SNR \cdot \left|H\left(e^{-j\omega T}\right)\right|^2} + \left|\Phi\left(e^{-j\omega T}\right)\right|^2 = \lambda$$

(7.30)

This equation can be solved a number of ways including finding the value of λ such that the area of the shaded areas in Figure 7.10 satisfies the transmitted energy constraint. Again, the symbol rate and/or carrier/center frequency will need to be varied until the transmit filter(s) are (each) full band. DFE designers need cognizance of the symbol rate(s) and carrier/center frequencies optimization. Many a DSL designer has been surprised by poor DFE performance, usually because this optimization was forgotten.

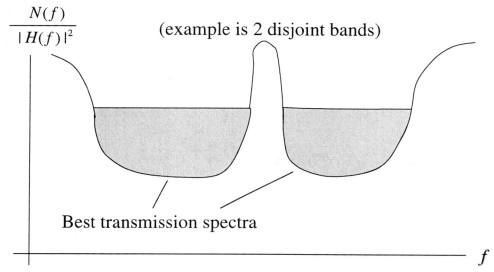

Figure 7.10   Water-filling spectrum.

The MMSE is computed according to the MMSE equation in Section 7.1.2.2 with the understanding that the optimized transmit filter is now used in all dependent quantities. This MSE can be shown to lead to the expression [12]

$$SNR_{mmse-dfe,U} = 2^{2\overline{C}} - 1$$

(7.31)

where $\overline{C}$ is the capacity or maximum theoretical limiting data rate for the channel at the selected symbol rate. *When the symbol rate(s) and/or carrier/center frequencies are optimized at the best possible symbol rate, this expression implies that a DFE-equalized system can perform at the highest possible data rates when good codes are used.* This does *not* mean that single-carrier QAM/CAP systems will perform at this level, except under very limiting circumstances that almost never occur in DSL practice because the occurrence of multiple symbol rates and carrier/center frequencies occurs commonly for water filling in DSL situations. However, with only a few bands, say 2 to 10, this performance level can be achieved.

7.1.3.4 Optimization with FIR Filters

One could take the expressions in Section 7.1.2.6 for the finite-length equalizers and attempt to maximize them over a stationary transmit filter with unit gain constraint. This problem remains to be addressed in terms of a simplification that is easily used. Brute force optimization may otherwise be necessary. Part of the reason for the difficult mathematics is that a stationary transmit filter is fundamentally inconsistent with finite-length filter implementations. Some work in this area has been undertaken by Al-Dhahir [10] and by Henkel [11].

Cioffi and Forney [12] note this problem and relax the transmit filter to be cyclostationary over a period (block symbol) of transmission, leading to a time-varying DFE structure that converges to all the same infinite-length results as filter length (and packet size) go to infinity. In this alternative case, it is much easier to determine optimum transmit filters. However, this approach has yet to be used in DSLs and does require a greater complexity than the methods of Section 7.2, which obtain the same performance [12].

7.1.4 Partial-Response Detection

Any bandlimited channel may be approximated by a discrete channel with finite duration over ν successive symbol periods, conveniently indexed as time samples 0 to ν. Typically, ML detection with ISI makes use of the WMF or MS-WMF (Section 7.1.2.2) of an infinite-SNR DFE (this is sometimes called a "zero-forcing" DFE). The resultant causal sampled ISI channel has D-transform representation $P(D)$ and the WMF output can be written

$$Y(D) = X(D) \cdot P(D) + N(D)$$

$$(7.32)$$

where $P(D)$ is the minimum-phase equivalent of the channel and the noise $N(D)$ is white with variance σ^2/η, where η is the value of γ in canonical factorization when the SNR is infinite (and the factorization still exists[6]). Such a minimum-phase equivalent channel can only be formed when the channel autocorrelation can be factored, which essentially occurs as long as $R(e^{-j\omega T})$ is not zero or infinite over any frequency range except for a countable number of bands of zero width (infinitesimally small notches). Thus $P(D) = 1 + p_i \cdot D + \dots + p_\nu \cdot D^\nu$. Such channels of finite length are called **controlled-ISI** channels. When the values p_k are also integers, then the channel is a **partial-response** channel. Often with partial-response channels, the DFE filtering is simplified significantly to just a lowpass filter, and the channel is considered to be equal to some presumed simple integer-p_k model with the noise sequence consequently having larger variance to accommodate the inaccuracies in filtering and modeling. In DSL transmission, only

---

6.   For finite SNR, one can show the factorization always exists. However, for infinite SNR, the Paley-Wiener integral $\int_{-\pi}^{\pi} \left| ln\left(R(e^{-j\omega})\right)\right| d\omega < \infty$ must be finite [9], meaning that the channel has no "dead-zones" where $R(\omega) = 0$ over a band or is not infinite itself. When this integral is large, as is often the case in DSLs, the near-singularity causes large precision to be required in the implementation of the receiver DFE filters.

a few partial-response channels have found application, the **duobinary channel** $P(D) = 1 + D$, which models a lowpass channel, the **DC notch** channel, $P(D) = 1 - D$, often used to model channels with transformer coupling, and the **modified duobinary channel** $P(D) = 1 - D^2$, which has a notch at DC and is also lowpass. These channels are almost always used in DSL with binary PAM inputs. The corresponding noise-free outputs are easily found to be -2, 0, or $+2$. A **binary differential encoder** has the simple property of encoding an input bit of 1 into a sign change from the last transmitted input to the channel and an input bit of 0 into no change. Such a differential encoder is illustrated in Figure 7.11 along with a DC-notch partial-response channel. On this channel, a simple (suboptimum) decoder then maps a $+2$ or a -2 output into a bit decision of 1 and a 0 output into a bit decision of 0. The duobinary channel uses the same encoder but the opposite mapping of channel outputs. No error propagation occurs with such a decoder. When the $1 - D$ channel shaping is intentionally introduced in the transmitter (anticipating the DSL is transformer coupled and thus does not pass DC), one obtains the AMI code discussed earlier. AMI is used in the simplest and earliest DSLs, like T1 circuits.

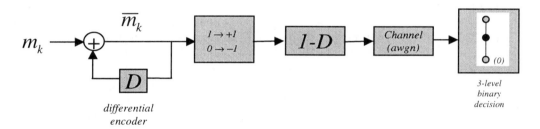

Figure 7.11 Illustration of partial response encoding and decoding.

For any partial-response channel with additive Gaussian noise with variance σ^2 per sample, the probability of symbol error is upper bounded by

$$P_e \leq 2 \cdot \left(1 - \frac{1}{M^{\nu+1}}\right) \cdot Q\left[\frac{d_x}{2\sigma}\right]$$

$$(7.33)$$

where d_x is the distance between points in the constellation at the channel input (meaning all amplification in the channel of energy, $\|p\|^2$, is lost). This tight upper bound on the probability of error for the suboptimum detector in Figure 7.11 is often well below the theoretical best performance for an ML detector. This formula also presumes use of the equivalent of a modulo-M differential encoder, sometimes called a partial-response precoder [13] that permits symbol-by-symbol detection.

7.1.5 Maximum-Likelihood Detection (Viterbi Algorithm)

The maximum-likelihood detector for a channel that has intersymbol interference is not an equalizer. Viterbi detection allows the input to be viewed not as individual unrelated inputs to a channel, but rather as an entire sequence of inputs smeared by a channel. One might draw an analogy with a television set where a viewer might have great difficulty viewing the picture if he/she had to look at each horizontal trace individually and then somehow integrate them in his/her brain. If instead, the entire picture is viewed in its entirety, more of the intent of the picture is easily decoded. In the same way, a maximum likelihood detector looks at the entire sequence of messages and then makes a decision on what was transmitted, rather than looking at individually messages separately and without view of other adjacent messages.

For partial response or controlled-ISI channels, the matched-filter receiver operation is lossless for detection of the sequence x_k. Furthermore, the feedforward filter of a DFE is invertible, so that a maximum likelihood detector can scan the entire sequence of feedforward filter output samples and see which input sequence is most likely to have produced that output, i.e., which input sequence causes a noiseless output that is closest to the noisy output in terms of the sum of squared differences over all time. This differs from the usual decision-feedback situation where instantaneous decisions are made at each time without delay or scanning the entire sequence. Clearly such a ML detector requires infinite computation, but when ν is finite, the amount of computation per symbol period for a ML detector is fixed and also finite. The method used for recursively computing that input sequence most likely to have produced the output uses a mathematical search method known as dynamic programming and is most commonly called the **Viterbi algorithm,** or just Viterbi detection in data transmission, after Viterbi [14].

Viterbi detection is described in terms of the mathematical quantities associated with the M^ν state transition diagram or "trellis diagram" for the channel. A state transition diagram and corresponding trellis diagram appear in Figure 7.12 for the DC-notch channel. In Figure 7.12, there are two states at any time corresponding to $x_{k-1} = \pm 1$. For the $1 - D$ AMI channel the outputs are ±2 and 0. Each transition between states has a label of channel, output/channel input. The trellis is a time-indexed state diagram. The sequences of branches trace all possible sequences of channel outputs. This channel has only two states corresponding to binary PAM inputs ($M = 2$) and $\nu = 1$. More sophisticated controlled-ISI diagrams may have more states and more transitions between states. Every allowed output sequence follows a connected path through the trellis. An accumulated sum of squared differences between channel output and sequence is kept for each path. At any time, only one such sum need be maintained for each of the states, which is the minimum sum for all paths into that state. An illustration of the Viterbi detector for the binary PAM channel is illustrated in Figure 7.13.

In Figure 7.13, at each step the two branches into each state are compared in terms of total sum squared difference between the corresponding channel output values and the noise-free values specified in the trellis, called the **metric**. The path with the smallest metric is kept as a **survivor**, while the larger is discarded because any sequence emanating from that state will have shortest distance only if this survivor is extended. In the diagram at time 2, a **merge** occurs in

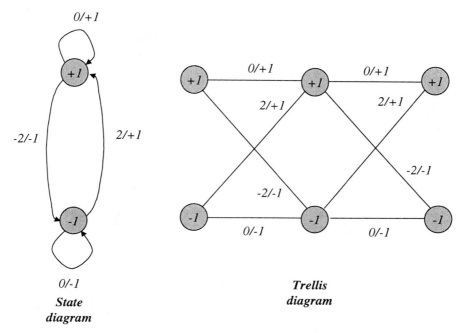

Figure 7.12 Illustration of state and trellis diagrams.

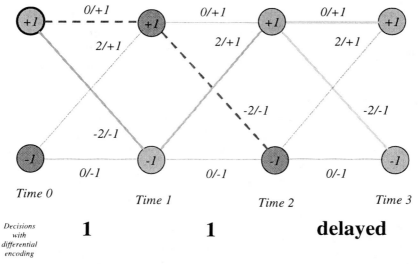

Figure 7.13 Example of Viterbi detection.

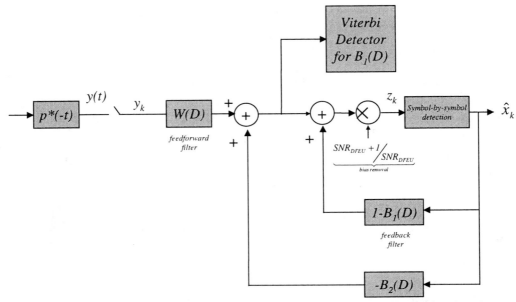

Figure 7.14   DFE with separated feedback section for Viterbi decoding (or approximated sequence detection).

which all survivors into time 3 and beyond come from the +1 state, so the decoder then knows all bits previous to this merge, which in the example is 1 bit, with differential encoding presumed on this $1 - D$ (AMI) channel.

The computation of probability of error for ML sequence detectors uses the same formula that we have always used, but calculation of minimum distance and/or the number of nearest neighbors can be intensive [1]. Computer search procedures are required to compare all possible pairs of sequences (some reduction is possible, see Ref. [15]) and compute distance, then counting the number of neighbors on the average at that closest difference. Such procedures are beyond the scope of this book. However, suffice it to say that matched-filer bound (MFB) performance (i.e., performance as if there were no ISI) can be attained on any channel with two taps, on any channel with three taps where the first and last have opposite signs, and on a number of practical four-tap channels like the channel $1 + D - D^2 - D^3$. While MFB performance may not be attained, significant gain can be achieved on almost all channels through maximum likelihood sequence detection (MLSD) or equivalently, a Viterbi detector.

Some enhanced-performance HDSL systems [16] use Viterbi detection with 4^ν states where the first ν taps of the feedback section of a DFE are intentionally separated as shown in Figure 7.14. Those first ν taps are described by partial-response polynomial $B_1(D)$ as shown, while the feedback taps for greater delays are used to shape more exactly the feedforward output to the correct settings used by the Viterbi detector. Improvements of 3 to 4 dB in performance can be measured on typical DSLs. Such gain can be obtained more consistently with another approach called trellis coding, described later in Section 7.3.

There are many approaches to approximate the sequence detection implemented in the Viterbi detector. Some methods [17] attempt to eliminate all but the best few survivors to reduce computational burden. Others, called reduced state sequence detection [18], [19] use partitioning methods and effectively several feedback sections that correspond to different survivors, while a third class uses fixed-depth tree search [20] for the approximation to full ML-based sequence detection.

Suboptimal sequence detection methods are presently receiving intense scrutiny in the coding research community at time of writing. The reader with further interest may desire to find the results of this research in decoding algorithms that appear concurrent with or after the publication of this book.

7.2 Multichannel Line Codes

The concept of multichannel transmission is one often used against a formidable adversary — "divide and conquer" — where the adversary in this case is the difficult transmission characteristic of the DSL twisted pair. Multichannel methods transform the DSL transmission line into hundreds of tiny transmission lines, each of which is easy to transmit on. The overall data rate is the sum of the data rates over all the easy channels. The most common approach to the "division" is transmission in nonoverlapping narrow frequency bands. The "conquer" part is that a simple line code on each such channel achieves best performance without concern for the formidable difficulty of intersymbol interference, which only occurs when wide-bandwidth signals are transmitted.

Multichannel line codes also have the highest performance and are fundamentally optimum for a channel with intersymbol interference [97]. A key feature of multichannel transmission for DSLs is the adaptation of the input signal to the individual characteristics of a particular phone line. This allows considerable improvement in range and reliability, two aspects of an overall system design that can dominate overall system cost. Thus, multichannel line codes have become increasingly used and popular for DSLs.

7.2.1   Capacity of the AWGN Channel

The capacity of a transmission channel is a theoretical upper bound on the data rate that can be reliably transmitted [21]. For an AWGN channel (with no ISI), this maximum data rate in bits per real dimension is

$$\bar{c} = \tfrac{1}{2} log_2 (1 + SNR)$$

(7.34)

where SNR is the ratio of transmitted energy per symbol to the noise power spectral density, or $SNR = \mathcal{E} / \sigma^2$ (again recalling the number of dimensions on noise and signal should be the same). To compute the maximum data rate in bits per second, one need only multiply the capacity above by the number of dimensions per symbol and the symbol rate

$$c = \bar{c} \cdot \frac{N}{T}$$

(7.35)

For a line code to obtain this maximum data rate at small P_e, infinite complexity needs to be used. Most practical coding/modulation methods can be characterized at a given fixed probability of symbol error by a gap Γ that quantifies the effective loss in SNR with respect to the capacity. Thus, the data rate achieved by the code is

$$\bar{b} = \tfrac{1}{2} log_2 \left(1 + \frac{SNR}{\Gamma} \right)$$

(7.36)

The smaller the gap, the better the code. QAM and PAM methods as presented above achieve a gap of 9.8 dB at a probability of symbol error 10^{-7}. Good coding methods (see Sections 7.4 and 8.1) can reduce this gap to 3 to 5 dB. Some extremely powerful codes not considered here can reduce the gap to about 1 to 2 dB (see Turbo Codes [17], [22] or Concatenated Codes [23]).

7.2.2   Basic Multichannel Transmission

Multichannel transmission methods achieve the highest levels of performance and are used in ADSL and VDSL. The equalizers of Section 7.1 only partially mitigate intersymbol interference and are used in suboptimum detection schemes. As the ISI becomes severe, the equalizer complexity rises rapidly, and then performance loss usually widens with respect to theoretical optimums. The solution, as originally posed by Shannon in his famous mathematical theory of communication [24], is to partition the transmission channel into a large number of narrowband AWGN subchannels. Usually, these channels correspond to contiguous disjoint frequency bands and the transmission is called **multicarrier** or **multitone** transmission. If such multitone subchannels have sufficiently narrow bandwidth, then each has little or no ISI, and each independently approximates an AWGN. The need for complicated equalization is replaced by the simpler need to multiplex and demultiplex the incoming bit stream to/from the subchannels. Multicarrier transmission is now standardized and used because the generation of the subchannels can be easily achieved with digital signal processing. Equalization with a single wideband carrier may then be replaced by no or little equalization with a set of carriers or "multicarrier," following Shannon's optimum transmission suggestion, and can be implemented and understood more easily. The capacity of a set of such parallel independent channels is the sum of the individual capacities, making computation of theoretical maximum data rates or use of gaps for practical rates easy.

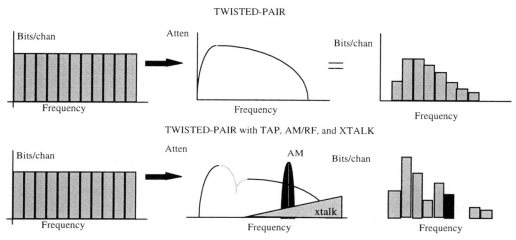

Figure 7.15   Basic multitone concept.

The basic concept is illustrated in Figure 7.15. There, two DSL transmission line characteristics are posed, each of which would have severe ISI if a single wideband signal were transmitted. Instead, by partitioning the transmit spectrum into narrow bands, then those subchannels that pass through the channel can be loaded with the information to be transmitted. Note, the receiver has a matched filter to each transmit bandpass filter, thus constituting an easily implemented maximum-likelihood receiver (without need for Viterbi sequence detection, even on a channel with severe spectral filtering). Better subchannels get more information, while poor subchannels get little or no information. If the subchannels are sufficiently narrow, then no equalizer need be used.

The set of signal-to-noise ratios that characterize each of the subchannels is important to performance calculation. It is assumed that there are N subchannels each carrying

$$\bar{b}_n = \tfrac{1}{2} log_2 \left(1 + SNR_n \middle/ \Gamma \right)$$

(7.37)

bits/dimension. The average number of bits is the sum of the numbers of bits carried on each channel divided by the number of dimensions (assumed here to be N) *as*

$$\bar{b} = \frac{1}{N} \sum_{n=1}^{N} \frac{1}{2} log_2\left(1 + \frac{SNR_n}{\Gamma}\right) = \frac{1}{2} log_2\left[\prod_{n=1}^{N}\left(1 + \frac{SNR_n}{\Gamma}\right)\right]^{1/N} = \frac{1}{2} log_2\left(1 + \frac{SNR_{geo}}{\Gamma}\right)$$

(7.38)

where SNR_{geo} is the **geometric signal-to-noise ratio**, or essentially the geometric mean of the 1 + SNR/Γ terms,

$$SNR_{geo} = \Gamma\left\{\left[\prod_{n=1}^{N}\left(1 + \frac{SNR_n}{\Gamma}\right)\right]^{1/N} - 1\right\}$$

(7.39)

The entire set of parallel independent channels then behaves as one additive white Gaussian noise channel with SNR_{geo} essentially equal to the geometric mean of the subchannel SNRs.[7] SNR_{geo} can be compared against the unbiased SNR of equalized passband and baseband systems directly. This SNR_{geo} can be improved considerably when the available energy is distributed nonuniformly over all or a subset of the parallel channels, allowing a higher performance in multichannel systems.[8] The process of optimizing the bit and energy distribution over a set of parallel channels is known as loading and is studied in the next subsection.

7.2.3   Loading Algorithms

The process of assigning information and energy to each of the subchannels is called **loading** in multichannel transmission. Each subchannel has a transmit energy $\mathcal{E}_n$ and a number of bits b_n. Using the gap approximation for QAM transmission on each of these subchannels, and recalling that each subchannel is actually two dimensional and has gain g_n noise power spectral density σ_n^2, is

$$b_n = log_2\left(1 + \frac{\mathcal{E}_n \cdot g_n}{\Gamma \sigma_n^2}\right)$$

(7.40)

The total data rate over M such parallel channels is then

---

7.   This geometric mean statement becomes very accurate for moderate to high SNR on all subchannels.
8.   Under certain severe restrictions, the DFE SNR can be made equal to the geometric SNR when MMSE-DFEs are used, the used subchannels are all next to each other in frequency (i.e., no gaps), and the same energy distribution is used by the DFE (see Refs. [3],[12]).

$$b = \sum_{n=1}^{M} b_n$$

(7.41)

which is maximized for a fixed total energy $\mathcal{E} = \sum_{n=1}^{M} \mathcal{E}_n$ by the so-called **water-filling solution:**

$$\mathcal{E}_n + \frac{\Gamma \cdot \sigma_n^2}{g_n} = \text{constant}$$

(7.42)

That is, the sum of the transmitted energy and the noise normalized to the channel gain (increased by the gap) must be constant. As illustrated in Figure 7.10, this can be interpreted in terms of the channel noise-to-signal ratio plot as pouring water from above into the curve with an amount of water equal to the amount of total energy. Such water would rest at a level that is the constant in the above equation. An algorithm for computing the maximum bit rate for a system using a code with gap Γ is given below:

Water-Filling Algorithm:

1. Set initial noise-to-signal ratio sum $NSR(i) = 0$; $i = 0$ and sort subchannels in terms of smallest to largest.

2. Update number of used subchannels $i = i + 1$

3. Compute $NSR(i+1) = NSR(i) + \frac{\sigma_i^2}{|P_i|^2}$

4. Set $\lambda(i) = \frac{\mathcal{E} + \Gamma \cdot NSR(i)}{i}$

5. Is $\mathcal{E}_i = \lambda(i) - \Gamma\left(\frac{\sigma_i^2}{|P_i|^2}\right) < 0$

   a) No, go to step 2.

   b) Yes, compute $\mathcal{E}_j = \lambda(i-1) - \Gamma\left(\frac{\sigma_j^2}{|P_j|^2}\right)$ $\quad b_j = \log_2\left(1 + \frac{\mathcal{E}_j \cdot |P_j|^2}{\Gamma \cdot \sigma_j^2}\right)$ $\quad$ for $j = 1,...,i-1$

The water-filling energy distribution can be approximated by a flat distribution on virtually all DSLs with minuscule loss in performance, as long as the correct transmission band is used (i.e., the transmitter turns off the right tones), as shown by P. Chow [25]. J. Tellado[9] has found a direct method/approximation for calculating the correct transmission band from the SNRs that avoids

---

9. J. Tellado, "Selected Topics on Energy Loading for DMT," EE400 class presentation, Stanford University, December 1997.

water-filling. The approximation notes that the problem

$$\max_I \sum_{n \in I} \log_2 \left(1 + \frac{\varepsilon |P_n|^2}{|I| \cdot \Gamma \cdot \sigma_n^2} \right)$$

(7.43)

where I is a discrete set of subchannel indices and $|I|$ is the number of indices in the set, is approximated by the integral

$$\max_\Omega \int_\Omega \log_2 \left(1 + \frac{\varepsilon |P(\omega)|^2}{|\Omega| \cdot \Gamma \cdot S_n(\omega)} \right) d\omega$$

(7.44)

By ordering the SNRs, taking a derivative with respect to the set size Ω, and approximating the term inside the log as 1 over the band used, the optimum bandwidth is then approximated by

$$\frac{|P(\Omega^*)|^2}{S_n(\Omega^*)} = SNR(\Omega^*) = \frac{\Gamma \cdot (e - 1)}{\varepsilon} \cdot \Omega^*$$

(7.45)

which can be solved by a single pass over the ordered SNRs to find at which point the SNR is equal to the expression. In discrete form, the solution is then

$$SNR(N^*) = \frac{\Gamma \cdot (e - 1)}{\varepsilon} \cdot N^*$$

(7.46)

The ratio is a constant that can be stored (e is the base of the natural logarithm) based on total energy ε. Thus, the complexity is essentially the ordering, which can be executed in an average of $N\log_2(N)$ operations using binary sort.

Water-filling has two drawbacks. First, the bit distribution is not necessarily a set of integers, potentially a severe complication for simple implementations. Second, the bit rate is maximized at given margin, and in many applications what is desired is maximum performance (lowest probability of error) at a given data rate. These drawbacks are eliminated in the loading algorithms to follow.

When integer bit distributions are not of concern, the margin can be maximized at fixed data rate according to the following **margin-adaptive loading** criterion:

$$\min_{E_n} \sum_{n=1}^{N} \varepsilon_i$$

$$st : \sum_{n=1}^{N} \log_2 \left(1 + \frac{\varepsilon_n \cdot |P_n|^2}{\Gamma \cdot \sigma_n^2} \right) = b$$

(7.47)

Finding the minimum energy necessary to achieve a given bit rate is a water-filling problem where the energy (water)-filling level is raised or lowered until the rate equals the desired fixed data rate corresponding to b bits per N (two-dimensional) subchannels. The optimum number of subchannels to be used is then the same for larger amounts of energy with all energy evenly distributed among those used subchannels to raise the margin uniformly to its maximum level. Equivalently, a Lagrangian can be formed and differentiated to find the solution

$$\varepsilon_n = K_{ma} - \Gamma \frac{\sigma_n^2}{|P_n|^2} \text{ where } \varepsilon_n \geq 0 \text{ and } \log_2 (K_{ma}) = \frac{1}{N^*} \cdot \left[b - \sum_{n=1}^{N^*} \log_2 \left(\frac{|P_n|^2}{\Gamma \cdot \sigma_n^2} \right) \right]$$

(7.48)

where N^* is the number of used channels (that is the ones with nonzero energies in the solution). Clearly this is another water-filling problem, except that the total "water" level, K_{ma} is determined by the bit rate. The maximum margin is then

$$\gamma_{\max} = \frac{\varepsilon}{\sum_{n=1}^{N^*} \varepsilon_n}$$

(7.49)

This simple solution seems to have first been noted by Campello [31]. This formula can be programmed as follows:

Margin-Adaptive Water-Filling Algorithm:

1. Sort the $SNR(n) = \dfrac{|P_n|^2}{\sigma_n^2}$ in order from largest to smallest, $n = 1, \dots, N$. Initialize $N^* = N$

2. Calculate $K_{ma} = 2^{\frac{1}{N^*} \left[b - \sum_{n=1}^{N^*} \log_2 \left(\frac{|P_n|^2}{\Gamma \sigma_n^2} \right) \right]}$ $\quad \varepsilon_{N'} = K_{ma} - \Gamma / SNR(N^*)$

3. If $\varepsilon_{N^*} \leq 0$, and $N^* \to N^* - 1$ repeat step 2.

4. Set $\varepsilon_n = K_{ma} - \dfrac{\Gamma}{SNR(n)}$, $b_n = \log_2 \left(1 + \dfrac{\varepsilon_n \cdot SNR(n)}{\Gamma} \right)$ $\quad n = 1, \dots, N^*$

5. Compute max margin $$\gamma_{max} = \frac{\mathcal{E}}{\sum\limits_{n=1}^{N^*} \mathcal{E}_n}$$

Maximum-margin water-filling, however, does not address the important issue of subchannel granularity, namely, integer bits on each subchannel are not necessarily guaranteed.

The **granularity** of a bit distribution is defined as the smallest unit of additional information that can be encoded on to a subchannel, β, which is usually $\beta = 1$ bit for a two-dimensional QAM subchannel. However, through the use of multidimensional coding, values like $\beta = 1/2$ or even $\beta = 1/4$ can be implemented with relatively minor complexity increase. Very fine (small) granularity does not usually improve a system's design significantly. Thus, $\beta = 1$ is usually sufficient, although pathological cases do exist where finer granularity would be of benefit. The developments of this chapter will use arbitrary β, but the reader will usually want to assume it is one bit, $\beta = 1$. The number of bits per symbol is then

$$b = B \cdot \beta \tag{7.50}$$

meaning there are B units of information. Each subchannel thus may have B_i units of information, $i = 1,...,N$. Each subchannel is presumed to have an encoding function that requires $\mathcal{E}_i(B_i)$ units of energy, a monotonically increasing function of B_i with $\mathcal{E}_i(0) = 0$. The incremental energy to increase from B_i to $B_i + 1$ units of information is written $e_i (B_i)$. The encoding functions need not be the same from subchannel to subchannel (thus the gap may vary and is tacitly contained within the energy/information formula). The previous water-filling and Chow's algorithms assumed a constant relationship

$$\mathcal{E}_i \equiv \frac{\left(2^{2b_i\beta} - 1\right) \cdot \sigma_i^2}{\Gamma} \tag{7.51}$$

on all subchannels, but the more general formulation now allows for different codes on different subchannels and thus more exact constellation-specific formulae to be used. All reasonable energy formulas are also convex functions, a necessary restriction for some of the following algorithms to be globally convergent.

There are two dual loading problems of interest, expressed in terms of the vector of energies $\mathcal{E} = [\mathcal{E}_1 \ \mathcal{E}_2 \ ... \ \mathcal{E}_N]$ and the vector of information distribution $B = [B_1 \ B_2 \ ... \ B_N]$:

(P1) Problem 1 — Rate-Adaptive DSL (RADSL):

$$\max_{E} B = \sum_{i=1}^{N} B_i$$

$$subject \ to: \ \sum_{i=1}^{N} E_i = E$$

(P2) Problem 2 — Margin-Adaptive DSL (MADSL):

$$\min_{B} E = \sum_{i=1}^{N} E_i$$

$$\text{subject to: } \sum_{i=1}^{N} B_i = B$$

There are several possible solutions, two of which are suggested here, **Chow's method** [26], [25] and **the Greedy algorithm** [27].

Combinations of the RADSL and MADSL for dual-service applications have recently been defined and solved.

7.2.3.1 Chow's Algorithms

Chow's algorithms can begin with the step of water-filling or any energy/bit distribution. This section assumes that each subchannel is two-dimensional, and that the SNR_n have again been sorted from largest to smallest (water-filling arranged NSR from smallest to largest, which is the same ordering). However, these algorithms originally suggest an on/off energy distribution that is determined by the following steps.

Chow On/Off Energy-Distribution Calculation:

1. Set $B_i = 0$ and $i = 1$
2. Set $\mathcal{E}_n = \mathcal{E}_{total}/i$ and $SNR_n = \mathcal{E}_n |P_n|^2 \Big/ \sigma_n^2$ for $n = 1, ..., i$
3. Compute $B(i) = \sum_{n=1}^{i} log_2 \left(1 + \dfrac{SNR_n}{\Gamma} \right)$
4. If $B(i) < B(i-1)$, then save old bit distribution and energy, that is

   a) $\mathcal{E}_n = \dfrac{\mathcal{E}_{total}}{i-1}$ for $n < i$, and $\mathcal{E}_n = 0$ for $n \geq i$

   b) $B_n = log_2 \left(1 + \dfrac{\mathcal{E}_n |P_n|^2}{\sigma_n^2 \cdot \Gamma} \right)$

   Otherwise, $i \leftarrow i + 1$ and GOTO 2

To solve (P1), Chow's algorithm executes the following additional steps:

5. Round B_n to the nearest integer and recompute energy as $\mathcal{E}_n = \dfrac{2^{B_n \cdot \beta} - 1}{|P_n^2|} \cdot \Gamma \cdot \sigma_n^2$ for all subchannels

6. Rescale energy on each subchannel's energy by $\left(\mathcal{E}_{total} \Big/ \sum_n \mathcal{E}_n \right) \cdot \mathcal{E}_n$

To solve (P2), Chow's algorithm executes instead the following steps:

5. Compute the geometric SNR: $SNR_{geo} = \left\{ \left(\prod_{n=1}^{N} \left[1 + \frac{SNR_n}{\Gamma} \right] \right)^{1/N} - 1 \right\} \cdot \Gamma$

6. Set the current margin at $\gamma_m = \left(\frac{SNR_{geo}/\Gamma}{2^{b/N} - 1} \right)$

7. Compute the updated bit distribution $B_n = \left(\frac{1}{\beta} \right) \cdot log_2 \left(1 + \frac{SNR_n}{\Gamma \cdot \gamma_m} \right)$

8. Round B_n to the nearest integer and recompute energy as $\mathcal{E}_n = \frac{2^{B_n \cdot \beta} - 1}{|P_n^2|} \cdot \Gamma \cdot \sigma_n^2$ for all subchannels.

9. If $\sum_n B_n \neq B$, then look at subchannels for which rounding error is greatest and round the other way until the data rate is correct, recomputing energies as in step 9.

10. Rescale energy on each subchannel's energy by $\left(\frac{\mathcal{E}_{total}}{\sum_n \mathcal{E}_n} \right) \cdot \mathcal{E}_n$

The reader should note that Chow's algorithms only approximately solve problems (P1) and (P2). The Greedy algorithms of the next section solve them exactly.

7.2.3.2 Greedy Algorithms

Where Robin Hood is famous for robbing from the rich and giving to the poor, the best strategy in multichannel transmission is the inverse — robbing the poor and giving to the rich. This may sound like the tax collector, and indeed that is what Greedy algorithms do for loading.

Greedy algorithms are well known in mathematics, a special case of which appears to have been first applied to multicarrier transmission by Hughes-Hartog [29]. The approach of this section follows a general approach recently derived by Campello [27], which is considerably more complete than the method of Hughes-Hartog. Greedy algorithms essentially allocate bits to the subchannels incrementally in a way that always chooses a subchannel for the next bit to be allocated according to the least cost in energy. The reallocation of energy is analogous to the benevolent tax collector taking money and reapportioning it to those who can best use it to improve the collective good of all (in this case maximum data rate or best performance). Thus, the best subchannel gets the first bit, and then a table of incremental next energies for the next bit has to be computed and the process continued until all bits have been allocated for **margin-adaptive loading** and until no more energy is available for **rate-adaptive loading.**

Three concepts facilitate the use of Greedy algorithms.

Efficiency of B:

A bit distribution vector B is said to be **efficient** if

$$\max_n \left[e_n \left(B_n \right) \right] \leq \min_m \left[e_m \left(B_{m+1} \right) \right]$$

(7.52)

where $e_n(B_n)$ is the incremental energy cost to add the nth information unit on subchannel n. In other words, any other bit distribution for the same total B would take more energy.

The following algorithm, which starts with any vector $\boldsymbol{B}$ whose components sum to B, is guaranteed to converge in a finite number of steps (under the previous restrictions on the encoding function of montonicity and convexity) to an efficient bit distribution:

Campello's Efficiency (EF) Algorithm:

1. $m \leftarrow \arg\left[\min_{1 \leq i \leq N}\left(e_i\left(B_i + 1\right)\right)\right]$

2. $n \leftarrow \arg\left[\max_{1 \leq j \leq N}\left(e_j\left(B_j\right)\right)\right]$

3. While $e_m\left(B_m + 1\right) < e_n\left(B_n\right)$ do

   a) $B_m \leftarrow B_m + 1$

   b) $B_n \leftarrow B_n - 1$

   c) $m \leftarrow \arg\left[\min_{1 \leq i \leq N}\left(e_i\left(B_i + 1\right)\right)\right]$

   d) $n \leftarrow \arg\left[\max_{1 \leq j \leq N}\left(e_j\left(B_j\right)\right)\right]$

Campello has proved that simple rounding of a water-filling distribution leads to an efficient bit distribution. An additional measure is **E-tightness** (ET), which is defined as an information distribution vector $\boldsymbol{B}$ whose components satisfy

$$0 \leq \mathcal{E}_{total} - \sum_{n=1}^{N} e_n\left(B_n\right) < \min_m\left[e_m\left(B_m + 1\right)\right]$$

(7.53)

E-tightness simply implies that no greater number of information units can be carried with the given amount of energy. To obtain an E-tight distribution, one starts with any bit distribution and executes the following.

Campello's E-Tightness (ET) Algorithm:

1. Set $S = \sum_{n=1}^{N} e_n\left(B_n\right)$
2. While $\mathcal{E}_{total} - S < 0$ or $\mathcal{E}_{total} - S \geq \min_{1 \leq i \leq N}\left[e_i\left(B_i + 1\right)\right]$
3. IF $\mathcal{E}_{total} - S <$ THEN

   a) $n \leftarrow \arg\left[\max_{1 \leq j \leq N}\left(e_j\left(B_j\right)\right)\right]$

   b) $S \leftarrow S - e_n\left(B_n\right)$

   c) $B_n \leftarrow B_n - 1$

ELSE

a) $m \leftarrow \arg \left[\min_{1 \leq i \leq N} \left(e_i \left(B_i + 1 \right) \right) \right]$

b) $S \leftarrow S + e_m \left(B_m + 1 \right)$

c) $B_m \leftarrow B_m + 1$

The dual of E-tightness is **B-tightness**, which trivially means the sum of the bits in the bit distribution adds to a desired total B.

Dual B-Tightness (BT) Algorithm:

1. Set $\tilde{B} \leftarrow \sum_{n=1}^{N} B_n$

2. While $\tilde{B} \neq B$

3. IF $\tilde{B} > B$

   a) $n \leftarrow \arg \left[\max_{1 \leq j \leq N} \left(e_j \left(B_j \right) \right) \right]$

   b) $\tilde{B} \leftarrow \tilde{B} - 1$

   c) $B_n \leftarrow B_n - 1$

   ELSE

   a) $m \leftarrow \arg \left[\min_{1 \leq i \leq N} \left(e_i \left(B_i + 1 \right) \right) \right]$

   b) $\tilde{B} \leftarrow \tilde{B} + 1$

   c) $B_m \leftarrow B_m + 1$

We may now solve the loading problems (P1) and (P2) easily and exactly.

Campello's Solution to Rate-Adaptive Loading (P1):

1. Choose any B.
2. Make B efficient with EF algorithm.
3. E-tighten B with ET algorithm.

Campello's Solution to Margin-Adaptive Loading (P2):

1. Choose any B.
2. Make B efficient with EF algorithm.
3. B-tighten B with BT algorithm.

Steps 1 through 2, and 4 through 5 can be replaced with rounding of a water-filling distribution if that is easier to compute and all subchannels use the same gap-based energy formula. Several sorting steps are implied in both tightening algorithms and the efficiency algorithm.

Sorting is a relatively complex procedure, generally requiring $N\log_2(N)$ operations for each sort. Advanced loading algorithms, especially those that might be used for dynamic rate adaptation and thus would have to execute in real time rather than just once initially, focus on elimination of the sorting steps. Campello has done additional work [27] using the data structure of a heap, a tree-structured list where each parent node has a higher value (incremental energy) than any of its children. It is possible to execute loading (on the average) in order N steps using heaps and other data structure partitioning methods. In these algorithms, SNRs are sorted into two unordered sets: a set where all exceed a pivot SNR, and a set where all are less than this pivot SNR. By iterating the pivot value, average computation can be significantly reduced.

7.2.3.3 Dual Bit Mapping

Multiple time-indexed bit loading for synchronized cyclostationary crosstalk has recently been used to improve ping-pong systems' performance for VDSL (see Ref. [30]) and for Japanese ADSL (see appendixes of Refs. [28] and [31]).

7.2.4 Channel Partitioning

To this point, the existence of a set of parallel independent subchannels that represent an original transmission channel has been postulated. In practice, such subchannels can never be completely independent. **Channel partitioning** methods attempt to construct the set of parallel channels so they are largely independent. **Intermodulation distortion** occurs when partitioning is not perfect and the subchannels interfere with each other. Intermodulation distortion is a form of frequency-domain ISI.

Channel partitioning consists of modulation and demodulation with basis functions. A set of N orthonormal basis functions is chosen so that after transmission through the known channel with transfer function $H(f)$ (with additive white Gaussian noise), those functions remain orthogonal. The set of functions that achieve such partitioning may require infinite complexity and delay for realization with most DSL channels, but are known as the **eigen-functions** of the channel or sometimes as the **Karhuenen-Loeve** functions. These functions $\varphi_n(t)$ satisfy the equation

$$\int_0^T \varphi_n(s) \cdot r(t-s)\,ds = \lambda_n \varphi_n(t)$$

$$(7.54)$$

where $r(t) = p(t) * p^*(-t)$, and λ_n are the eigen-values of the channel. A bank of matched filters in the receiver will produce sampled outputs that are the desired N parallel independent channels. The noise at the output of each channel will be Gaussian and independent of all other subchannels, and the gains will be the eigenvalues λ_n. Figure 7.16 illustrates such a system, sometimes referred to as **modal modulation.** The determination of the nonunique eigen-functions (the eigenvalues themselves form a unique set) can be very difficult. As the symbol period spans all time, a valid choice of the eigen-functions is a set of infinitesimally narrow sinusoids, that is, a multitone transmission system with very narrow tones.

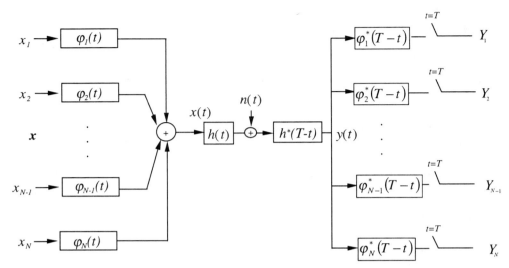

Figure 7.16   Illustration of modal modulation.

Figure 7.16 illustrates that modal modulation is a special form of the basis modulator described in Section 6.1. N basis functions are used, and the receiver is indeed optimum maximum likelihood. The matched filter outputs are capitalized because these matched filters can be shown to be taking a "transform" of the channel output. The amount of energy and information carried by each mode or subchannel can be optimized for best transmission in a channel-dependent way. Further, the basis functions themselves are a function of the channel instead of just generally selected to represent a class of modulation schemes. In DSLs, the basis functions and matched filters are implemented digitally in discrete time at a sampling rate sufficiently high to allow the continuous-time eigen-functions to be approximated. If each subchannel is excited with independent information, then detection proceeds with each receiver matched filter output independently. No equalization or sequence/block detection is needed — "symbol-by-symbol" (read "mode-by-mode") detection is optimum on the bandlimited channel when modal modulation is used.

Most modern multichannel transmission systems are implemented in discrete time as in Figure 7.17. A vector X represents the subchannel inputs, which are modulated into a corresponding vector x that is formatted for transmission over the discrete version of the channel. Usually, the transform is unitary or orthogonal so that it preserves the energy and independence of the individual samples. The corresponding receiver transformation accepts the vector of channel output samples y and produces an output vector Y that has each component as a noisy estimate of the corresponding input component in X. The selection of the transform and also the use of some extra samples called the guard period (described later) distinguishes the channel partitioning methods from one another, as described in the sections to follow.

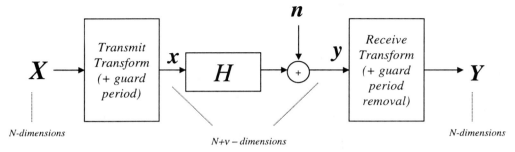

Figure 7.17 General multichannel partitioning for discrete-time implementation.

7.2.4.1 Vector Modulation/Coding

Because the symbol period must be finite, a packet of time-domain samples is often augmented by a **guard period**. The guard period occurs in time at the beginning of a symbol transmission and carries no new information on these samples because samples placed there get corrupted at the channel output by transients from the previous symbol. The receiver often ignores guard period samples (although not always).[10] The channel impulse response (including any transmit filter, so perhaps more accurately called a "pulse" response) must have negligible energy outside a time interval of nT', where $1/T'$ is the sampling rate. Thus, v is the length of the channel in sampling periods. The symbol period is thus $T = (N+v)T'$, where N is the number of samples after the guard period. The excess bandwidth is thus $\alpha = v/N$. The transmitted samples are indexed $x_{-v}, \ldots, x_0, x_1, \ldots, x_{N-1}$. The corresponding channel outputs $y_0, \ldots, y_{N-1}$ lead to a vector/matrix channel description for each transmitted packet/symbol

$$y = Hx + n \tag{7.55}$$

where the highest time index is presumed at the top of the vector and the lowest time index at the bottom of the vector. The matrix H is

$$H = \begin{bmatrix} h_0 & h_1 & \cdots & h_v & 0 & \cdots & 0 \\ 0 & h_0 & h_1 & \ldots & h_v & 0 & \cdots \\ 0 & 0 & \ddots & \ddots & \cdots & \ddots & 0 \\ 0 & \cdots & 0 & h_0 & h_1 & \cdots & h_v \end{bmatrix} \tag{7.56}$$

a Toeplitz convolution matrix. Input vectors are of length $N + v$ samples, while outputs are of length N samples, meaning that the receiver essentially ignores the guard band samples because

---

10.   Clearly, these samples do carry information and we will return to its recovery later in this section.

they contain interference from previous symbols (equalization can be used to reduce the guard-band length, as discussed later).

If the guard period is long enough so that successive transmissions are independent, the N output samples are essentially sufficient to recover all information about the transmitted packet. y and n are $N \times 1$ column vectors

$$y = \begin{bmatrix} y_{N-1} \\ \vdots \\ y_0 \end{bmatrix} \quad \text{and} \quad n = \begin{bmatrix} n_{N-1} \\ \vdots \\ n_0 \end{bmatrix} \tag{7.57}$$

H is an $N \times (N + v)$ matrix; x is an $(N + v) \times 1$ column vector

$$x = \begin{bmatrix} x_{N-1} \\ \vdots \\ x_0 \\ x_{-1} \\ \vdots \\ x_{-v} \end{bmatrix} \tag{7.58}$$

The equivalent of modal modulation with discrete independent packets is **vector modulation**, which is defined by the **singular value decomposition (SVD)** of the matrix H,

$$H = F\Lambda M^* \tag{7.59}$$

where F and M are unitary matrices, $FF^* = I$, $MM^* = I$, and Λ is a unique (other than ordering) "diagonal" matrix of nonnegative real singular values. When H is real, F, M, and Λ can be selected as real. The basis functions for discrete-time implementation in the transmitter are found through the matrix multiplication $x = MX$, where X is the vector corresponding to the transmitted symbol. The bank of matched filters is implemented according to $Y = F^* y$. Clearly, the combined action of the transmit and receive filters

$$Y = F^* y = F^*(Hx + n) = F^* HMX + N = \Lambda X + N \tag{7.60}$$

leaves a set of parallel independent (if the noise is white) subchannels with gains given by the singular values λ_n:

$$Y_n = \lambda_n \cdot X_n + N_n \tag{7.61}$$

where the noise on each subchannel is independent (because $F^*(\sigma^2 I)F = \sigma^2 I$). When the noise is not white, the channel is first converted to a white-noise equivalent channel by the transformation $y \leftarrow R_n^{.5} y$ where the noise autocorrelation matrix factors into its matrix square root $R_{nm} = \sigma^2 R_n^{.5}(R_n^{.5})^*$. With the channel matrix now becoming $H_{white} = R_n^{-.5} H = F \Lambda M^*$, the development then proceeds identically to the white noise case (although the guard-period length, v, may need adjustment).

As the block length goes to infinity, it can be shown that the set of singular values become the magnitudes of the channel Fourier transform. That is, vector modulation converges to multitone and modal modulation as the symbol length becomes infinite.

7.2.4.2 Discrete Multitone (DMT)

The computation of singular value decomposition is very complex, but can be executed infrequently because DSL channels do not change rapidly. However, the matrix multiplies F and M can have large complexity (essentially as complex as the equalizers studied in Section 7.1). A method by which to approximate these complex filters by simpler operations is to exploit the knowledge that these matrices essentially tend to Discrete Fourier Transforms. This is the observation in the standardized method for ADSL, **discrete multitone (DMT)** [98] and also in many wireless transmission applications that use **orthogonal frequency division multiplexing (OFDM)**.[11]

In DMT, the guard period must contain a **cyclic prefix.** The samples in the prefix must repeat those at the end of the symbol, that is $x_{-i} = x_{N-i}$ $i = 1,...,v$. When a cyclic prefix is used, the matrix H becomes what is called a square "circulant" matrix (the last v output samples of each transmitted packeted are ideally ignored in DMT). Circulant matrices have the property they may be decomposed as

$$H = Q^* \Lambda Q \tag{7.62}$$

where Q is a matrix corresponding to the discrete Fourier transform (DFT) and Λ is a diagonal matrix containing the N Fourier transform values for the sequence h_k that characterizes the channel. The klth element of Q starting from the bottom right at $k = 0$, $l = 0$ and counting up is

$$Q_{kl} = \frac{1}{\sqrt{N}} e^{-2\pi \frac{kl}{N}} \tag{7.63}$$

---

11.   The difference between OFDM and DMT is that DMT uses loading while OFDM uses a fixed number of bits per subchannel (because the channel is varying too rapidly or is broadcast so that the transmitter cannot know the channel and "load" to it).

The DFT is a heavily used and well-understood operation in digital signal processing, and a variety of structures exist for its very efficient implementation in $N\log_2(N)$ operations rather than the usual N^2 for most matrix multiplication (see Section 7.2.8). Thus, DMT is a very efficient multichannel partitioning method. Even when H is real, the matrices in the DMT decomposition are complex.

The $N \times N$ circulant matrix is written

$$
H = \begin{bmatrix}
h_0 & h_1 & \cdots & h_v & 0 \\
0 & h_0 & h_1 & \cdots & h_v \\
h_v & 0 & h_0 & \cdots & h_{v-1} \\
\vdots & \ddots & \ddots & \ddots & \ddots \\
0 & 0 & h_v & \cdots & h_0
\end{bmatrix}
\tag{7.64}
$$

which the reader can verify implements circular convolution on the packet of inputs. The transmit symbol vectors are created by $x = Q^*X$, where X is a frequency-domain vector of channel inputs. Each element of X is complex and can be thought of as a QAM signal. When the channel is real and thus the symbol x must also be real, the frequency-domain input must have conjugate symmetry, which means that

$$
X_n = X^*_{N-n}
\tag{7.65}
$$

meaning there are $N/2$ complex subchannels, not N. The energy of the cyclic prefix guard period is clearly wasted, thus effectively reducing the amount of power available for transmission by $N/{N+v}$ on the average. This power penalty is in addition to the excess bandwidth penalty of the cyclic prefix, but accepted for the extremely efficient implementation of DMT. If the guard-period length v can be made small with respect to the packet length N, then this penalty is small. The receiver generates the outputs of the set of parallel channels by forming:

$$
Y = Qy = Q(Hx + n) = QHQ^*X + N = \Lambda X + N
\tag{7.66}
$$

Again, a set of parallel independent channels when the input noise is white. When the input noise is not white, the output noise vector tends to white as long as the block size is long enough, so noise prewhitening need not be used in reasonable DMT implementations.

As block length goes to infinity, DMT becomes multitone and optimum, but for finite lengths, its performance only approximates that of vector modulation. Nevertheless, for reasonable block lengths, DMT is near optimum if the guard period is long enough to avoid overlapping packets at the channel output. DMT for ADSL is described in more detail in Section 7.2.6.

Vector modulation uses real matrices, while DMT uses complex matrices. Thus, even though both converge to the same performance at large block length (or large symbol size), they are not the same. However, the subchannel gain magnitudes are the same in vector modulation and DMT. DMT differs from vector modulation in the limit only on the phase of the subcarriers. In vector modulation, all subchannels have zero phase, while in DMT the phases are arbitrary. The phases of DMT do not admit easy decoder implementation, so what is called an **FEQ** (frequency-domain "equalizer") often appears at the output of the receiver FFT (after multiplication by Q). The FEQ is not really an equalizer in this case, but a gain/phase adjustment through multiplication by a complex number that inverts the channel (and thus rotates the phase to zero). Then all subchannels may use a common decoder with common assumed spacing of points in the constellations on each of the subchannels. The basic FEQ is shown in Figure 7.18. It does not improve performance; it simplifies decoder implementation only. The error is used in the dynamic loading (or "bit swapping") described later.

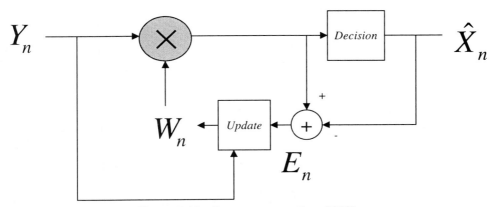

Figure 7.18   Frequency equalizer (FEQ).

The update of the FEQ is best implemented with a **zero-forcing algorithm** to avoid bias in this one-tap equalizer

$$W_{n,k+1} = W_{n,k} + \mu \cdot E_{n,k} \cdot \hat{X}_{,k}^{*}$$

(7.67)

7.2.4.3   Other Bases for Multichannel Modulation
The slight difference between DMT and VC, even when H is the same cyclic matrix in both, raises the question of just how many forms of multichannel transmission can be constructed (in concept, an infinite number). Some have other useful properties and are investigated in this section.

Discrete Hartley — Fan's Real DMT   The discrete Hartley transform applies when H is cyclic and real. In this case, a decomposition of the channel matrix of interest is

$$H = A\overline{\Lambda}A \tag{7.68}$$

where A is a real involution matrix $A^2 = I$. That is, A is its own inverse. In this decomposition, $\overline{\Lambda}$ is not diagonal, but rather a real "cross" matrix related to the complex Λ of DMT by the following:

$$\overline{\Lambda} = \Re\{\Lambda\} + J\Im\{\Lambda\} \tag{7.69}$$

where J is the reflection matrix

$$J = \begin{bmatrix} 0 & \cdots & 1 & 0 \\ \vdots & \cdot^{\cdot^{\cdot}} & \vdots & \vdots \\ 1 & \cdots & 0 & 0 \\ 0 & \cdots & 0 & 1 \end{bmatrix} \tag{7.70}$$

that takes any vector and reverses the order of the first $N - 1$ components. Note $J^2 = I$. Thus the channel is partitioned into two-dimensional real subchannels (similar to the complex subchannels in DMT) with each dimension being real. The corresponding pairs of subchannels can be easily inverted to get the channel input, essentially corresponding to an FEQ after the receiver multiplication by the matrix A. The matrix A can be found as

$$A = \Re\{Q\} + J\Im\{Q\} = \Re\{Q\} - \Im\{Q\}, \tag{7.71}$$

which corresponds to the Hartley transform, as first noted by John Fan [32]. The klth element of A is (starting with j,k at zero in the lower right corner and counting up) is

$$A_{kl} = \frac{1}{\sqrt{N}}\left[cos\left(2\pi \frac{kl}{N} \right) - sin\left(2\pi \frac{kl}{N} \right) \right] \tag{7.72}$$

Wavelet Transforms and Partitioning   Wavelet transforms are perhaps best interpreted in terms of a set of basis functions/vectors. The diagram in Figure 7.19 applies, except that not all basis vectors are excited on any given symbol interval. Instead, the process is periodic in that the set of basis functions (with equal numbers of basis functions in each set) is divided into g subsets of an equal number of basis functions, where g is known as the **genus** of the wavelet transform. These subsets form orthonormal sets of basis functions, and the functions in each of the subsets are orthogonal to the functions in all of the other subsets when translated by any integer multiple of the M-sample subsymbol period. There are many choices for such sets, which are beyond the scope of this book (see Refs. [33], [34]). Tzannes [35] has investigated these functions for transmission on DSLs, sometimes using the terminology discrete wavelet multitone transmission (DWMT).

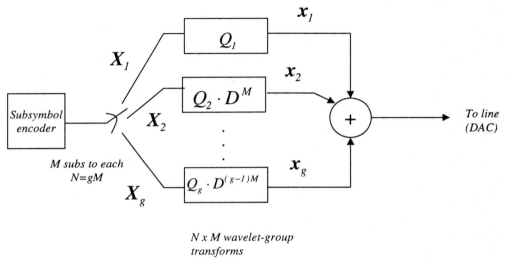

N x M wavelet-group transforms

Figure 7.19 Wavelet-based transform (overlapping groups of samples corresponding to sets of subchannels).

In addition to the orthogonality constraints of the set of wavelets, three additional restrictions are often imposed:

1. Restrict the "out-of-band" energy of the basis functions (note this presumes a frequency-domain interpretation of the basis functions, which is true in most cases).

2. Choose basis functions that will remain orthogonal as they pass through the channel.

3. The functions can be implemented with a fast transform of $N\log_2(N)$ operations.

The first constraint can simplify analog filter design and can reduce emissions, which could be helpful for crosstalk and radiation purposes. Inevitably, such design leads to increase in latency. The second constraint of course recognizes that a set of orthogonal functions will not remain orthogonal after filtering unless they are the Karhuenen-Loeve functions of modal modulation. The designer then attempts (perhaps adaptively as a function of the channel) to select a set of basis functions that approximate the KL functions or can be compensated adaptively to reduce intermodulation distortion. The third constraint is to match the complexity of DMT.

Inevitably, the approximation of the KL functions leads to a level of intermodulation distortion at the receiver, which is often mitigated with the use of an equalizer after the receiver transform. Without the assistance of a cyclic prefix, such intermodulation distortion can be large and span many subchannels. Thus the equalizer may add significant complexity. The "equalizer" will actually be different and essentially a generalization of the FEQ of DMT in that each receiver transform output will be adjusted by terms constructed from other subchannel outputs to eliminate the intersubchannel interference (intermodulation distortion). Such equalization differs for each subchannel, and thus a better name might be interference cancellation. Some call this an "MFEQ" for "multichannel FEQ."

The selection of basis functions for DSLs, both fixed and adaptive, as well as the corresponding complexity of the interference cancellation after the receiver transform, remain active areas of research. Alternative approaches to the wavelet are to use the TEQ of the next section or to filter the DMT signals after the transmitter IFFT. With digital complexities rapidly becoming negligible for DSLs, the area of alternative transforms is likely to become one of great interest over the next decade, especially when constraint 1 above is considered.

Transmitter Windowing   Windowing is a very low complexity alternative to transmit filtering or wavelets that uses basic spectral-estimation/filter-design signal processing to reduce the side effect of the slow decrease in the energy with frequency from the main lobe of a DMT basis function. In the transmitter, a DMT basis function is essentially the multiplication of a rectangular window of length $N+\nu$ samples by a sinusoid. Such windowing causes the frequency spectrum of the basis function to follow the form

$$\phi(f) \propto \frac{sin(\pi [f - f_n] T)}{[f - f_n]} \tag{7.73}$$

which decays linearly with frequency deviation from the DMT tone frequency. Fortunately, this function goes to zero at all other tone frequencies, and thus there is no inter subchannel interference (when the cyclic prefix is long enough). Unfortunately, there may be other constraints than performance of the transmission link; for instance, the out-of-band spectrum decay may be important for spectral compatibility with other DSL services. Thus, at the transmitter side, it may be of value to impose the additional constraint of faster decay with frequency, at least to ease analog filter design.

An easy solution is to use a transmit window. A window is a time-function W_k for each symbol where the transmitted sequence (possibly including any guard bands) is multiplied by the window so that

$$x_k \leftarrow x_k \cdot w_k \tag{7.74}$$

This operation is very low complexity, with only one multiply per sample. The operation of windowing corresponds to convolution of the DFTs of the window W_n and the channel input X_n. Such convolution can dramatically reduce the energy level of sidelobes with careful selection of the window. A number of common windows (Hamming, Hanning, etc.) can be found in almost any digital signal processing book [2]. However, the windowing does cause interference between tones that can be rejected (basically by convolution with the inverse of the window when the channel is flat) in the receiver. The cyclic prefix may be windowed, but if it is, the orthogonality of the DMT subchannels at the channel output is lost. However, if the windowed symbol is not cyclically prefixed, then some of the benefit of the window is lost. Isaksson et al. [36] have studied this problem and found some acceptable windows, as have Spruyt et al. [37]. Use of a transmit window virtually forces the use of the GDFE of the next section in the receiver to reduce the effects of inter subchannel interference or, more generally intermodulation distortion.

Another approach is to use nonlinear windowing instead. This is a method conceived by Bingham [38] that actually places some energy on tones nominally zeroed. The energy is a function of the exact symbols on adjacent tones that are nominally used. By forming a linear combination of nearby used tones' symbol inputs and placing the corresponding result on nominally zeroed tones, the sidelobe energy can be reduced by a factor of approximately 10 dB. This type of window is considered nonlinear because it depends on the symbol inputs. This method requires yet less computation than a full window. The coefficient values depend on the number of zeroed tones used as well as the number of used tones that are input to the linear combination. The coefficients in the linear combination may themselves become functions of the input data symbols.

Bingham's canceler interpolates the energy using sinc functions. The basic observation is that the power spectral density component caused by a DMT tone at frequency n/T is

$$PSD_n(f) = |X_n|^2 \cdot \text{sinc}^2\left(\left[f - \frac{n}{T}\right]T\right)$$

(7.75)

Tones at integer multiples of $1/T$ away will have zero interference, resulting in no intermodulation distortion. However, signal energy is zero only at these frequencies and can be relatively high at other frequencies. The energy of the sinc function's sidelobes is maximum at frequencies solving the transcendental equation $\tan(\pi f) = \pi f$, which may be approximated by the frequencies $(2k+1)/2T$, that is, odd multiples (3 or greater) of half the symbol rate. The amplitude at $3/2T$ is the largest of the sidelobe contributions, and has amplitude given by

$$I_{n+1.5} = \sum_{k=n-M}^{n} X_k \cdot \text{sinc}(k - [n+1.5])$$

(7.76)

Thus, by placing a small amplitude on tone $n+1$ that is opposite in sign and equal in magnitude at the same frequency can significantly reduce the energy. Thus, for tone $n+1$,

$$X_{n+1} = -\frac{I_{n+1.5}}{\text{sinc}(.5)} = \frac{-\pi I_{n+1.5}}{2} = -1.571 I_{n+1.5} \neq 0 \tag{7.77}$$

Typically, this amplitude is a small number since the contributions of other tones are small and the inverse of $\text{sinc}(0.5) = 2/\pi = .637$ does not amplify the function much. Bingham shows about a 10 dB improvement from this method when M is just a few tones [38]. This method causes no intermodulation distortion, but clearly has limited gain.

Receiver Windowing In receiver windowing, the window is applied at the receiver. In the receiver, the use of the window is to reduce the effect of narrowband interference. Sinusoidal or narrowband noise cause the equivalent channel impulse response corresponding to $H(f)/S_n^5(f)$ to be very long, essentially much longer than any practical cyclic prefix. In this case, intermodulation distortion will occur unless the frequency of the sinusoidal noise fortuitously happens to be exactly or nearly on one of the DMT subchannel center frequencies. Application of a window can dramatically reduce the effect of a sinusoidal or narrowband noise. The improvement is evident when one recalls that convolution of a window function (in the frequency domain) with the narrowband transform of a sinusoid will reduce the sidelobes of that sinusoid's spectrum dramatically (thus preventing energy from spreading to all DMT tones and just affecting a few).

The consequence of receiver windowing is again the induced intermodulation distortion. Again, the GDFE method of the next section needs to be used.

7.2.5 Equalization for Multichannel Partitioning

Multichannel methods like DMT or vector coding make use of the previously discussed guard band of length v samples. Sometimes this guard band does not exceed the length of the channel response, and so an equalizer can help performance significantly. The purpose of the equalizer is to limit the equalized sampled-time channel response to the length of the guard period. One can always increase N in a multichannel system and thus increase v with negligibly small performance loss, but larger N means more delay and more memory, which may not be acceptable. Equalizers with multichannel designs that are implemented in the time domain are often called **time-domain equalizers** (**TEQs**). TEQs will follow a design procedure very close to a DFE, with one subtle but performance-significant difference, the feedback section of a TEQ need not be causal, nor minimum phase.

The basic idea parallels the DFE development with error signal again given by

$$e_k(\Delta) = \boldsymbol{b}^* \boldsymbol{x}_{k-\Delta} - \boldsymbol{w}^* \boldsymbol{y}_k \tag{7.78}$$

where $\boldsymbol{b}^*$ is now the $1 \times (N_b + 1)$ vector $\boldsymbol{b}^* = \begin{bmatrix} b_0^* & b_1^* & \dots & b_v^* \end{bmatrix}$ that models the length-v channel, whose only constraint is that its norm be nonzero, $\boldsymbol{x}_{k-\Delta} = \begin{bmatrix} x_{k-\Delta} & \dots & x_{k-\Delta-N_b} \end{bmatrix}$ is the vector of delayed channel inputs, $\boldsymbol{y}$ is the $N_f \cdot q$ vector of channel output samples (there is no matched filter, but rather a lowpass filter, whose output is sampled at q/T, a speed sufficiently in excess of twice the usable channel bandwidth to capture useful information), and $\boldsymbol{w}^*$ is the corresponding vector of TEQ coefficients. Again for the bandlimited AWGN channel, one can always find a matrix H such that a packet of channel output samples $\boldsymbol{y}$ is given (as earlier in Section 7.2.4.1) by

$$y = Hx + n \tag{7.79}$$

for any input packet $\boldsymbol{x}$ and noise $\boldsymbol{n}$ with given autocorrelation matrices R_{xx} and R_{nn} (zero mean). This assumes the input packet's spectrum is known or approximated, meaning that loading effects may be included. This model also allows the output $\boldsymbol{y}$ to be sampled at a rate differing from the channel input sampling rate, thus generalizing the model of Section 7.2.4.1, and allowing for fractionally spaced equalization [99]. The vector $\boldsymbol{x}$ need only have a length sufficient to compute the samples in the matrix $\boldsymbol{y}$, which in turn depends on N_f and q (the oversampling factor). Thus, in terms of known quantities H, R_{xx}, and R_{nn}, the designer can again compute two new matrices:

$$R_{xy,\Delta} = E\begin{bmatrix} \mathbf{x}_{k-\Delta} \mathbf{y}^* \end{bmatrix} = R_{xx,\Delta} H^* \tag{7.80}$$

where $R_{xx,\Delta}$ is an $(N_b + 1) \times (N_f + v)$ matrix of autocorrelation coefficients and

$$R_{yy} = H R_{xx} H^* + R_{nn} \tag{7.81}$$

(note R_{xx} in the above equation is a square $N_f + v$ matrix, $E[\boldsymbol{x}\boldsymbol{x}^*]$). The TEQ differs from the DFE in that the choices for Δ allow effectively the feedback section (which is never implemented and only used for designing the settings of the TEQ) to be noncausal. There is no subtraction of trailing ISI in a TEQ, as this equalizer is intended to focus ISI into an interval that has the same length as the guard period of the associated multichannel transmission system. The MMSE finite-length stationary TEQ is then computed according to

$$w^* = b^* \cdot R_{xy,\Delta} \cdot R_{yy}^{-1} \tag{7.82}$$

for any value of $\boldsymbol{b}^*$. For the TEQ, the value of $\boldsymbol{b}^*$ that minimizes MMSE is the eigen-vector corresponding to the smallest eigen-value of the matrix

$$R_{TEQ}(\Delta) = \tilde{R}_{xx,\Delta} - R_{xy,\Delta} \cdot R_{yy}^{-1} \cdot R_{xy,\Delta}^* \qquad (7.83)$$

This eigen-value/eigen-vector can be recognized as what is known as a Rayleigh quotient in mathematics, computed efficiently through the use of a Lanczos algorithm [100].

Note that $\tilde{R}_{xx,\Delta} = E\,[x_{k-\Delta}\,x_{k-\Delta}^*] \neq R_{xx,\Delta}$ unless $N_b = N_f + v - 1$, but both are different size autocorrelation matrices for the same input symbol sequence. R_{xx} is yet again a different size autocorrelation matrix for the same sequence. The eigen-values problem may need to be solved for all reasonable values of Δ to find the TEQ with smallest MMSE over Δ. This TEQ equation assumes that MMSE is the best criteria (which it turns out is not true) but generally provides an excellent solution if the sampling rate and number of taps in the filter w is sufficiently large. Al-Dhahir [39] and Henkel [11] have independently studied the more difficult problem of maximizing product-SNR for the set of parallel channels, with Henkel appearing to have the best-behaved algorithm. Zervos also has studied the discrete modeling problem via the more exact and precise basis function approach for a finite rank approximation to a stochastic space in [40].

The number of taps for a given level of TEQ performance will always be less than that required for a DFE because there are fewer restrictions on b, but of course there is also the complexity of an additional multichannel system with the TEQ. One algorithm for adaptively setting the TEQ is the **Chow iteration** after Jacky Chow, as described in Ref. [41]. Performance for the TEQ system is still calculated by the geometric SNR, except that the TEQ now affects the SNR on each subchannel. Each subchannel SNR becomes a ratio of the known channel input energy on that subchannel to the mean-square distortion on that subchannel's output in the receiver. Computation of this new SNR requires inclusion of the TEQ response in the channel pulse response so that the new channel response used is $P(D) \rightarrow P(D) \cdot W(D)$ and the new "noise" power spectral density is given by

$$R_{ee}(D) = B(D) \cdot R_{xx}(D) \cdot B^*(D^{-*}) - 2 \cdot \Re\{B(D)R_{xx}(D) \cdot P^*(D^{-*})\} +$$
$$W(D) \cdot [R_{nn}(D) + P(D) \cdot R_{xx}(D) \cdot P^*(D^{-*})] \cdot W^*(D^{-*}) \qquad (7.84)$$

where the input power spectral density may not be flat because it may have been optimized for transmission. By substituting $D = e^{-j\frac{2\pi n}{N}}$ into the channel and MSE power spectral density, one can obtain the N signal to distortion ratios necessary for performance evaluation through calculation of the geometric SNR for this set of equalized subchannels, as described in Sections 7.2.2 and 7.2.3.

An alternative form of equalization, sometimes called a **frequency-domain equalizer (FEQ)** (also discussed in Section 7.2.4.2) can be used in combination with, or separate from, the TEQ. This type of equalizer is implemented after the transform in the receiver. If the guard period is sufficiently large (greater than the length of the channel pulse response), this type of equalization trivializes to a single coefficient on each subchannel output. This coefficient's only purpose is to scale and rotate the constellation on each subchannel so that a common decision device (or decoder) can be used on all subchannels, as in Section 7.2.4.2. Such a one-tap FEQ has no performance improvement but may simplify decoder implementation.

The FEQ becomes more complex when the guard period is not sufficiently long and there is intermodulation distortion. Essentially in this case, the combination of the FFT in the receiver and the FEQ will become the MMSE block estimator, which suggests that vector coding be used. In some cases, designers use only adjacent tones (or a few adjacent tones), assuming all intermodulation distortion is limited to narrow bands, to simplify the FEQ structure, sometimes called a **modified FEQ, or just MFEQ.** There is, however, a better general solution in the next section.

7.2.5.1 Generalized Decision-Feedback Equalization

Restricting again momentarily to the case of ISI only within a symbol, a special case of what is called **generalized decision feedback equalization (GDFE)** [12] can be applied to multichannel transmission systems. Again noting the noncausality of b in the TEQ formulation, intermodulation distortion in the transform output of a multichannel receiver is often a stronger function of adjacent frequencies than it is of distant frequencies. A receiver may choose to decode subchannel outputs in any order for a single packet. This means that a decision made earlier can be used to assist subsequent decisions. The basic GDFE structure is shown in Figure 7.20 where the transform-output subchannels are reordered by a **shuffle matrix** that chooses an order so that the subsequent signal processing optimizes the geometric signal-to-distortion ratio. ADSL uses a shuffle matrix (in fact a good one ordered in terms of SNRs for additional purposes of secure operation with dual latency), but this shuffle matrix is different and appears only in the receiver. The matrix operation V follows the shuffling of indices and consists of an index-varying N_v-coefficient sum-of-products operation of weighting adjacent (or more generally any set of N_v) subchannels. The corresponding triangular matrix A uses N_A previous-index decisions to subtract intermodulation distortion from those previous-index subchannels. An unordered form of the GDFE has been investigated by Wiese [42].

The overall GDFE problem can be formulated mathematically by denoting the index of subchannels as i and presuming that an ordering of N such indices is denoted S_i. There are $N!$ such possible orderings, denoted by the set of orderings S. The SDR for each channel is a function of the channel, any TEQ, the settings for V and A, along with the choice of ordering, thus, simply stated the best GDFE satisfies

$$\max_{S_i \in S; V, A} \Gamma \left\{ \left[\prod_{i \in S_i} \left(1 + \frac{SDR_i}{\Gamma} \right) \right]^{1/N} - 1 \right\}$$

(7.85)

where SDR_i is the ratio of the subchannel symbol energy to the mean-square distortion (noise plus whatever interference from other tones and other symbols). This task appears formidable mathematically even though relatively simple to express. A search of all possible orderings, for each of which the MMSE settings for A and V would need to be computed along with the SDR, is intensive for even moderate N.

To eliminate such a search, the author suggests the following ordering:

1. Select any ordering.
2. Compute the set of differences in SDR with no GDFE compared with the best possible (infinite-length) transmission system at the center frequencies of a multitone system. Call this set of differences $\{\delta_i\}$ for the selected ordering.
3. Reorder the subchannels with the smallest δ_i first, then the next smallest, and so on until the last corresponds to the largest SDR difference. This new order is used for A and V to minimize mean-square distortion as follows:

Method 1 — Matrix MSE Approximation: The N-dimensional vector of errors E corresponding to the reordered index is

$$E = VZ - A\hat{X} \tag{7.86}$$

where Z is the $1 \times N$ transform output vector and $\hat{X}$ is the vector of "previous" subchannel decisions, which is assumed to be correct and henceforth equal to X, the vector of subchannel inputs for this symbol, again reordered. For any channel, one can always find a matrix H (a permuted form of the matrix H nominally used without reordering and discussed for DMT and vector coding) such that

$$Z = HX + N \tag{7.87}$$

for any input and noise with given autocorrelation matrices R_{xx} and R_{nn} (zero mean). Thus, in terms of known quantities H, R_{xx}, and R_{nn}, the designer can compute two new matrices:

$$R_{xz} = E[XZ^*] = R_{xx}H^* \tag{7.88}$$

and

$$R_{zz} = HR_{xx}H^* + R_{nn} \tag{7.89}$$

Further, one computes the Cholesky factorization of the **GDFE discrete key-equation matrix**

$$\left[R_{xx} - R_{xz} \cdot R_{zz}^{-1} \cdot R_{xz}^{*}\right]^{-1} = GSG^{*} \tag{7.90}$$

with G an upper triangular, monic (ones down diagonal) matrix and $S = diag\left\{S_{0}, ..., S_{N_{b}}\right\}$ a diagonal matrix with smallest element in the upper left corner and positive values non-decreasing down the diagonal.

The MMSE GDFE then satisfies

$$V^{*} = A^{*} \cdot R_{xy} \cdot R_{yy}^{-1} \tag{7.91}$$

for any triangular value of A^{*}, including $A = I$ for the MMSE-FEQ settings, and the best A is determined by $A = G$. The MMSE values are the diagonal elements of the matrix S. (This matrix should not be confused with the set of all possible orderings.) The GDFE, like all decision-aided systems, is subject to error propagation. This can be eliminated through the use of index-varying Tomlinson-like or flexible-like precoders in the transmitter (which would then need to know both G and the reordering). One concern with this method is the growing size of the feedback section, which may be a bit too complex. Usually the longer feedback filters have zeroed coefficients so by inspection the feedback filter length can be reduced. Also, the feedforward section has a fixed (index-varying) length of N coefficients (or Nq when fractionally spaced), which may also be too complex. A more complete development of the GDFE with explanations can be found in Ref. [43]. Method 2 addresses these complexity problems.

Method 2: A second method proceeds similarly, except that the DFE problem is solved individually for each subchannel. The number of feedforward filter coefficients and "feedback" coefficients is arbitrary and can be set by complexity constraints. The method for computing a finite-length DFE in Section 7.1.2.6 follows with the matrix H being set for each subchannel according to the permutation and the channel response values. The solutions for the individual feedforward and feedback filters then follow the same equations as in Section 7.1.2.6. This approach can clearly be extended to use selected decisions from previous symbols.

The GDFE methods are sufficiently general to be used with all multichannel transmission methods. With overlapping or wavelet methods, a set of g GDFEs (that is, a number of GDFEs equal to the genus) needs to be used for each of the g phases. Clearly decisions from previously decoded symbols can be added also (as may be more appropriate for overlapping transmission systems like wavelets) to effect additional improvement.

7.2.5.2 Block Decision Feedback

Kasturia's block decision feedback and block precoding methods [44], [45] can be used to eliminate interference between blocks when vector coding is used. An extension to DMT was recently developed by Cheong [46] (see Section 7.2.5.3). These methods can be combined with the first type of GDFE to avoid intrasymbol interference or intermodulation distortion altogether.

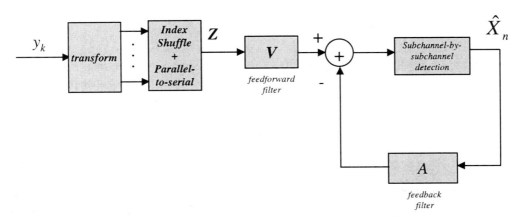

Figure 7.20   Generalized decision feedback equalizer (GDFE).

7.2.5.3   Precoding Methods — Cheong's Precoder

A desirable characteristic for DMT transmission is that the transmit signal appear periodic in the guard period. Decision feedback methods instead try to force the guard period essentially to look like it was zeroed. If instead of zeroing, the input could be altered to add a cyclic prefix as well as remove ISI, then intermodulation distortion would be eliminated for DMT. Unfortunately, the latter step of **cyclic reconstruction** cannot be executed in a receiver because *current* decisions are necessary in addition to previous decisions. However, well-designed systems do not actually implement decision feedback in the receiver (to avoid error propagation) but instead use a precoder in the transmitter to create an identical effect (without error propagation) as the feedback section, as has now been discussed several times (e.g., Tomlinson-Harishima and Laroia precoders). Cheong's precoder [46] instead notes that in the transmitter, current and past decisions are available if the precoder is implemented with an additional symbol period of delay. A cyclic prefix can be "faked," and then intermodulation distortion is eliminated with a shorter cyclic prefix than otherwise necessary. The **cyclic reconstruction** in the transmitter does cause a transmit power increase that is larger than that of the earlier precoders, like Tomlinson-Harishima and Laroia. However, the low complexity of the FFT in DMT is now maintained in the receiver. The transmitter complexity increases by the precoder complexity, which may be non-negligible. Considerable improvement for fixed complexity can be attained, as in Cheong [46].

7.2.5.4   Training Methods

The straightforward adaptive computation of TEQs or GDFEs is difficult because of the large dynamic range of quantities, typically leading to long convergence times and excessive precisional requirements. There are some methods, particularly those popularized by Chow [41], and some modifications [47], but these methods can also have convergence problems. The best methods today would simply compute the various matrices above from estimates of the channel response (defining the matrix H) and noise autocorrelation (defining the matrix R_{nn}), and then execute well-conditioned matrix operations. Fortunately, these equalizers need not be updated often in DSL.

7.2.6   ADSL T1.413 DMT

The ANSI T1.413 (ITU g.dmt) ADSL standards [63] specify use of DMT. This particular downstream system employs a sampling rate of $1/T' = 2.208$ *MHz*, a block size of $N = 512$ with conjugate symmetry, meaning 256 tones from 0 to 1.104 MHz, and a cyclic prefix length of $v = 32$ samples. Every 68 symbols, a constant known **synch symbol** with $512 + 32 = 544$ samples is inserted, so that the actual symbol rate is $2.208/(544 \times 69/68) = 4000$ Hz. The width of a tone is 2.208 MHz/512 = 4.3125 kHz. The excess bandwidth is thus 0.3125/4 = 7.8%. The data rate is any multiple of 4 kbps, but since information is transmitted in bytes, the actual standard allows any multiple of 32 kbps to be transmitted. The downstream transmit power spectral density is − 40 dBm/Hz on average with tolerable variation over the tones of ± 2.5 dB. The maximum transmit power is thus about 20 dBm.

This T1.413 upstream transmission system is also DMT and employs a sampling rate of $1/T' = 276$ kHz, a block size of $N = 64$ with conjugate symmetry, meaning 32 tones from 0 to 138 kHz, and a cyclic prefix length of $v = 4$ samples. Every 68 symbols, a constant known synch symbol with $64 + 4 = 68$ samples is inserted, so that the actual symbol rate is $276/(68 \times 69/68) = 4000$ Hz. The width of a tone remains 276 kHz/64 = 4.3125 kHz. The excess bandwidth is thus 0.3125/4 = 7.8%. The average upstream power spectral density is −38 dBm/Hz with again up to 2.5 dB tolerable deviation, leading to a maximum transmit power of 14 dBm. Recently, ADSL Issue 2 allows the use of 32 to 64 tones upstream, with the additional tones often being used when ISDN coexists on the same line (tones 33 to 64 are above the ISDN band). All 64 tones are sometimes used to reduce the asymmetry between downstream and upstream data rates.

In both directions, T1.413 mandates the potential use of two paths through the transmission link, known as the **fast** and **interleave** paths. The interleave path bits may have been interleaved over more than one symbol and thus may incur greater delay through the link. The fast path bits have not been interleaved beyond symbol boundaries. The tones indices in T1.413 are adaptively shuffled so that the tones with the smallest numbers of bits are encoded first with bits from the fast buffer. Then subsequently tones with increasingly larger numbers of bits are encoded from the remaining bits in the fast buffer and then from the interleave buffer until all have been encoded. This **tone-shuffle** interleaving leaves the transmission link more robust to impulse noise and to clip noise than a straightforward allocation of bits. The tones with largest numbers of bits are usually most affected by impulses and by clipping and so get extra burst-error protection in the interleave path of external error correction (see Section 7.4).

A diagram of T1.413 transmission appears in Figure 7.21.

The detailed state of DMT in evolving ADSL standards (ADSL-LITE [31] and ADSL-HEAVY [48]) is likely to change, but the basics of this chapter will, hopefully, provide a fundamental foundation to understanding and design of any DMT/DSL system.

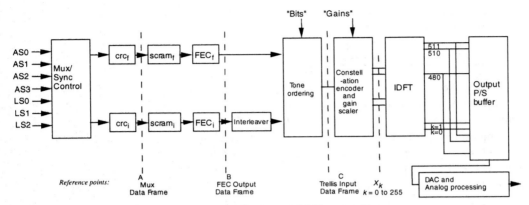

Figure 7.21   ANSI T1.413 DMT transmitter.

7.2.7   Clipping and Scaling (Peak-to-Average Issues)

Entropy is an information-theoretic quantity that is a generalization of data rate. In DSL, designers often use signal constellations where each point is equally likely to occur. If there are M equally likely points for each symbol, the entropy and the data rate are related easily by

$$R = \frac{H}{T} = \frac{log_2(M)}{T} \quad \text{bits/second} \tag{7.92}$$

where T is the symbol period or time it takes to send each constellation point and H is defined below.

When the points in the input signal distribution are not equally likely, as sometimes occurs with one-dimensional distribution of the samples entering the line with transmission methods like DMT, the entropy generalizes to

$$H = \sum_{i=1}^{M} - p_i \cdot log_2\left(p_i\right) \tag{7.93}$$

which reduces to $log_2(M)$ if all points are equally likely with $p_i = 1/M$.

If viewed in the time domain as one-dimensional signals, the probability distribution of multichannel signals approaches a Gaussian distribution through the (ab-) use of central-limit theorem arguments.[12]

---

12.   Actually, no discrete distribution ever truly becomes Gaussian, and there are always peak limits unlike the Gaussian, but the approximation is good enough for the tutorial purposes of this document.

A **clip is defined** to occur when the transmit signal sample exceeds the maximum implemented value for the transmitter (often set by a digital-to-analog converter's maximum value). This random event's probability of course depends on the clip level. For a Gaussian distribution and a probability of clip of 10^{-7}, the clip instantaneous power level is 14.2 dB higher than the average power of the signal. For a clip probability of 10^{-5}, this value reduces to 12.5 dB, and for 10^{-3}, the value is 10 dB. A reduction of a very large peak-to-average ratio (PAR) to a smaller one can reduce the complexity of a transmitter, especially reducing the power consumption of linear analog driver circuits.

As an example, let us take an ADSL signal, as per T1.413. The probability of a clip is about 10^{-8} with the specified 15 dB PAR. Thus, at a sample rate of 2.208 MHz, clips occur then at data rate $10^{-8}(2.208 \times 10^6) \cdot b_{clip}$ where b_{clip} is the number of bits necessary to convey the size and position of the clip to the receiver, typically less than 20 bits. Thus, it would then take about 0.5 bits/second to eliminate clips with PAR = 15 dB, a very low data rate loss. This 0.5 bits/second is the **entropy of the clip**. Equivalently, if the clip occurs, in reality only 0.5 bits/second of data rate should be lost.

This study of entropy summarizes the motivations of Wulich [49], and Ochiai and Imai [50], Davis and Jedwab [51], and Patterson [52], who all investigate the use of various coding methods that preclude the occurrence of data sequences that cause high PAR in OFDM transmission. Such methods, unfortunately, seem to exhibit higher rate loss than necessary, largely to simplify implementation. Further study is necessary before any of these methods are applicable to DMT ADSL.

Indeed, ADSL modems could increase PAR significantly without noticeable reduction in the data rate. Even at a PAR of 10 dB, the loss in data rate, or the entropy of the clip, is still below 50 kbps, which is much less than the downstream data rates of 1.5 to 8 Mbps.

A practical drawback is that to keep the additional data rate reduction small, a large buffer (memory) may be necessary, along with the consequent delay increase. If a transmitter first needs to compute a preliminary output to detect a clip and then redo this computation with a slightly altered bit stream to encode the clip instead of allowing it, the transmitter complexity can double. A similar doubling can occur at the receiver in first decoding the information to learn of the clip, reconstruct the large sample, and then decode again.

Fortunately, a number of low-complexity and low-delay methods have been developed and are individually discussed in the next few subsections. The two Tellado methods, tone reduction [53] and tone injection [54], [55] (parts of which were independently found by Gatherer and Polley [56] and by Kschischang, Narula, and Eyuboglu [57]) are those that gain the most PAR reduction and appear strong candidates to be implemented in DSLs of the future (see Sections 7.2.7.4 and 7.2.7.5). An upcoming book [66] has a chapter by M. Friese of Deutsche Telekom devoted to PAR reduction methods (see Chapter 4).

7.2.7.1 Dynamic Clip Scaling

Dynamic clip scaling [59] is an early proprietary clip-mitigation method used in some DMT DSLs. A single tone is reserved for indication of a scale factor for the entire symbol. This tone is typically modulated with 4QAM so to be independent of gain in decoding and is added last after the transmitter IFFT output has been checked for the presence of a clip. The four phases for this tone represent four possible scalings, typically 0 dB, –1 dB, –2 dB, and –3 dB; one of the last three factors is applied to transmit when a clip is detected. Usually 0 dB is sent, but when clips and consequent scaling occur, depending on the magnitude, –1, –2 , or –3 is sent. The receiver first decodes this same tone (usually added at a level where it will not cause another clip elsewhere because its amplitude is too small) and then scales the signal back to its original level before decoding.

Such scaling does reduce noise immunity, but the small effective increase in noise is less than the consequent noise caused by a clip that would have otherwise occurred. This method clearly reduces the data rate by approximately $1/N$ (typically small) and requires the transmitter and receiver to coordinate through the use of the additional tone (sometimes combined with the pilot).

7.2.7.2 Random Reencoding

A number of random reencoding methods have been suggested and are discussed below. All these methods note that since the occurrence of a clip has low probability, it is very unlikely that a rephased encoding of the same information will also clip.

Selected Mapping (SLM Method)   At least nine authors appear to have simultaneously published the same method in three papers in October 1996 [60], [61], [62], which has been dubbed by two other authors (Muller and Huber [64]) as the **selected mapping method (SLM)**. Basically all suggest a set of known rotations be each applied to the known transmit sub-symbol inputs (essentially an extra IFFT or more, instead of one, when used) in a corresponding set of encoders. The encoder with the lower peak is transmitted, and again a tone and/or part of the bit stream is reserved to tell the receiver which encoder was used. The receiver then knows which decoder to use. This method is fairly simple, except for the extra complexity of the additional transmitter IFFT(s), and essentially reduces the probability of a clip to the nth power of the probability for one encoder, where n is the number of encoders, resulting in a few dB improvement. The SLM method's improvement is limited by secondary peaks, i.e., the second largest peaks in a symbol.

Friese and Muller/Huber Methods   Friese [65], [58], and Muller and Huber [64] introduced a method they call **partial transmit sequences (PTS)** that separates the DMT/OFDM inputs into a set of groups of tones. Each set of tones has a fixed phase rotation applied to each tone within it, selected from a set of possible rotations. The partial IFFTs are performed for each group, and then one by one the resultant time-domain outputs are added in a way that keeps peaks as low as possible. The selected rotation for each of the groups needs to be carried to the receiver. This method shows a modest improvement over SLM. Friese investigates a number of optimization criteria, including optimization of the continuous-time PAR, in his work and in [66].

Verbin's Method   Another approach due to Verbin [67] instead makes use of the observation that the sum of two cyclically prefixed sequences will produce at the channel output a sequence that corresponds to the sum of the two corresponding sets of parallel channels. Verbin then selects from up to 7 possible sums, each corresponding to essentially an orthogonal energy-preserving transformation, of the even and odd parts of the nominal channel input. The one with the smallest peak is transmitted (and again the receiver must be told which corresponding receiver transformation to use via a control channel like a single tone or the pilot). This method obtains about 2 dB improvement. To avoid complicating gain issues with the FEQ implementation, Verbin selects orthogonal transformations that correspond to permuting the order of the subchannels at the output, but otherwise do not lead to interference between DFT outputs. This restriction is not necessary, but makes the receiver simpler. This method also causes a reduction in the probability of a peak roughly as the nth power of the clip probability, where n is the number of transforms. However, the occurrence of secondary, tertiary peaks limits gain to about 2 to 3 dB as Verbin shows [67]. This method is simpler than recomputing an entire IFFT for introduction of randomness.

7.2.7.3 Gatherer/Polley Method

The Gatherer/Polley (GP) method [56] represents a significant improvement on dynamic clip scaling and on all the randomized reencoding methods. First, this method does not require coordination between transmitter and receiver, a significant benefit because standardization is less relevant in this case. Second, no extra IFFT or transforms are necessary although there is hidden complexity in a search procedure.

The GP method notes that tones are often unused in DMT (intentionally or because DMT turns them off). Some energy may be allocated to unused tones as desired as long as spectral mask constraints (and any power constraints) are met. A signal may be constructed corresponding to use of unused tones. Offline optimization of the possible inputs is done to find a time-domain **signature waveform** that has a large peak in it and is otherwise small at other time instants. Addition of the signature waveform to the transmitted DMT signal corresponds to insertion of energy on unused tones and does not increase intermodulation distortion. When a large peak is detected, the signature waveform is added to the transmit signal with negative amplitude to the peak to reduce the level to under the desired clip level. The signature waveform must be cyclically rotated into the position of the peak, but that corresponds only to a linear-phase change on the inputs of the unused tones (meaning it can be ignored by the receiver). Better signature waveforms require more unused tones and allow increasingly lower settings for the clip level. Gatherer and Polley have been able to reduce PAR by 3 dB using this method.

One limitation is the possibility of generating new peaks elsewhere, and so the signature waveform may have to be added several times until acceptable peak level is obtained, thus limiting the reduction in PAR. Another caution is that peaks can be shifted to the receiver if only subchannels that would not make it through the transmission channel are used to generate the signature waveform. Indeed simple analog lowpass filtering can lead to loss of the PAR reduction because the GP method optimizes only at sampling instants and not the entire filtered signal. This effect can also lead to clipping in the receiver.

GP also note that a DMT receiver may elect to underuse certain tones. When underused, the FEQ error signals on these tones can be used to fit a straight line to phase error. The slope of this line determines the position of a clip, while the amplitudes of the error signal determined the clip strength. Then, the clip may be subtracted at the receiver. This method again requires no coordination and is receiver based, a unique contribution of Gatherer and Polley as all other previous methods appear transmitter based.

7.2.7.4 Tellado's Tone-Reduction Method

Tellado's method [53] improves upon the GP method to obtain a significantly higher PAR than had been previously achieved. The basic idea is the addition of a known constant symbol offset to the transmitted waveform to change its PAR. The general problem is

$$\min_{\hat{C}} \left\| x + Q\hat{C} \right\|_{\infty} << \left\| x \right\|_{\infty} \tag{7.94}$$

The subscript of infinity simply means maximum value, and the vectors are DMT symbols of length N samples. The idea is that the constant to be added is actually determined in the frequency domain and then transformed through the FFT Q to the time-domain. Of particular interest are restrictions on the constant $\hat{C}$ so that the implementation is facilitated and/or the value of the added offset need not be known by the receiver. The GP method can be viewed as a very special case of this criterion where the additive constant is restricted to only unused tones and the problem is solved by an approximate algorithm.

Tellado noted that the unused tones are often not good choices for the constant and found, however, that by using a few good tones (typically 5 to 20% or less), an additive constant that led to large PAR reduction (as much as 6 to 10 dB) could be achieved. The use of good tones for PAR reduction instead of data transmission is known as **tone reduction.** Tellado also notes when $\hat{C}$ is restricted to tones that otherwise will not carry data, whether intentionally silenced to get large PAR reduction or just not used anyway, the well-known simplex algorithm applies: PAR reduction in this case is an instance of what computer scientists call linear programming. Figure 7.22 shows the improvement for this case.

The level of PAR reduction is dramatic (actually causing the PAR for the high-performance DMT method to be lower than other so-called "simple" transmission methods that do not perform as well), but the complexity is high for the full PAR reduction. Tellado then studied various approximation methods for solution of quadratic criteria that iteratively solve a related problem. The minimization of a so-called signal-to-clip power ratio, which was also tacitly considered by Gatherer and Polley, leads to a method that can converge close to the performance of the simplex method, but with less complexity. The algorithm that arises turns out to be equivalent to the heuristic interpretation of Gatherer and Polley that the time-domain waveform is optimized to have a peak in the location of the peak of the otherwise unaltered DMT symbol. By

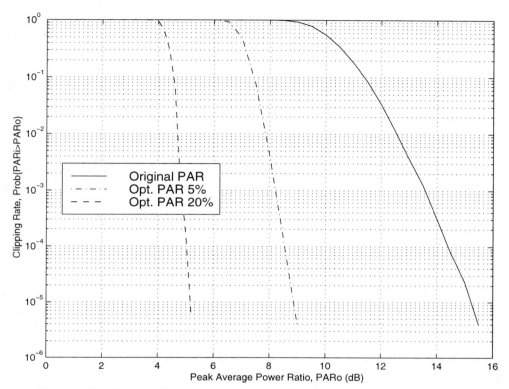

Figure 7.22 Best solution for PAR reduction with tone reductions of 5% and 20%.

finding waveforms that have single high peaks and energy only at a few frequencies (that will not be used for data transmission), the algorithm consists of rotating the peak into position (rotation does not change the tones on which the energy of the additive constant is nonzero), scaling, and subtracting. Tellado's observation that the use of good tones dramatically improves performance still holds for this method. Tellado also observed that the occurrence of secondary peaks (the next largest sample) often limits the performance of the GPF method to a 3 dB PAR reduction. The complexity of the approximate methods is roughly N multiply-accumulates per iteration, and about 10 iterations gains most of the PAR reduction as in Refs. [53] and [54]. Space constraints prevent further discussion of this method here. However, both ADSL-LITE [28] and ADSL-HEAVY [31] standards of the ITU, as well as ADSL Issue 2 [63], had last-minute adjustments that allow vendor-discretionary use of Tellado's and/or the GP method.

7.2.7.5 Tellado's Tone-Injection Method

The original criterion above in Tellado's tone-reduction method need not be restricted to tones that cannot or do not carry data. The optimization becomes far more difficult but can indeed be solved in general for any number of tones used to construct peak-reduction signals. The peak-reduction signals are injected on the tones used along with the data. One method for simplifying the distinction of the injected constant and the data is to ensure that the added constant is much larger than any message (as long as this is not done on too many tones too often, the energy and PSD increases are not measurable). The receiver then knows the value of the large constant that may have been added and simply translates its decision regions by the known constant when a large signal is detected. This method was also noted by Kschischang, Narula, and Eyuboglu [57], who concentrated only on additive constants that maintain a lattice structure (which reduces the amount of PAR reduction from the levels independently found by Tellado in [55]). This method results in no bandwidth loss but usually more complicated algorithms for PAR reduction than tone reduction. (The reader is referred to Refs. [100] and [101] for more details.) This method also can be altered most easily to minimize PAR at the output of the analog transmit filter, whereas the Tellado tone reduction does not appear to alter easily to fix this important problem.

7.2.8   Fast Fourier Transforms for DMT

Fast Fourier transform (FFT) methods reduce the computation in an N-point DFT from N^2 complex multiples to $N \log_2(N)$ complex multiples, when N is a power of two, as is assumed throughout this subsection. Most modern DSPs are becoming very efficient in the implementation of the FFT algorithm, furthering the popularity and simplicity of the DMT multichannel technique in particular. There are many FFT algorithms, but the basic ones are reviewed here. In the case of DMT, a further reduction is possible in the FFT for the receiver by observing that inputs are real (not complex) and similarly in the IFFT for the transmitter by observing that outputs are real. Those simplifications are also described here.

The DFT and its associated inverse are described by the transform pair

$$X_n = \sum_{k=0}^{N-1} x_k \cdot e^{-j\frac{2\pi kn}{N}} \quad \text{and} \quad x_k = \frac{1}{N} \cdot \sum_{n=0}^{N-1} X_n \cdot e^{j\frac{2\pi kn}{N}} \tag{7.95}$$

The $1/N$ scale factor in the inverse is often ignored because scaling is more often determined by dynamic-range constraints of the implementing processor. The basic FFT algorithm exploits the structure of the DFT through a recursion relating a size N DFT to two size $N/2$ DFTs, allowing $\log_2(N)$ stages of the recursion to construct a DFT from size 2 DFTs, which are trivial. To simplify notation,

$$W_N = e^{-j\frac{2\pi}{N}} \tag{7.96}$$

and

$$X_n = \sum_{k=0}^{N-1} x_k \cdot W_N^{kn} = \underbrace{\sum_{k=0}^{\frac{N}{2}-1} x_{2k} \cdot W_N^{2kn}}_{even} + \underbrace{\sum_{k=0}^{\frac{N}{2}-1} x_{2k+1} \cdot W_N^{(2k+1)n}}_{odd}$$

$$= \underbrace{\sum_{k=0}^{\frac{N}{2}-1} x_{2k} \cdot W_{N/2}^{kn}}_{N/2 \, \text{DFT of even}} + W_N^n \cdot \underbrace{\sum_{k=0}^{\frac{N}{2}-1} x_{2k+1} \cdot W_{N/2}^{kn}}_{N/2 \, \text{DFT of odd}}$$

$$= G_n + W_N^n \cdot H_n \tag{7.97}$$

which is the desired recursion of the original DFT in terms of two half-size DFTs. The process then iterates $\log_2(N)$ times to complete the FFT algorithm [68]. The essence of the recursion is illustrated in Figure 7.23. Each half-size DFT is in turn divided into a pair of quarter-size DFTs. The time-domain inputs are successively partitioned into even and odd sets, which (over all FFT stages) results in an index order in the time domain that can be determined over an N-point block to be simply the binary index written in reverse order in terms of bits. This ordering is sometimes known as **bit-reverse addressing**, implemented by an instruction option often included on most programmable DSPs. Each stage of the DFT consists of N complex multiply-accumulates in a straightforward implementation. However, as also shown in Figure 7.23, the complex multiples can be grouped into execution of two by two sections known as **butterfly operations**. A butterfly can simplify to six real operations, rather than the usual eight. Indeed there are $(N/2) \log_2(N)$ butterflies in an FFT, leading to $3N \log_2(N)$ real multiples.

Further simplification ensues when the time-domain sequence is real, as in DMT for DSLs. For any complex sequence of length $N/2$

$$y_k = \Re\{y_k\} + j\Im\{y_k\} \tag{7.98}$$

basic complex algebra leads to

$$\Re\{y_k\} \leftrightarrow \text{Even}\{Y_n\} = \tfrac{1}{2}\left[Y_n + Y_{N/2-n}^*\right] \tag{7.99}$$

and

$$j\Im\{y_k\} \leftrightarrow \text{Odd}\{Y_n\} = \tfrac{1}{2}\left[Y_n - Y_{N/2-n}^*\right] \tag{7.100}$$

One can form a length $N/2$ complex sequence from a length N real sequence via

$$y_k = x_{2k} + jx_{2k+1} \quad k = 0,\ldots,\tfrac{N}{2}-1 \tag{7.101}$$

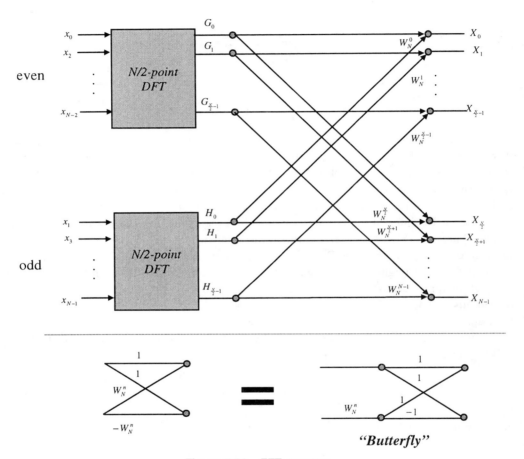

Figure 7.23 FFT structure.

The DFT of the length N sequence x_k is again X_n. From the derivation of the FFT above, then one notes

$$X_n = \text{Even}\big(Y_n\big) + W_N^n \cdot \text{Odd}\big(Y_n\big) \tag{7.102}$$

which is itself a size-$N/4$ butterfly operation. Thus, there are $(N/4)\log_2(N/2)$ butterflies for the size-$N/2$ complex FFT that generates Y_n plus the final $N/4$ butterflies to generate X_n from the even and odd parts of Y_n, totaling to $(N/4) \log_2(N)$ butterflies, or $1.5N \log_2(N)$ real instructions/operations to implement, a savings of exactly 2 with respect to the full complex implementation.

An IFFT can always be implemented by an FFT using the formula

$$IDFT(X_n) = \frac{1}{N} \cdot \left[DFT(X_k^*)\right]^*$$

(7.103)

For the IFFT in the transmitter of DMT, a different formulation of the computations is used for this inverse transformation, which leads directly to a simplification in the complex-to-real case. The (scaled by N) IFFT can be written

$$x_k = \sum_{n=0}^{\frac{N}{2}-1} \left[X_n \cdot W_N^{-kn} + X_{n+\frac{N}{2}} \cdot W_N^{-k\left(n+\frac{N}{2}\right)}\right]$$

$$= \begin{cases} \displaystyle\sum_{n=0}^{\frac{N}{2}-1} \left[X_n + X_{n+\frac{N}{2}}\right] \cdot W_{\frac{N}{2}}^{-nr} & k = 2r \quad \text{(even time samples)} \\[3ex] \displaystyle\sum_{n=0}^{\frac{N}{2}-1} W_N^{-n} \cdot \left[X_n - X_{n+\frac{N}{2}}\right] \cdot W_{\frac{N}{2}}^{-nr} & k = 2r+1 \quad \text{(odd time samples)} \end{cases}$$

(7.104)

Two half-length frequency-domain sequences

$$G_n = X_n + X_{n+\frac{N}{2}}$$
$$H_n = \left(X_n - X_{n+\frac{N}{2}}\right) \cdot W_N^{-n}$$

(7.105)

can be inverse transformed to generate the even and odd time-domain samples. This type of decimation in output IFFT appears in Figure 7.24.

The computation required for this IFFT is almost exactly the same as for the FFT and is again $(N/2) \log_2(N)$ butterfly operations or equivalently $3N \log_2(N)$ real operations.

Again with DMT, the real nature of the IFFT output in the transmitter can be exploited to half the necessary computation of the IFFT. This is achieved by noting the conjugate symmetry of the DMT input, namely

$$X_{N-n} = X_n^*$$

(7.106)

The sequence

$$G_n = X_n + X_{\frac{N}{2}+n} = X_n + X_{\frac{N}{2}-n}^*$$

(7.107)

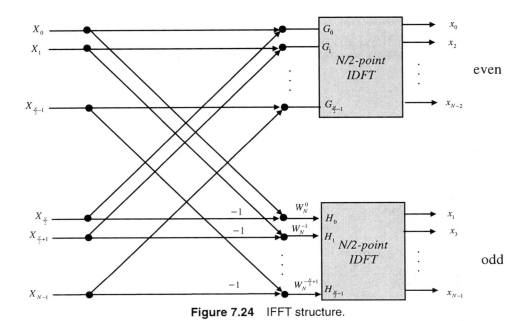

Figure 7.24   IFFT structure.

is thus even, and the sequence

$$O_n = X_n - X_{\frac{N}{2}+n} = X_n - X^*_{\frac{N}{2}-n}$$
(7.108)

is thus odd; the function $H_n = O_n \cdot W_N^{-n}$ is even. Because of the conjugate symmetry, G_n and H_n are reflections about $N/4$ and thus only $N/4$ values are distinct when x_k is real. Thus, both G_n and H_n are even and have corresponding real time domain IFFT outputs. Thus, the sequence

$$\tilde{X}_n = G_n + jH_n \quad n = 0,...,\tfrac{N}{2}-1$$
(7.109)

has an $N/2$-size IFFT that is $g_k + jh_k$ where both g_k and h_k are real. But from our earlier development, we know that

$$x_{2k} = g_k \text{ and } x_{2k+1} = h_k \text{ for } k = 0,...,\tfrac{N}{2}-1$$
(7.110)

Thus a half-size IFFT, plus the construction of G_n and H_n, which is equivalent to an additional $N/4$ butterfly operations, are executed, leading to $(N/4) \log_2(N)$ butterflies or $1.5N \log_2(N)$ real operations for this special IFFT that exploits the conjugate symmetry in DMT for DSLs.

The operations of butterfly depend on N being a power of two and lead to what are known as **radix-2** algorithms. By constructing factorization of an arbitrary N into prime factors, it is possible to derive alternative butterfly structures that may be useful or perhaps slightly more computationally efficient, depending on the desired dimensionality's (N's) actual value.

7.2.9 Multiplexing Methods for Multicarrier Transmission

Multiplexing methods were generally considered in Chapter 5 for DSLs, but this section investigates in more detail some specific implementation considerations for multichannel transmission, particularly DMT. Some particularly noteworthy work specifically on the co-existence of VDSL with ADSL appears in Jacobsen [69], [70] where the concepts of a CO-mix (different services emanating from the Central Office, or "CO") and CPE-mix (remote modems in similar physical positions at the customer premises, or "CPE") are introduced.

7.2.9.1 FDM

Frequency-division multiplexing (FDM) is one means by which to avoid self-NEXT in DSL transmission, namely the upstream and downstream transmission bands are nonoverlapping. However, simple silencing of tones with DMT is often not sufficient to ensure that out-of-band energy is low enough to eliminate crosstalk. Specifically, the spectrum drops only with the square of the $\text{sinc}(fT)$ function that arises from the rectangular windowing nominally used with DMT. Bingham's canceler or windowing can be used to reduce the out-of-band spectrum another 10 dB, but this may not be sufficient. Thus, analog filtering is often added to ensure a sufficiently low spectrum. Care must be exercised in the design of the analog filters to prevent significantly lengthening the response of the channel and thus causing greater need for TEQ-like equalization.

7.2.9.2 TDM (Ping-Pong)

If self-NEXT should be avoided (often the case in ADSL, VDSL, etc.), then another multiplexing method is time-domain or ping-pong transmission. This method is used in Japanese ISDN (with AMI) and in the VDSL Alliance[13] proposal [30]. In the latter, a number of large manufacturers use a **superframe** of 20 DMT symbols with $1/T = 43.125$ kHz, nominally. In this proposal, two of the symbols are silent and used for "line turnaround" (allowing transients to decay before reversing the direction of transmission). Thus, for example, as in Figure 7.25, symmetric transmission is implemented with nine downstream symbols, a silent symbol, nine upstream symbols, and another silent symbol. Clearly, self-NEXT is not of concern because a transceiver does not listen while it is transmitting. All lines in a cable are presumed to use the same symbol-frame boundaries. Asymmetric transmission is often implemented as 16 down, 1

---

13. The VDSL Alliance is a group of companies writing a DMT-based VDSL standards draft. There is also a VDSL Coalition writing a similar CAP/QAM document.

silent, 2 up, and 1 silent. It is thus easy to implement various levels of symmetry and asymmetry with ping-pong, whereas FDM requires analog filters that may be difficult to design.

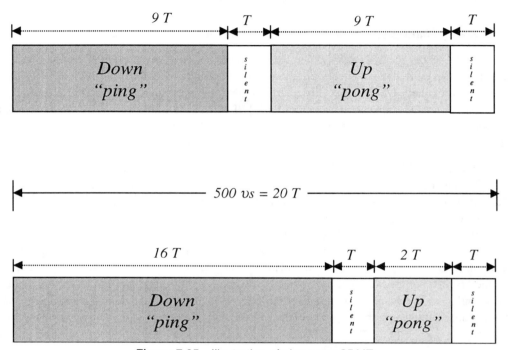

Figure 7.25   Illustration of ping-pong SDMT.

7.2.9.3 Echo Cancellation

Echo-canceled DMT presumes that self-NEXT is tolerable and that the bandwidth can be shared in both directions. Echo canceling is used in ISDN, HDSL, and ADSL. For DMT in ADSL, some simplifications for asymmetric echo cancellation occur that allow a very low complexity in software implementation. If the self-NEXT is tolerable, echo cancelers are usually cheaper in ADSL than the analog filters for FDM, but ADC precision is increased by 1 to 2 bits, which leads to an interesting modem cost trade-off.

Echo cancellation with DMT is fortunately simplified with respect to echo cancellation for other modulation methods like PAM, QAM, or CAP through the use of **circular echo synthesis (CES)** of Ho et al. [71], which appears in Figure 7.26. The normal DMT transceiver functions of FFT and IFFT are augmented by the shaded boxes with echo cancellation. Echo impulse responses are typically longer than those of the forward transmission path on a twisted pair, mainly because signal energy bounces from the far-end transceiver as well as leaking across the hybrid circuit. The guard period of DMT is thus usually insufficiently long to ensure that the

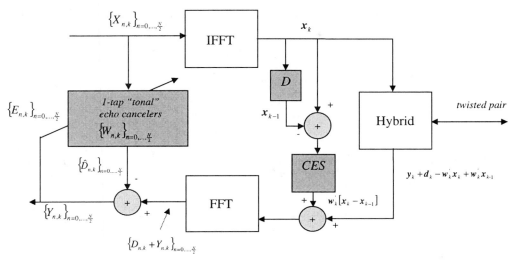

Figure 7.26   Circular echo synthesis (CES): echo cancellation with DMT.

echo signal appears periodic. Melsa and Younce [72] show how an equalizer like a TEQ can jointly serve to reduce transmit response length for the channel and echo paths to be confined to the guard period, as an alternative or addition to the CES. Thus the signal returning from the hybrid is the sum of the far-end signal symbol y_k for the kth symbol block time, a circular echo component d_k, and the noncircular echo component, $\mathbf{w}_k[\mathbf{x_k} - \mathbf{x_{k-1}}]$. The desirability of a circular echo component is that it can be constructed entirely in the frequency domain with on the order of N multiplies, rather than the usual time domain complexity of N^2 multiples to implement a convolution on a long filter. Recognizing that noncircular echo components are caused only by the tail of the echo response that exceeds the guard-period duration in length, these echo components are the only ones that need be addressed in the time domain. So-called **frame synchronous** operation [71] can occur at one end of the transmission system, meaning DMT symbol periods are loop-timed at that end as well as the actual sampling clock. This configuration minimizes the CES complexity (but may increase it at the other end which is not frame synchronous). These few echo components typically require on the order of no more than one-quarter the complexity of a full time-domain echo canceler. The CES is the simple operation of using the known echo response $\mathbf{w}_k$ only to compute echo at those instants that deviate from a circular echo. Then, the echo canceler can be completed with the comparatively negligible N multiply-accumulates in the frequency domain. This second part of the echo canceler occurs after the receiver FFT. This is a simple adaptive echo canceler, typically implemented with the LMS algorithm:

$$W_{n,k+1} = W_{n,k} + \mu E_{n,k} \cdot X^*_{n,k}$$

$$(7.111)$$

A method to minimize the total computation at both ends of transmission was developed by Jones in [101].

The time-domain CES coefficients are then obtained periodically by IFFT. Since updating is not required for every symbol, the complexity may be distributed in time to achieve a sufficiently low real-time computational requirement.

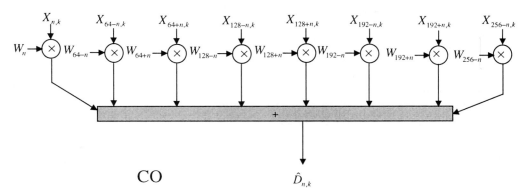

Figure 7.27   Asymmetric decimating DMT echo canceler for ADSL ATU-C.

For ADSL, as Ho et al. [71] further noted, the asymmetry can be exploited to reduce the computational complexity of echo cancellation further. At the CO side, the ADSL Transmission Unit at the Central Office (ATU-C) has a transmit ADSL-downstream sampling rate that is exactly eight times higher than the receive sampling rate. This means that the echo signal is aliased seven times on to a single tone n as illustrated in Figure 7.27. There is no simplification in the frequency-domain. However, the CES need only compute one of eight outputs using the echo-canceler coefficients derived from the ATU-C transmit signal. The same error signal must be used for all eight corresponding **echo coefficients**, which suggests caution in echo-canceler updating if one of the tone inputs (or more) have significantly less energy transmitted than the others. However, all should be updated or some leakage of echo-tap energy should be used. Figure 7.28 shows the mirror-image interpolating DMT echo canceler for the ADSL Transmission Unit-Remote (ATU-R). In this case each upstream transmit tone produces eight echoes on corresponding tones as shown. In this case, the CES reduces to 8 sub-CESs in parallel, each with the same 276 kHz upstream input repeated. This again reduces complexity by a factor of eight.

7.2.9.4 Isaksson's Zipper

A recent method of duplexing, specific to DMT, is known as the "zipper" [73]. Zipper uses synchronization of DMT frames in both transmit and receive directions through both loop timing and the use of a cyclic suffix (in addition to a cyclic prefix). By exactly synchronizing both symbol clocks in phase, then downstream tones on DMT will not interfere with different-index upstream tones. Thus, the assignment of tones to upstream and downstream is completely flexi-

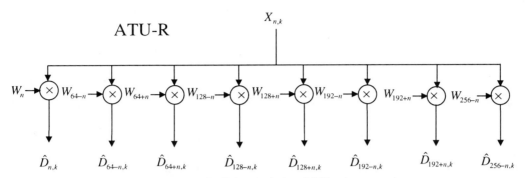

Figure 7.28 ATU-R interpolating DMT echo canceler.

ble, allowing even the possibility of every other tone being assigned to opposite directions, hence the name zipper. The cyclic suffix bandwidth penalty may or may not be of consequence and depends on the maximum delay of the forward transmission path. Also, analog dynamic range of outbound signals interlaced with inbound signals at either end forces ADC requirements to be the same as in echo cancellation. The zipper method may find increasing use in future DSLs.

7.2.10 Narrowband Noise Rejection

Narrowband noise is defined as having a bandwidth that is significantly less than the width of a multicarrier subchannel. Examples of narrowband noise are amateur and AM radio signals with respect to VDSL (where tone width is 43 kHz while amateur and AM radio signals have bandwidths of 2.5 kHz and 10 kHz, respectively). Narrowband noise under this condition appears qualitatively more deterministic than random. Under this situation only, it is possible to construct narrowband interference at one frequency from a measured amplitude at another frequency.[14] One method is **Wiese's Canceler** [42]:

The basic concept in cancellation is to model a sinusoid approximately as the function

$$n(t) \approx t \cdot e^{j2\pi f_o t} \tag{7.112}$$

---

14.   The reader is cautioned that more generally, synthesis of a truly random process at one frequency from another frequency is not possible. However, it is possible for a set of deterministic waveform shapes parameterized by a few random variables to be reconstructed at any frequency, once the parameters are estimated.

The Fourier transform of such a linearly time-varying exponential generally has the form

$$N(f) = \frac{A}{f - f_o} + \frac{B}{(f - f_o)^2},$$

(7.113)

which depends on three parameters, A, B, and f_o. Once these parameters are known, the noise at any frequency can be computed. Thus, RF cancellation reduces to estimation of these parameters from a DMT symbol and synthesis of the RF noise (to be then subtracted) at those used frequencies where it is significant. The parameters are estimated from frequencies that are intentionally not used (often because transmission at the same frequencies as RF interference is prohibited by a spectral mask associated with the DSL service). The best example is amateur radio frequencies with VDSL. A used frequency has index n so $f = \frac{n}{T}$, the interference is at frequency $f_o = \frac{(m+\delta)}{T}$, and

$$m < \delta < m + 1$$

(7.114)

where m and $m + 1$ are unused frequencies just below and just above the interferer. Then,

$$X_m = -\frac{A}{\delta} + \frac{B}{\delta^2}$$

(7.115)

and

$$X_{m+1} = \frac{A}{1 - \delta} + \frac{B}{(1 - \delta)^2}$$

(7.116)

which can be solved for A and B if X_m and X_{m+1} are known

$$\begin{bmatrix} A \\ B \end{bmatrix} = \begin{bmatrix} -1 & 1 \\ 1 - \delta & \delta \end{bmatrix} \begin{bmatrix} \delta^2 \cdot X_m \\ (1 - \delta)^2 \cdot X_{m+1} \end{bmatrix}$$

(7.117)

If windowing is also used, the effect of the intermodulation of the window needs to be removed from the values X_m and X_{m+1} (which is simple deconvolution since these frequencies are not used for transmission). Wiese [42] found simple division by the window complex amplitude at these frequencies to be sufficient. Usually, instantaneous values are used for the measurements on the unused tones because of the time variation of the interferer being such that the model used is valid individually only over one symbol. The use of averaging is still an area of research. The canceler also needs a value for δ, which amounts to determining the center frequency of the RF interference from measured values of DMT subchannel outputs. There are many approaches to the estimation of the location of a frequency in noise, see [74] and [75]. Weise [42] found the instantaneous approximation

$$\delta = \frac{\left|\Re\{X_{m+1}\}\right| + \left|\Im\{X_{m+1}\}\right|}{\left|\Re\{X_{m+1}\}\right| + \left|\Im\{X_{m+1}\}\right| + \left|\Re\{X_m\}\right| + \left|\Im\{X_m\}\right|} \tag{7.118}$$

to be sufficient to obtain 50 dB of cancellation for VDSL.

This type of RF cancellation is implemented digitally after conversion with an ADC. Unfortunately, RF signal levels may be so high that saturation of the conversion device can occur. Thus, some level of analog cancellation may also be desirable.

Analog RF Cancellation: Figure 7.29 generically depicts an analog RF noise canceler [76] that estimates RF noise and subtracts it from the differential phone line signals. The estimate of such RF noise is based on a reference RF noise signal that should be correlated with only the RF noise and not with the VDSL signals. The difficulty in such cancellation is the generation of this reference RF signal so that it is uncorrelated with DSL signals and does not contain other independent noise. Any reference signal containing significant VDSL signal components would result in degradation of the DSL link performance.

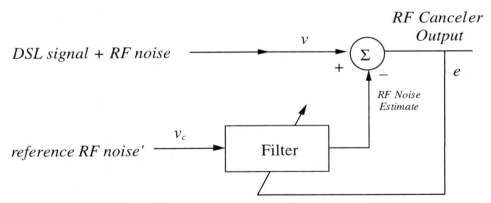

Figure 7.29 Illustration of RF cancellation.

Amateur radio signals change their frequency position often. This time variation imposes the additional constraint that any RF canceler be adaptive. Most adaptive filters will critically use the correlation between the reference and the RF canceler output to update the RF canceler settings. Ideally, the RF canceler output is the VDSL signal, complicating the above poor-reference problem. The synchronized DMT (SDMT) proposal of Section 7.2.9 describes a time-division "ping-pong" multiplexing of upstream and downstream DMT signals, consisting of a multisymbol superframe. Two of the DMT symbols in the superframe are silent for the purpose of facilitating the reversal of transmission direction on the phone line. Transients are much shorter than a full symbol length, leaving only RF noise and other disturbances on the line during the last part of these silent symbol periods. The analog RF canceler updates during these silent intervals when both the reference and the RF canceler output are free of DSL signals, which would have been severe noise to the RF canceler adaptive updating. The consequent clean updating can swiftly adjust RF canceler filter parameters to synthesize an exact replica of the RF noise in the differential signal. Updating is then off during any "nonsilent" DSL signal transmission on the remaining symbols of the superframe.

Radio frequency (RF) noise couples into phone lines because the two copper wires in the twisted pair act like antennae and are not quite identical. This "ingress" effect is primarily caused by an imbalance in common-mode impedance to ground for the pair. The twisting of the pair helps to balance the coupled components so that the differential-mode RF noise voltage between the wires has only a small fraction of the common-mode noise voltage induced in either element. However, the balance of the twisted pair, which measures the ratio of differential to common-mode (or vice versa) transfer is not perfect. At very low frequencies, common-mode imbalance can be as low as −60 dB. At the frequencies of interest for VDSL, the common-mode imbalance may only be −30 dB over most of the used band.

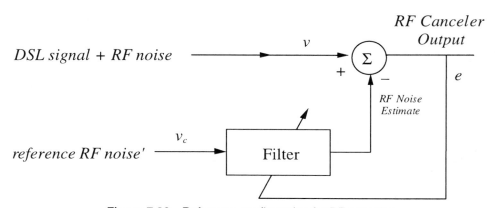

Figure 7.30   Reference configuration for RF canceler.

Figure 7.30 shows the configuration of the RF canceler under analysis. The line transformer differential output is the DSL signal + RF noise, v, and the center-tap signal (reference to chassis ground) is the reference RF noise, v_c. Typically, for RF noise, $v_c \gg v$. The RF canceler output is e, and the canceler filter is $w(f)$. The canceler filter may in the simplest implementations be just a single constant complex gain, in which case $w(f) = w$. All signals in this paper's analysis could be considered either time-domain signals (constant $v(f) = v$) or as functions of frequency (in the latter case the analysis as derived here applies independently to each frequency), and so we do not use frequency or time arguments on the various signals to simplify notation.

The differential signal is

$$v = s + k_c \cdot n \tag{7.119}$$

and the reference signal is

$$v_c = k_d \cdot s + n + u \tag{7.120}$$

u represents any extra uncorrelated noise that only appears in the reference signal, which may be broadband — in theory, u should be zero, but in practice it can be nonzero. Ideally, $k_c=k_d=0$, but in practice they are constants small in magnitude ($k << 1$), but not zero. The error is

$$e = v - w \cdot v_c = (1 - w \cdot k_d) \cdot s + (k_c - w) \cdot n - w \cdot u \tag{7.121}$$

from which a good setting for w would be correctly inferred as $w = k_c$, leading to only the very small noise $-k_c u$ in e. The minimum mean-square error (MMSE) setting for w minimizes the average squared value of the error signal. All signals shall have zero mean (no DC component) and the variance (power or power spectral density) of s is E_s while the variance of the noise n is σ^2. Then basic calculus can be used to determine that the MMSE setting for w is

$$w = \frac{k_d \cdot E_s + k_c \cdot \sigma^2}{k_d^2 \cdot E_s + \sigma^2 + \sigma_u^2} \tag{7.122}$$

which is not equal to k_c in general. The corresponding MMSE is thus

$$MMSE = \frac{E_s \cdot \sigma^2(1 + k_c^2 \cdot k_d^2 - 2 \cdot k_c \cdot k_d) + \sigma_u^2(E_s + k_c^2 \cdot \sigma^2)}{k_d^2 \cdot E_s + \sigma^2 + \sigma_u^2} \tag{7.123}$$

While these settings are those that would be attained by minimizing the mean square error with adaptive algorithms for this problem, transmission on the link would prefer $w = k_c$. Since $\sigma_u^2 << \sigma^2$ for RF noise of interest, then $w = k_c$ approximately occurs when $E_s = 0$ or when the RF noise is very large. $E_s = 0$ corresponds to no VDSL signal, hardly a situation desired while large noise is not guaranteed on every line and is itself undesirable from a transmission (not RF canceler) standpoint.

However, with SDMT, or ping-pong in general, the updating is sufficiently frequent to allow excellent RF rejection, as long as the updating interval occurs a few hundred to a few thousand times or more per second. The length of the filter in Figure 7.30 may need to be long to ensure narrowband cancellation at RFI frequencies, while simultaneously maintaining good balance and strong signal transfer at non-RFI frequencies.

7.3 Trellis Coding

Trellis codes use sophisticated examples of sequential encoders, usually with either PAM or QAM modulation. Straightforward use of PAM or QAM allows the detector and decoder to make a separate instantaneous decision on each symbol or message transmitted through the communication channel. Trellis codes map a series of groups of bits into a series of successive symbol transmissions to improve performance. This concept seems to disturb many DSL engineers at first. The difference between simple line coding and trellis coding might be compared to the difference between a letter and a word. There are 26 letters in the English alphabet, and an archaic means of speaking to another person might be to spell every word, rather than to say the word itself. If the listener were to hear any letter incorrectly, there is no way to correctly interpret the word without looking at the context (which he could not do with simple letters, or PAM/QAM, because he must decide which letter was sent before looking at the next letter). Trellis codes are analogous to introducing spoken language rather than spelling. Much more than 26 words can be formed, but the listener need not hear each letter correctly as long as he or she can infer the words from the context. Trellis codes similarly make it more reliable to decode a transmitted sequence of symbols, even though there is a much larger number of message possibilities that could have been conveyed.

Trellis codes are used to reduce the gap to capacity. At $P_e = 10^{-7}$, the **coding gain**, γ_c, of a trellis code is the amount by which the gap is reduced and typically measures between 3 and 5 dB for commonly used trellis codes. This improvement occurs with no increase in transmit power, but rather at the expense of the complexity associated with encoding several successive transmit symbols as a sequence of symbols and the consequent increase in decoder complexity to decide which of many possible sequences was most likely to have been transmitted. A very complete treatment of trellis codes appears in the classic articles by Forney [77] and [78], or the tutorial by the original trellis-code inventor, Ungerboeck [79]. Here, we will describe in detail only two codes and then enumerate through some tables the most popular related codes.

7.3.1 Constellation Partitioning and Expansion

The first step in trellis coding is the expansion of the constellation, typically by doubling the number of points. This text will investigate only one, two, and four-dimensional codes. The extra points actually reduce the distance (at fixed energy) in the constellation for a symbol, but the redundancy thus provided allows a more than offsetting increase in the distance between the closest of symbol sequences for the entire code. An example for $b = 3$ appears in Figure 7.31 where a four-state code for QAM appears in the form of a constellation and associated trellis. The labels on the trellis correspond to one of the four subsets into which the constellation with redundant points is partitioned, as indicated on the right in Figure 7.31, where the subset index is the second listed for a point in the constellation. Each subset has four points. To specify a constellation point, one follows the subset index (0, 1, 2, or 3) by the label of the point within a subset (also 0, 1, 2, or 3) in the example of Figure 7.31. So, the point 00 is then the 0th point in the 0th subset. A transmitted point is specified first by a subset index from the trellis encoder, determined by two output bits specifying that index. The two output bits (formed from one input bit and the encoder state) specify one of the four subsets. The remaining two input bits select a point within the selected subset for transmission. Let d represent the distance between points in the constellation with the extra points. An encoder circuit and generator matrix $G(D)$ also appear — the matrix $G(D)$ derives from the linear binary circuit in the obvious way with all addition implied as binary.

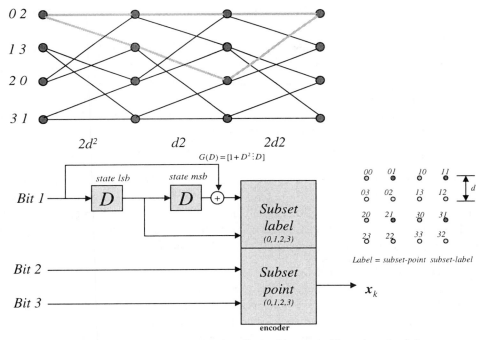

Figure 7.31 Constellation and trellis for four-state Ungerboeck code.

Through inspection of the trellis, one can determine that no two sequences (no matter what point is selected from the subset on each branch) can differ by less than $2d^2 + d^2 + 2d^2 = 5d^2$. An uncoded constellation would require only eight points and would then have distance $2d^2$. The decoder first decides which point in each subset is closest to the received sample, and then decodes the sequence of subsets with a Viterbi sequence detector. The first step of deciding the closest point within a subset has a distance of $4d^2$, which leads to a minimum distance improvement over uncoded of 3 dB, the coding gain for this code. Larger coding gain can be achieved by using more states (complexity) in the description of the code.

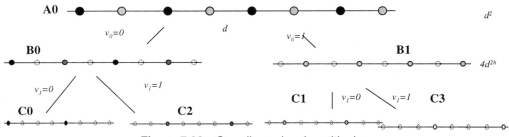

Figure 7.32   One-dimensional partitioning.

Figure 7.32 illustrates partitioning of one-dimensional constellations. Figure 7.33 describes two-dimensional partitioning constellations. At each step, that is from A sets to B sets, or from B sets to C sets, distance increases. Codes with greater than 3 dB of coding gain would need the C-level one-dimensional subsets or the D-level two-dimensional subsets. In each case, the number of points doubles in the constellation with respect to that needed for uncoded transmission, which means that the encoder uses an extra bit of redundancy to specify the subset. C-level partitioning leads to a rate ½ encoder and D-level partitioning leads to a rate $^2/3$ code, and so on.

Since the trellis encoder only deals with subsets, any integer-power-of-two-size constellation can be used in trellis coding as long as $b > 1$. Then the same code can be used at many different data rates and with a variety of constellation sizes that correspond to the different data rates. The only change is the trivial operation of encoding and decoding between the points within a particular subset, which corresponds to simple slicing in the receiver.

7.3.1.1   Four-Dimensional Codes

Trellis codes for symbol dimensionality greater than two usually lead to a lower complexity for the same gain and an improvement in constellation expansion. (Doubling constellation size across many dimensions leads to a smaller increase in lower-dimensional peak-to-average ratio, which is good when nonlinearity is of concern in the transmission path; the peak reduction occurs because fewer points occur in the lower-dimensional constituent dimensions.) ADSL uses a four-dimensional trellis code. The four-dimensional symbols are constructed by concatenation of two two-dimensional QAM symbols. The subset partitioning for this case is shown in Figure 7.34. Note that the distance increases by 3 dB for every two partitions in four dimensions. A C-level code has eight subsets and would thus have a rate $^2/3$ encoder.

The four-dimensional 16-state Wei code [77] is used in ADSL [63]. The trellis and corresponding encoder are detailed within [63]. This code has a nominal gain of 4.2 dB when nearest

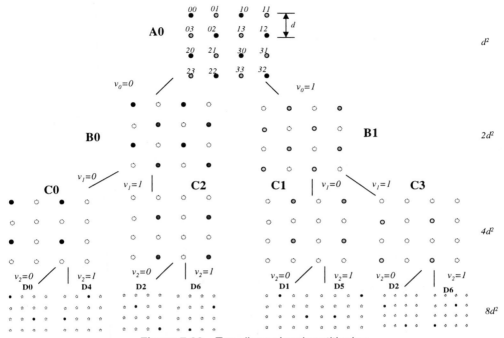

Figure 7.33 Two-dimensional partitioning.

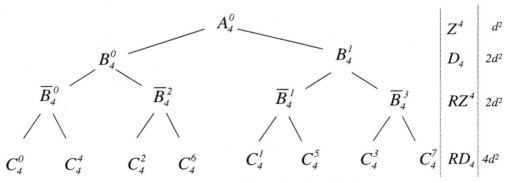

Figure 7.34 Four-dimensional partitioning.

neighbor effects are included. The eight four-dimensional subsets are comprised of concatenations of two-dimensional subsets as follows:

$$C_4^0 = \begin{bmatrix} C_2^0 & C_2^0 \end{bmatrix} \cup \begin{bmatrix} C_2^2 & C_2^2 \end{bmatrix}$$

$$C_4^4 = \begin{bmatrix} C_2^0 & C_2^2 \end{bmatrix} \cup \begin{bmatrix} C_2^2 & C_2^0 \end{bmatrix}$$

$$C_4^2 = \begin{bmatrix} C_2^1 & C_2^1 \end{bmatrix} \cup \begin{bmatrix} C_2^3 & C_2^3 \end{bmatrix}$$

$$C_4^6 = \begin{bmatrix} C_2^1 & C_2^3 \end{bmatrix} \cup \begin{bmatrix} C_2^3 & C_2^1 \end{bmatrix}$$

$$C_4^1 = \begin{bmatrix} C_2^0 & C_2^1 \end{bmatrix} \cup \begin{bmatrix} C_2^2 & C_2^3 \end{bmatrix}$$

$$C_4^4 = \begin{bmatrix} C_2^0 & C_2^3 \end{bmatrix} \cup \begin{bmatrix} C_2^2 & C_2^1 \end{bmatrix}$$

$$C_4^2 = \begin{bmatrix} C_2^1 & C_2^0 \end{bmatrix} \cup \begin{bmatrix} C_2^3 & C_2^2 \end{bmatrix}$$

$$C_4^6 = \begin{bmatrix} C_2^1 & C_2^2 \end{bmatrix} \cup \begin{bmatrix} C_2^3 & C_2^0 \end{bmatrix} \tag{7.124}$$

where $\cup$ means union and, for instance, subset C_4^1 points can be formed only by first sending a point in C_2^0 followed by C_2^1 or by C_2^2, C_2^3.

The trellis is terminated/initiated within a single DMT symbol. This requires three inputs to the trellis code where the nonparallel transition bits are nulled, reducing the data-carrying capacity of the symbol by 4 bits (16 kbps) in ADSL (see Ref. [63]). This loss is negligible downstream but can be appreciable for upstream transmission on very long loops. This loss suggests that trellis coding might best be avoided upstream on very long loops, although it is possible to essentially recover some of the equivalent energy loss through the decoder's gain.

7.3.1.2 Rotationally Invariant Trellis Codes

Voice-band modems also use trellis codes, but with a modification that renders decoding insensitive to arbitrary rotations by multiples of 90 degrees. This is necessary because carrier recovery loops in voice-band modems may recover from phase hits that cause an ambiguity in phase while operating. Voice-band modems operate over a carrier network where the symbol rate and carrier frequencies may be asynchronous, rendering the voice-band modem sensitive to such phase hits. In DSLs, the modems are directly connected to a twisted pair and the carrier and symbol rate are locked. As a result, there is no need for rotational invariance in DSLs, and it is usually not used. Furthermore, the trellis code has reinitialized every symbol in DMT ADSL to eliminate error propagation, while simultaneously making rotational invariance unnecessary.

7.3.1.3 Fundamental Coding Gain

The same trellis code may be applied to constellations with any integer number of bits (per 1D, 2D, or 4D symbol) with the only difference being the size of the subsets after partitioning. Since these subsets correspond to parallel transitions in the trellis code and the corresponding simple encoding and decoding, they can have any number of points without affecting the encoder or the Viterbi sequence detector. There will, however, be small differences in gain caused by the shape of the constellation.

Coding gain consists of two components, **fundamental gain** γ_f and **shaping gain** γ_s so that

$$\gamma = \gamma_f \cdot \gamma_s \qquad (7.125)$$

Fundamental gain is independent of the number of points in a constellation and instead measures the improvement in minimum distance relative to two-dimensional cross-sectional area of a decision region around a point. Fundamental gain then measures the number of points per unit volume, defined formally by

$$\gamma_f = \frac{d_{min}^2 \Big/ 2^{2\bar{b}+2\bar{r}} V_{\Lambda}^{2/N}}{d'^2_{min} \Big/ 2^{2\bar{b}} V_{\Lambda'}^{2/N}} \qquad (7.126)$$

where V_{Λ} is the volume associated with the decision region around any point in the N-dimensional symbol constellation based on some **lattice** Λ.[15] Λ' is the uncoded lattice with respect to which coding gain is computed, and r is the number of redundant bits in a symbol (typically 1 for trellis codes when the constellation size is doubled), and the overbar simply means divided by the number of dimensions. Energy should be proportional to total volume, which is roughly the number of points in the constellation times the volume of a decision region. However, because some points in a finite constellation are not surrounded on all sides by other points (those at the boundaries), then there is a shaping gain,

$$\gamma_s = \frac{2^{2\bar{b}+2\bar{r}} V_{\Lambda}^{2/N} \Big/ \bar{E}_{\Lambda}}{2^{2\bar{b}} V_{\Lambda'}^{2/N} \Big/ \bar{E}_{\Lambda'}} \qquad (7.127)$$

The fundamental gain is constant for a trellis code, independent of the constellation size, while the shaping gain is an adjustment factor for the shape of the constellation. The maximum value for shaping gain with respect to hypercubic (SQ QAM) constellations is 1.53 dB — typi-

---

15. A lattice (in this context) is an infinite set of N-dimensional vectors that is closed under addition, essentially the points in a constellation extended to infinity.

cally, shaping gain is very small. Fundamental gain can be as high as 6 dB for the most powerful trellis codes.

7.3.2 Enumeration of Popular Codes

The fundamental gains and parity polynomials for many popular codes have been tabulated and are shown in Table 7.1. The number of states, fundamental gain, effective fundamental gain (a reduction of 0.2 dB for every factor of 2 in nearest neighbors about the PAM minimum of 2 per dimension), and parity polynomial are listed for several codes. The convolutional encoder that describes the trellis is either rate $\frac{k}{k+r}$ for most codes, and since $r = 1$, the parity polynomial more compactly specifies the encoder. One infers the encoder matrix $G(D)$ by taking $H(D)$, dividing each entry by the last and calling the new parity matrix $[h(D) \quad 1]$. The corresponding encoder realization is described by $G(D) = [I \quad h^T(D)]$. For instance, in Figure 7.31, the parity matrix is

$$H(D) = 25 = \begin{bmatrix} D & D^2 + 1 \end{bmatrix} \rightarrow \begin{bmatrix} \dfrac{D}{D^2 + 1} & 1 \end{bmatrix}$$

(7.128)

and the encoder matrix is

$$G(D) = \begin{bmatrix} D^2 + 1 & D \end{bmatrix} \rightarrow \begin{bmatrix} 1 & \dfrac{D}{D^2 + 1} \end{bmatrix}$$

(7.129)

and the leftmost bit is the most-significant bit in the subset index. Many equivalent realizations all lead to the same set of codewords, with a systematic encoder realization often preferred.

Underlined entries indicate that a distance other than the minimum distance dominates probability of error at 10^{-6} because of a large number of next-to-nearest neighbors.

7.3.3 Shaping Effects

The shaping gain of a constellation can be improved separately from the fundamental gain of a trellis code. Such methods typically offer 1 dB of further improvement in performance. The two most popular methods are **trellis shaping** [81] and **constellation mapping** [82], [83]. Neither is used in DSLs because a significant complexity increase that characterizes their implementation provides only a small gain. In addition, the flexible precoder of Section 7.1.2.5 must be used if DFEs or GDFEs are also present, and this adds complexity and increases the peak-to-average ratio.

Table 7.1 Tabulation of Trellis Codes

One and Two-Dimensional Code Listing

2_ν	h_2	h_1	h_0	d^2_{min}	γ_f	(dB)	$\bar{N}_e$	$\bar{N}_1$	$\bar{N}_2$	$\bar{N}_3$	$\bar{N}_4$	γ_f	$\bar{N}_D$	N
One-Dimensional Codes														
4	-	2	5	9	2.25	3.52	4	8	16	32	64	3.32	12	1
8	-	04	13	10	2.50	3.98	4	8	16	40	72	3.78	24	1
32	-	10	45	11	2.75	4.39	12	28	56	126	236	3.87	96	1
64	-	032	135	14	3.50	5.44	8	32	66	84	236	4.38	204	1
128	-	052	341	14	3.50	5.44	4	8	14	56	136	4.63	384	1
256	-	336	755	16	4.00	6.02	14	0	108	0	484	4.87	768	1
512	-	0556	1461	16	4.00	6.02	2	0	44	0	248	5.10	1536	1
1024	-	1512	2461	16	4.00	6.02	2	0	4	28	68	5.26	3072	1
2048	-	2202	4105	16	4.00	6.02	2	0	0	16	12	5.44	6144	1
Two-Dimensional Codes														
4	-	2	5	4	2	3.01	2	16	64	256	1024	3.01	8	2
8	04	02	11	5	2.5	3.98	8	36	160	714	3144	3.58	32	2
16	16	04	23	6	3	4.77	28	80	410	1952	8616	4.01	60	2
64	036	052	115	7	3.5	5.44	20	126	496	2204	10756	4.78	228	2
256	274	162	401	7	3.5	5.44	2	32	124	552	2732	5.22	900	2

4D Trellis Codes

2_ν	h_4	h_3	h_2	h_1	h_0	d^2_{min}	γ_f	(dB)	$\bar{N}_e$	γ_f	$\bar{N}_D$	N
16	--	--	14	02	21	4	$2^{1.5}$	4.52	6	4.20	36	4
64	050	030	014	002	101	5	$5/2^{.5}$	5.48	36	4.65	524	4

7.3.4 Turbo Codes

Turbo codes were introduced by Berout [22]. They can exhibit higher gain than trellis codes, but have the drawback of applicability in each case to a fixed number of bits per symbol. These codes are based on the basic information-theoretical concept that codes selected at random (with very large number of states) will often be very good, which has now been verified by various tested turbo codes. The "turbo" comes from an analogy to turbo car engines where a suboptimal iterative decoder is made that estimates the probability of symbol (or bit) values rather than the actual values themselves. The algorithm iterates the probability distribution using extrinsic information based on one of two (or more) instances of a code within the transmitted bit stream (separated by interleaving), thus feeding back ("turboing") on itself. Good catenated coding systems (employing both trellis codes and block error correction) can achieve the same gains. In both cases, delay is increased significantly for ever-diminishing coding gain. ADSL uses a simple example of concatenation. Zogakis has shown that more careful concatenation can lead to yet higher coding gain [84] than achieved in ADSL with consequent complexity increase and delay enlargement. This area still remains active for DSL, and fundamental theory and some results in general for turbo codes suggest that DSLs with latency increase could operate within 1 dB (well over 90%) of the theoretical limits of capacity. The authors expect that revised versions of DSL standards will eventually use such methods as VLSI advances rapidly make the cost of such decoders negligible. The use of turbo codes with DMT requires special care [85].

7.4 Error Control

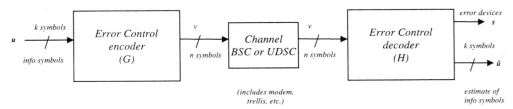

Figure 7.35   Error control coding.

Error control mechanisms work just outside the physical layer modulation to either detect and or correct bit errors in blocks of transmitted bits. All DSLs use some form of error control. ISDN, HDSL, T1 and early digital services all detect the presence of most bit errors within transmitted blocks of information. ADSL, VDSL, and ADSL-LITE additionally use forward error correction, which means they correct some of the errors after detecting their presence. A detailed treatment of error control is the subject of many fine textbooks (e.g., Refs. [86], [87], or [88]).[16] This section provides a basic description and analysis of error control in DSL, allowing for the reader to design error control within a DSL transmission system and achieve error-control objectives. However, detailed optimization of individual error-control circuits or justification for error-control choices made in DSL are beyond the scope of this book.

The designer of error-control systems models the entire channel-plus-modem (including trellis codes if used) by a simple probability of a bit being in error or not, p, as in Figures 7.35 and 7.36. The process of **error detection** in DSLs provides indication of whether or not bit errors have been made in a block of bits transmitted through the modem-plus-channel as in Section 7.4.6. Higher-level DSL management functions use the indication of detected errors to locate and service faulty DSL equipment. Error detection appends additional check bits to blocks of transmitted bits. These extra bits allow very reliable detection of errors. **Error correc-**

---

16.   [88] has interesting finite-field analogies to the FFT and cyclic prefixes also used in DMT, which may be of
        interest to a motivated reader

tion provides additional performance gain and can correct many bits that were in error through the modem plus channel. Error correction and error detection have a price in terms of additional bytes added for correction/detection purposes. A classic use of error correction in DSLs is to correct the occasional impulse noise disturbance that causes a burst of errors at the modem-plus-channel level, as described in Section 7.4.3.

Both error detection and error correction in DSLs to date only use of a powerful subclass of error-control coding methods known as **cyclic codes.** DSL transmission systems implement error control in nonoverlapping blocks (i.e., convolutional codes are not used). Thus, bits in error are corrected within a definite boundary without effect in other preceding or succeeding blocks. DSL error detection uses binary cyclic codes known as **cyclic redundancy checks or CRC methods**, as in Section 7.4.6. DSL error correction instead establishes blocks of bytes (groups of 8 bits) and corrects erroneous blocks, using the cyclic codes known as **Reed-Solomon** codes. Reed-Solomon code arithmetic executes in **Galois Field 256 (GF256)** and allows up to 16 errored bytes in a codeword of up to 255 bytes to be corrected. The power of RS error correction can be multiplied for impulsive error bursts by interleaving the errors into adjacent codewords as described in Section 7.4.3. The concatenation of error control coding and trellis coding for DSL is briefly discussed in Section 7.4.4. Section 7.4.5 provides insight into the somewhat sensitive issue of delay and code overhead, particularly for RS codes used in DSL.

Scrambling methods are simple rate 1 binary codes used with the expectation that long strings of zeros or ones in transmission occur more often than they should — the scrambler reduces the probability of such occurrence when this assumption is valid. Scramblers are described in Section 7.4.7.

7.4.1 Basic Error Control

The concept of a symbol remains in error control as in the general channel model of Section 6.1, where the channel conditional probability distribution is discrete (rather than continuous as with the AWGN channel earlier). Each symbol represents b successive bits. For DSL error control, either $b = 1$ ($M = 2$) or $b = 8$ ($M = 256$). For the case of $b = 1$, error control designs assume that channel bit errors are independent of one another and that the channel is modeled by the **binary symmetric channel** (Figure 7.36). The BSC model replaces the modem-plus-channel, where a bit error is made with probability p. This **crossover probability** is the $\overline{P_b}$ of the inner channel. The **universal discrete symmetric channel** (UDSC) model of Figure 7.37 models the modem-plus-channel for the case of $b = 8$. In the UDSC, the probability of a bit error is still p, but the probability of a byte error is then $p_s = 1 - (1 - p)^8 \approx 8p$, and the probability of receiving a byte correctly is then $(1 - p)^8 \approx 1 - 8p$. Any of the individual transitions in the UDSC has thus probability $\frac{8}{255} p$. In general, we will call the probability of a symbol error for either the UDSC or the BSC p_S, with $p_S = p$ in the BSC case.

A codeword consists of n consecutive symbols, of which k carry information and the remaining $r = n - k$ are extra check symbols added by the encoder. These extra parity or check bits/bytes (symbols) assist in the detection or correction of errors. DSL transmission systems

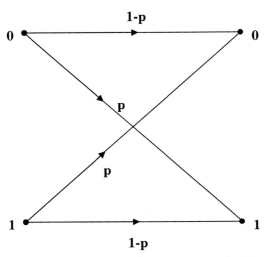

Figure 7.36   Binary symmetric channel (BSC).

keep the extra bytes to a minimum to avoid performance loss. In DSL, this number is typically below 15% additional bits for error control, which almost always ensures that more is gained from the code than is lost in the extra overhead bits.

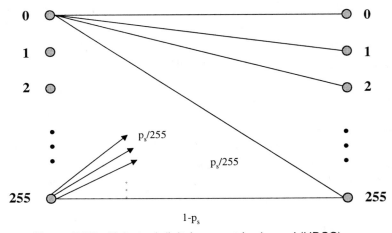

Figure 7.37   Universal digital symmetric channel (UDSC).

For error control, the **free distance,** d_{free}, is the number of symbol positions in which the two closest codewords differ. In general, the number of symbol positions in which two identical-length blocks of symbols, denoted by vectors v and $\bar{v}$, differ is known as the **Hamming distance,** $d_H(v,\bar{v})$. The related **Hamming weight** is the Hamming distance of any block from the all-zero symbol block, $w_H(v) = d_H(v,0)$. The free distance is thus the minimum Hamming distance between any two codewords. This quantity is analogous to d_{min}, the Euclidean distance between any two channel input vectors, for the AWGN channel. d_{free} measures the performance gain of the code. Higher d_{free} means more errors can be detected and/or corrected. If the closest codewords differ by d_{free} positions, then up to $d_{free} - 1$ errors can be detected and up to

$$\left| \frac{d_{free} - 1}{2} \right|$$ errors can be corrected. (Investigation of the probability distribution for the channel shows that picking the codeword that differs in the least number of positions from the received symbol sequence is indeed a ML detector for either the bits in a BSC or the bytes in a UDSC.)

For **linear codes**, the set of 2^k codewords is closed under n-dimensional addition.[17] That is, any two codewords sum to another codeword and in particular the set of codewords is the same as the set of possible errors that can be made, thus facilitating the structure and analysis of linear codes. DSLs use only linear codes. The row vector of n output symbols, $v = \begin{bmatrix} v_0 & v_1 & \dots & v_{n-1} \end{bmatrix}$, of the encoder are related to the row vector of k input symbols (bits or bytes), $u = \begin{bmatrix} u_0 & u_1 & \dots & u_{k-1} \end{bmatrix}$, by the linear matrix relationship.

$$v = u \cdot G \tag{7.130}$$

for a linear code, thus establishing the $\left(2^b\right)^k$ codewords as all the possible linear combinations of the rows of G, the rank-k $k \times n$ **generator matrix**. DSLs to date exclusively use systematic codes, which have the special generator structure

$$G = \begin{bmatrix} I_k & h^* \end{bmatrix} \tag{7.131}$$

meaning that the first k output symbols/bits are equal to the k input symbols. The reader should note that in DSLs, largely developed by digital signal processing persons, the time index is such that v_0 is the first bit transmitted in a block and v_{n-1} is the last. Coding theorists, on the other hand, almost all send v_{n-1} first and reverse the labeling in the blocks — this is largely semantics but causes enormous confusion and bugs in a variety of implementations. This DSL text will use the lowest time-index first notation and will describe where appropriate how to translate to the vast literature on primitive polynomials and cyclic-code polynomials to avoid confusion.

---

17. This addition is binary element-wise for binary codes and GF256 addition (see Section 7.4.2) for byte-oriented codes.

A second rank-$(n - k)$, $(n - k) \times n$ **parity matrix** H satisfies the equation $GH^* = 0$ and equivalently characterizes the code in that all codewords satisfy $v \cdot H^* = 0$. H itself is a generator matrix for a dual code in which every codeword is orthogonal to all the codewords of the original code generated by G. The parity matrix is very important in decoding. The free distance of the linear code can be found by the smallest number of columns of H that sum to zero. The parity matrix for a systematic linear code is trivially

$$H = \begin{bmatrix} h & I_{n-k} \end{bmatrix}$$

(7.132)

thus making the free distance equal to $d_{free} = n - k + 1$. Thus, a linear systematic code with even rank-$[(n - k) = 2t]$ parity matrix can correct t errors and detect $2t$ errors, a result often used in DSLs.

The generator essentially then describes the encoder and associated symbol processing, which can be implemented directly as a matrix multiply with arithmetic in the appropriate field (binary GF2 or byte-level GF256). However, a linear code can easily be simplified into an implementation involving rudimentary addition ("exclusive-or" for binary) and multiplication ("and" for binary) gates. GF256 addition is eight-dimensional vector addition of bytes, while GF256 multiplication is more complicated in general, but reduces to a circuit with a handful of gates for most multiples in coding where one of the operands to the multiplication is fixed and known as a quantity in G (or H). For systematic codes, only $(n - k)$ of the output symbols require logic, which usually simplifies the implementation.

The receiving decoder receives as an input the n-dimensional channel output vector y. Decoding is achieved (at least in concept) by computing an $(n - k)$ row vector **syndrome** for a received channel output vector of values $s = y \cdot H^*$. If there were no symbol errors in transmission, the syndrome is zero in all its $(n - k)$ positions. A nonzero syndrome indicates that at least one symbol error has occurred — such a situation corresponds to a **detected error.** The syndrome may also be zero if symbols are such that the received vector y is exactly equal to another codeword, which corresponds to an **undetected error.** The syndrome is also key to decoding through a ML receiver for a linear code. Given the syndrome, all ML receivers essentially solve the following equation

$$s = e \cdot H^*$$

(7.133)

for the value of e that has minimum Hamming weight. That particular value of e may be subtracted from y to obtain the best estimate of the transmitted codeword. The probability of symbol error in the receiver is then the probability that y is closer to another codeword than the one sent. Thus, with error control, it is possible to estimate both if an error has occurred as well as to correct it — the former is easier and has a higher reliability associated in the form of a lower probability of undetected error than the probability of a symbol error made by the receiver.

One conceptual method for decoding linear codes is to solve the syndrome equation in advance for all possible syndrome vectors and then store the results in a table of size $\left(2^b\right)^{n-k}$ locations, a **tabular decoder.** When $b=1$ for binary codes, this is usually a reasonable decoding strategy. However, for the Reed Solomon codes of ADSL, LITE and VDSL have $b = 8$, making table look-up far too consumptive of memory to be viable (at least in this straightforward manner).

In DSL, binary codes are typically used only for error detection with a nonzero syndrome indicating that an error has occurred. These binary codes are not used for correction. For error correction, RS codes with $b = 8$ are instead used.

The probabilities of undetected error and of symbol error (as well as the associated bit error) are important for any coded system. Fortunately, approximations are relatively easy to compute. Exact expressions require the knowledge of the Hamming weight distribution of all codewords in a code.

The probability of an undetected error sequence is upper bounded by

$$P_u \leq \sum_{i=d_{free}}^{n} \binom{n}{i} \cdot p_S^i \cdot \left(1 - p_S\right)^{n-i}$$

(7.134)

which unfortunately is not very tight. The weight distribution for a linear code is given by $W_i \quad i = 1,...n$, where W_i is the number of codewords with Hamming weight i. This distribution may be difficult to find, but when it is known, then the exact probability of an undetected error is

$$P_u = \sum_{i=d_{free}}^{n} W_i \cdot p_S^i \cdot \left(1 - p_S\right)^{n-i}$$

(7.135)

Unfortunately, this distribution does not appear to be known for the codes used in DSL.[18] The probability of detected error can be upper-bounded fairly tightly by the expression

$$P_d \leq 1 - \left(1 - p_S\right)^n$$

(7.136)

namely, 1 minus the probability that all the symbols are correctly detected by the inner channel. The exact expression, tacitly using again the weight distribution, is

$$P_d = 1 - \left(1 - p_S\right)^n - P_u$$

(7.137)

---

18.   At least the authors are not aware of it, nor was it discussed or presented in any of the public deliberations that led to the choice of these codes for DSL.

This latter expression is then the probability that an error did occur and it was detected, while P_u measures the more ominous event that errors occurred and went undetected, which is usually smaller for a reasonable code design. Both P_d and P_u measure probabilities of decoding error associated with a block of k transmitted information symbols (corresponding of course with coding to $n \geq k$ channel symbols). Of more interest usually are the probabilities of individual bit errors, which are upper- and lower-bounded through division by 1 and kb, respectively. Exact analysis requires a yet more detailed knowledge of the total number of input bit errors corresponding to all codewords of Hamming weight i, B_i, and replacing W_i by $B_i \big/ k$ in the exact expressions above. Some analyses instead use the more optimistic factor of $B_i \big/ n$, arguing that the error bits are actually distributed across a systematic codeword and that some of them may be parity bits that are not passed to the receiver. We prefer the use of the factor k here.

The exact formulas for symbol error and codeword error probability by the receiver are more involved than those for detection of an error. Fortunately, the codeword error bound remains simple, fairly accurate, and easy to compute

$$P_e \leq 1 - \sum_{i=0}^{\left\lfloor \frac{d_{free}-1}{2} \right\rfloor} \binom{n}{i} \cdot p_S^i \cdot (1-p_S)^{n-i} = \sum_{i=\left\lfloor \frac{d_{free}+1}{2} \right\rfloor}^{n} \binom{n}{i} \cdot p_S^i \cdot (1-p_S)^{n-i} \qquad (7.138)$$

More complete bounds occur in Chapter 10 of Ref. [86]. The probability of bit error is estimated by multiplying each term by an estimate of the number of bit errors and then dividing by the number of bits per codeword on the right-hand side, producing

$$\overline{P}_b \approx \sum_{i=\left\lfloor \frac{d_{free}+1}{2} \right\rfloor}^{n} \left(\frac{i \cdot 4}{8 \cdot k} \right) \cdot \binom{n}{i} \cdot p_S^i \cdot (1-p_S)^{n-i} \qquad (7.139)$$

Here we have assumed that a symbol is a byte and that a symbol error leads to 4 of the 8 bits in error on the average, while $8k$ is the number of information bits per codeword. Alternative expressions that are roughly the same appear in Refs. [89] and [90].

7.4.2   Reed-Solomon Codes

The table look-up that converts syndromes to **error-locating vectors** becomes far too complex for the implementation of powerful codes. Reed-Solomon (RS) codes are used in DSL because they provide coding gain against random errors (about 3 dB if correctly used) and also, with the interleaving of Section 7.4.3, allow correction of large error bursts caused by impulses. RS codes have symbols that are bytes, not bits, and thus are more complex than binary codes. They are, however, some of the most powerful error-correcting codes known and can be implemented with reasonable complexity. RS codes use an even number of parity bytes, $2t$, and correct up to t bytes as well as detect $2t$ bytes in error. When the channel can provide indications that channel-output symbols are likely to be in error, RS codes can then actually correct up to $2t$ such "erasures."

Reed-Solomon codes are special cases of what is generally called **cyclic codes**. In cyclic coding, additional codewords can be found by the $(n-1)$ possible circular shifts of any given first codeword. In some sense, this reduces the apparent complexity of the code space by a factor of n in some very realistic ways that lead to simpler decoding (see, e.g., Refs. [86–88]). We will focus here only on specifics necessary to understanding the RS codes of DSL.

7.4.2.1 GF256

All arithmetic in the RS codes for DSL is byte-wise in the finite field GF256. GF256 is an extension field of binary arithmetic and has well-defined addition and multiplication properties. Every symbol in a RS code for DSL is a byte, and thus the matrix arithmetic associated with those bytes/symbols being added or multiplied by other bytes/GF256 field elements in the generator matrix (or its equivalent) needs to be defined.

The 255 nonzero elements in GF256 can be viewed as successive powers of a primitive element, allowing enumeration of the field as

$$GF256 = \left\{0 \quad 1 \quad \alpha \quad \alpha^2 \quad \alpha^3 \quad \dots \quad \alpha^{254}\right\} \tag{7.140}$$

where the primitive element α is such that $N = 255$ is the smallest integer such that $\alpha^N = 1$. It can be shown that the element α can be viewed as a root of the **binary primitive polynomial**

$$\alpha^8 + \alpha^4 + \alpha^3 + \alpha^2 + 1 = 0 \tag{7.141}$$

This statement requires some qualification for the designer perhaps not familiar with Galois fields and coding: When arithmetic in the above equation is binary, then α is considered a place-holding dummy value — that is, α is not binary (not equal to 0 or 1), but just a polynomial variable. The primitive polynomial has in fact neither 0 nor 1 as a root and indeed cannot be factored with binary arithmetic. However, when viewed instead in GF256 with GF256 multiplication and addition (as yet to be defined), α and indeed $\left\{\alpha^2, \alpha^4, \alpha^8, \alpha^{16}, \alpha^{32}, \alpha^{64}, \alpha^{128}\right\}$ are roots of the equation, and it is factorable in GF256 into the product of factors of $\left(x - \alpha^i\right)$ terms for each of these eight roots. Then the equation tells us that

$$\alpha^8 = \alpha^4 + \alpha^3 + \alpha^2 + 1 \tag{7.142}$$

in GF256. (Any element in GF256 is its own additive inverse so the subtraction above is the same as addition, like with binary addition in GF256.)

The reproduction of α^8 in terms of lower powers of α according to the last equation means that all the elements of GF256 can be represented in terms of polynomials with binary coefficients in powers of α less than 8. That is, the designer can associate any byte with an element in GF256 by considering the individual bits in the byte to be the binary coefficients in a vector (or polynomial) with powers $\alpha^7, \alpha^6, ... \alpha^0 = 1$. Addition in GF256 is simple binary vector addition element by element to get a new byte or eight-dimensional binary vector/polynomial. Multiplication is conceptually simple and is polynomial multiplication, with substitution through the primitive polynomial solution in (7.142) for any powers of α exceeding 7. Generally, the implementation of such multiplication requires consideration of $2^{16} = 65,536$ combinations that might be stored in a look-up table of that many bytes (which may not sound too bad with the price of memory today), but there are many ways to simplify further the description of multiplication as used in RS encoders and decoders.

The most heavily used simplification for GF256 multiplication is that one of the coefficients is fixed in the multiply — the one associated with the known generator G or the known parity matrix H. Thus, the look-up table simplifies to 256 locations for each such multiply. Typically, further simplification is possible by simply investigating the multiplication: for instance, multiplication of the fixed element $\alpha^2 + 1$ (1010000) by any element in GF256 $\alpha^i = a_7 \cdot \alpha^7 + a_6 \cdot \alpha^6 + ... + a_0$ produces

$$\left(\alpha^w + 1\right) \cdot \alpha^i = \left(a_7 + a_5\right) \cdot \alpha^7 + \left(a_6 + a_4\right) \cdot \alpha^6 + \left(a_7 + a_5 + a_3\right) \cdot \alpha^5 + \left(a_7 + a_6 + a_4 + a_2\right) \cdot \alpha^4$$
$$+ \left(a_7 + a_6 + a_3 + a_1\right) \cdot \alpha^3 + \left(a_6 + a_2 + a_0\right) \cdot \alpha^2 + \left(a_7 + a_1\right) \cdot \alpha^1 + \left(a_6 + a_0\right)$$

$$(7.143)$$

that the reader will note is implemented with a handful of binary adders. Thus, multiplication of a fixed value in GF256 by any other value is easy. Only a few such multiplications (per symbol) are necessary in the encoder of a GF256 code.

7.4.2.2 Cyclic Encoders

Any system cyclic code can have its codewords described by the remainder of the division kth-degree message polynomial

$$m_0 \cdot Z^{k-1} + m_1 \cdot Z^{k-2} + ... + m_{k-1}$$

$$(7.144)$$

(where each element m_i $i = 0,...k-1$ is an information symbol) by an $(n - k) = r$th-degree generator polynomial

$$g(Z) = g_0 \cdot Z^{r-1} + g_1 \cdot Z^{r-2} + ... + g_{r-1}$$

$$(7.145)$$

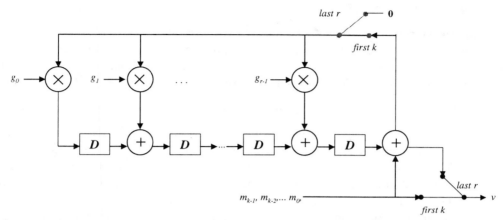

Figure 7.38   Encoding circuit for systematic cyclic code with generator *g*.

The codeword consists of the symbols m_0 (sent first) through m_{k-1} followed by the r successive symbols of the remainder with the symbol corresponding to Z^{r-1} sent first. For RS codes, all arithmetic in the polynomial division is in GF256 and $r = 2t$, where t is the number of bytes that can be corrected. The circuit in 7.38 implements the polynomial division exactly and in fact contains the remainder after k clock cycles in the delay elements. This remainder can be clocked from the circuit in r clock cycles with the feedback path zeroed. Note that each polynomial multiplication in the circuit is of the type where one of the GF256 elements is fixed and known, allowing for the simple implementation with a few gates.

Note that the division is described in terms of shift registers corresponding to D in the notation of this book. Coding theorists describe the polynomials for the generators in terms of "advances" instead of delay elements, so one can infer the division by $g(Z)$ by substituting $D = Z^{-1}$ in the circuit transfer function. (A caution to the reader: the editors of the ADSL and LITE standards have used the notation D inconsistently in their documents. When referring to cyclic codes for detection or correction, their D is an advance, while elsewhere their Ds are delays. This rather annoying practice pervades the industry and has been the source of more than one designer's despair upon debugging his circuits.)

7.4.2.3   Computing Syndromes and Decoding

The subject of decoding is well studied and somewhat complex for RS codes. The detection of errors is achieved by recomputing the parity bits in the receiver and comparing to those transmitted. If they are not equal, an error is declared. This difference in parity bits is the syndrome for the code. RS cyclic codes have a wealth of structure that allows enormous reduction in complexity in converting the syndrome and all its cyclic shifts into an error-locator polynomial that tells the position of errored bytes. Additional work allows the magnitude of the error to be computed and thus corrected. (For more information, see Refs. [86–91]).

7.4.3   Interleaving Methods

Interleaving is a reordering of transmitted bytes over a block of L codewords so that adjacent bytes in the transmitted data stream are not from the same codeword. Bytes are reordered at the receiver. Errors caused by impulsive disturbances in a DSL are concentrated in a **burst** of bit/byte errors. The duration of the burst can exceed the correctable number of errors $t = \dfrac{d_{free} - 1}{2}$ of the code. Thus, interleaving distributes the consequent bit/byte errors of the burst over a longer interval in time. Specifically, with good interleaving, the receiver deinterleaver leaves no deinterleaved data segments containing more errors than can be corrected. Since impulse noise (see Section 3.6) tends to occur with relatively long interarrival times, the errors in the burst can be distributed into adjacent codewords that have no such errors, effectively "evening" the load of error-correcting work among the codewords.

Block interleaving is the simplest type of interleaving. Figure 7.39 illustrates a block interleaver. Each successive codeword is written into a corresponding register/row in the interleaver. The number of codewords stored is called the **depth** L of the interleaver. If a codeword has N bytes and the interleaver depth is L, then LN total bytes must be stored in each of two transmit memories of a block interleaver as shown for $L = 3$ and $N = 4$. As $N = 4$ bytes at a time are written into each row of one of the transmit memory buffers, $L = 3$ bytes in each column are read. Bytes occur every T' seconds, making the byte clock $\frac{1}{T'}$. The interleaver input clock is thus $\frac{1}{N}$ of the byte clock rate. Thus, one codeword of $N=4$ bytes is written into a each row of the write memory buffer. Over $LN = 12$ byte periods, the entire write buffer is full. The transmit-memory output clock is $\frac{1}{L}$ of that same byte clock. The read-memory buffer is read $L = 3$ bytes from each column for each period of the interleaver output clock. The interleaver starts/completes writing of the write buffer at exactly the same two points in time as it starts/completes reading the read buffer. Every LN byte-clock period, the read and write memories are swapped.

The deinterleaver in the receiver accepts bytes from a decoder and writes them in terms of L bytes per column successively. After all LN bytes have been stored, the codewords are read in horizontal or row order. Again two memories are used with one being written while the other is read. The end-to-end delay is (no worse than) $2LN$ byte times and correspondingly there is a total of $4LN$ RAM locations necessary (at receiver and transmitter), half of which are in the receiver and the other half in the transmitter.

A burst of B errored bytes[19] is distributed roughly evenly over L codewords by the deinterleaving process in the receiver. If this is the only burst within the total LN receiver bytes, then the receiver FEC can correct approximately L times more errored bytes. Larger L means more memory and more delay, but greater power of the FEC as long as a second burst does not

---

19.   The value for B is computed as $\left(\frac{R}{8} \right) \cdot \tau_{burst}$ where τ_{burst} is the length of the impulse/error burst in seconds and R is the data rate of the DSL.

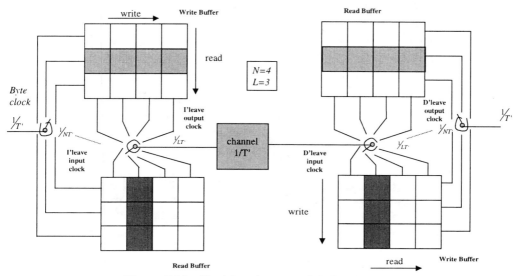

Figure 7.39   Block interleaver and de-interleaver.

occur within the same *LN* bytes.[20] The minimum distance of the code is thus essentially multiplied by *L* as long as errors are initially in bursts that do not occur very often. Such bursts correspond exactly to impulse noise bursts in DSLs.

Block interleavers, while easy to understand and describe, are not very efficient in terms of their use of memory nor the consequent delay they introduce. Convolutional interleavers halve the delay and reduce memory by a factor of 2to 4 for the same distribution of burst errors. DSL uses only convolutional interleaving.

To understand convolutional interleaving, the bytes within a codeword are numbered from $i = 0,..., N-1$. Again, a burst of errors is to be distributed over *L* (the depth) codewords, so each codeword also has an index of $l = 0,..., L-1$. In order for any burst to be distributed over *L* codewords, it is sufficient that *N* and *L* be coprime (i.e., have no common factors). If a desired *L* and *N* are not coprime, DSLs insert a dummy byte. For instance, in ADSL, $L = 2^d$ (a power of two) and *N* can be even. Then the dummy byte brings the codeword length to $N + 1$ (an odd number), which then is coprime with *L*. (The dummy byte need not be transmitted and costs nothing in data rate as long as the receiver knows to reinsert it.)

The *i*th byte in a codeword is always delayed in the convolutional interleaver by

$$\Delta_i = i(L-1) \qquad (7.146)$$

20.   For interarrival times of impulses, see Chapter 3, Section 6.

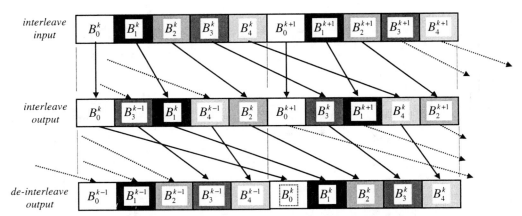

Figure 7.40   Illustration of byte ordering for convolutional interleaving and corresponding deinterleaving; $L = 2$, $N = 5$.

byte clocks as shown in Figure 7.40. Thus, the first byte $(i=0)$ in a codeword is not delayed at all while the last byte is delayed by $(N-1) \cdot (L-1)$ byte clocks. The deinterleaver reorders the bytes by delaying each byte by a varying amount that causes the total delay for any byte through both transmitter and receiver to be constant at $(N-1) \cdot (L-1)$.

The memory requirement of the straightforward implementation of the convolutional interleaver or de-interleaver is that of storing L, N-byte codewords, or LN bytes. Thus, the convolutional interleaver requires no more than half the memory of a block interleaver and furthermore has slightly less than half the delay. Error bursts are nevertheless distributed essentially evenly across L codewords with convolutional interleaving, and so the free distance appears multiplied by L, the same as in block interleaving. An additional benefit of the convolutional interleaver is that the first byte of a codeword need not be stored in the "upper corner" of the memory and can indeed be anywhere within the LN buffer. Thus, convolutional interleavers do not require synchronization to the codeword boundaries (while block interleavers do for even distribution of bursts). For these reasons, all DSLs using FEC also use convolutional interleaving. Block interleaving is not used.

The special structure of convolutional interleaving allows yet further reduction in memory, which is here called **Tong's interleaver** [92]. This concept follows directly from Figure 7.40 by noting that after each block of N bytes in the interleaver output, approximately N/L bytes need no longer be stored because they have already been transmitted. The memory locations thus vacated are available for storage of new incoming bytes. After L N-byte codeword lengths, there are no bytes remaining to be stored. By complicated addressing translations, the storage requirement can then be almost reduced to $LN/2$ bytes (see Ref. [92]).

7.4.4   Concatenated Coding and Multilayer Coding

Trellis codes and FEC can be concatenated, as shown in Figure 7.41. The inner trellis code provides a level of performance improvement. Trellis decoding with the Viterbi algorithm (or most other decoding methods) is subject to bursts of bit errors when the incorrect sequence is selected. FEC, sometimes with no or little interleaving, can correct these bursts, enhancing the system performance. As the incidence of bursts increases with decreasing margin on any coded transmission system, a threshold will be crossed where the bursts overwhelm the correction capability of the FEC, resulting in even larger bursts. At such low (or negative) margins, the secondary, tertiary, and generally error events corresponding to distances greater than the d_{min} of the trellis code become negligible and usually correspond to longer bursts than those at minimum distance. Thus, the system performance degradation is more rapid — in practice, the system tends to operate error-free right up to the threshold and then suddenly almost completely fails past the threshold for concatenated coding systems. Nevertheless, up to the threshold, the concatenated system has better performance.

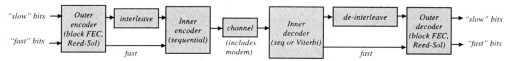

Figure 7.41   Concatenated coding for DSLs.

For this reason, ADSL allows the optional inclusion of both trellis coding and FEC. The consequent effective coding gain for the four-dimensional 16-state Wei code and the RS code of ADSL result in approximately 5.5 dB of coding gain. The FEC alone results in only 2 to 3 dB (and also corrects impulses as discussed above) while the trellis alone would have about 4.2 dB. Note that the coding gains do not add to the concatenation, but the aggregate coding gain is greater than either alone. The gain and its attainment are dependent on the line and data rate (the reader is referred to Ref. [93] for further information and also Ref. [84] for a more general treatment). Section 7.4.5 contains some specific simulation results that help to ascertain an appropriate level of interleaving in ADSL and LITE.

It is likely that this gain will be increased through more careful code-concatenation in DSLs of the future.

7.4.5   ADSL Special Case

For ADSL, a RS FEC is mandatory and an even number, $R = 2t$, of check bytes satisfying $0 \leq R \leq 16$ with polynomial description

$$C(Z) = c_0 \cdot Z^{R-1} + c_1 \cdot Z^{R-2} + \ldots + c_{R-2} \cdot Z + c_{R-1}$$

$$(7.147)$$

are appended to a codeword consisting of K bytes with polynomial description *(K<256)*:

$$M(Z) = m_0 \cdot Z^{K-1} + m_1 \cdot Z^{K-2} + ... + m_{K-2} \cdot Z + m_{K-1} \qquad (7.148)$$

where m_0 is the first transmitted byte and c_0 is the first transmitted byte of the appended parity bytes. A data byte has 8 bits $[d_7 \quad ... \quad d_1 \quad d_0]$ and is associated with GF(256) element $[d_7 \cdot \alpha^7 + ... + d_1 \cdot \alpha + d_0]$ and α is a root of binary polynomial $x^8 + x^4 + x^3 + x^2 + 1$. The check-byte polynomial satisfies

$$C(Z) = M(Z) \cdot Z^R \text{ modulo } G(Z) \qquad (7.149)$$

The generator is $G(Z) = \prod_{i=0} (Z + \alpha^i)$, and the arithmetic is performed in GF(256). The number of parity bytes is 0,4,8 with an option of 16. See Figure 7.38 for an implementation of this circuit.

Convolutional interleaving is allowed with a choice of interleave depths between $L = 0$ and 64 in ADSL. (The number of interleave bytes, number of parity bytes, and data rate [number of information bytes] are negotiated during initial training.) When the number of information bytes K is odd, the interleave depth L and $N = 2t+K$ are coprime, and thus full dispersion of error bursts is achieved. When K is even, a dummy byte is added at the end of each codeword for interleaving at the transmitter to make K effectively odd (the dummy byte is not transmitted, but reinserted at the receiver for deinterleaving). In ADSL and LITE, an additional parameter called S is specified, which is the number of DMT symbols per codeword. S is restricted to be one-half or one of the integers 1, 2, 4, 8, or 16 as negotiated during training in ADSL and to 1, 2, or 4 in LITE.

ADSL also uses dual latency, as discussed earlier. There are two different FEC paths through the modem, both using the same general RS code but likely with different numbers of check bytes and codeword sizes. The interleave path uses interleaving with the programmed interleave depth, while the fast path uses no interleaving ($S = 1$). The fast path guarantees an end-to-end latency of less than 2 ms, while sacrificing impulse-noise and other burst-error immunity.

ADSL and LITE negotiate four quantities during training (or retraining) between the ATU-C and the ATU-R: R_F, R_I, S, and L.[21]

R_F is the number of parity bytes in the fast buffer.

R_I is the number of parity bytes in the interleave buffer.

---

21.   Denoted by a capital letter I in the standards.

S is the number of 250 µs symbols per codeword in the DMT signal.

L is the interleave depth.

These four quantities augment the selected data rates implied by:

B_F — the number of information bytes per symbol in the fast buffer.

B_I — the number of information bytes per symbol in the interleave buffer.

Ignoring extra overhead bytes that are inserted as described in detail in [63], the codeword length is then

$$N_F = B_F + R_F \qquad\qquad (7.150)$$

for the fast buffer and

$$N_I = B_I \cdot S + R_I \qquad\qquad (7.151)$$

The interleave buffer bytes are convolutionally interleaved with depth L. The burst length normally considered a maximum for ADSL and LITE designs is 500 µs. This value is low because impulses in the field can often be several ms in length. Thus, ADSL and LITE allow optional interleaving to great depths. To understand the depth issue, consider a 500 µs burst. This corresponds to 375 bytes at 6 Mbps and about 94 bytes at 1.6 Mbps. If the number of parity bytes is 16, then 8 bytes can be corrected, meaning that an interleave depth of 47 is needed at 6 Mbps and of 12 at 1.5 Mbps. For 6 Mbps, this creates a codeword length of 188 bytes plus 16 parity for $204 = N$ and $S = 1$. The delay is about 12 ms, end to end and the coding overhead is about 8 to 9%. For 1.5 Mbps, set $S = 4$, and again use $k = 188$ bytes and $N = 204$ with also a delay of 12 ms. A impulse of 2 ms would lead to a delay of 48 ms and a large memory.

Thus, forward error correction, while effective against many impulses, may use much memory for interleaving and introduce corresponding long delay to eliminate impulses alone. (Fortunately, the DMT line code itself tends to reduce impulse height in amplitude by about a factor of 10, offering some additional relief.)

RS codes also provide coding gain, but only over a limited range of percentage parities. Figures 7.42 and 7.43 illustrate the coding gain for RS coding with 1.6 Mbps and 6.4 Mbps ADSL on four loops with DMT [94].

Clearly there is some room for optimization of both coding gain and interleave depth to achieve best performance against both. However, lower overhead percentage does increase delay. Coding gain becomes yet more sharp with respect to overhead percentage when trellis coding joins RS error correction in ADSL. Figures 7.44 and 7.45 repeat the earlier figures for the case of concatenated coding gain. Clearly the gain is higher, but the sensitivity to choice of

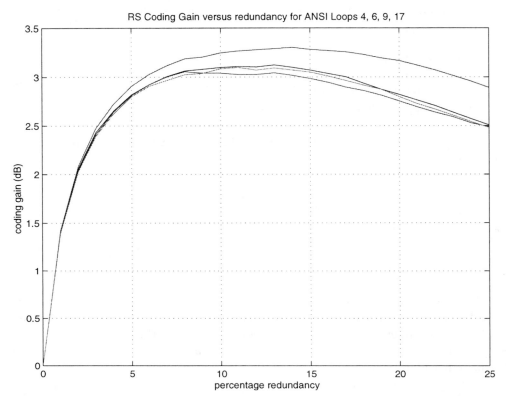

Figure 7.42   Coding gain for Reed-Solomon at 1.6 Mbps (no trellis) versus parity overhead
percentage.

correct overhead percentage is also evident. The lower overhead percentages then impose longer
interleave delays.

In the case of video services, long latency is not much of a problem (but bit errors are a
serious problem in compressed video), so latencies of 20 to 100 ms in interleaving are common
to ensure good picture quality over ADSL. Voice services are very tolerant to bit errors at 64
kbps, but are somewhat sensitive to delay. The network spec is 1.25 ms for voice; ADSL and
LITE can do at best 2 ms. In practice, a few milliseconds is tolerable on voice connections with
usually negligible effect elsewhere. However, the IP protocol used in Internet data traffic is sensitive to delay. At 1.5 Mbps, only about 10 to 12 ms is tolerable without reduction in throughput
caused by acknowledgment packet delay. At 3 Mbps, this number reduces to about 5 to 6 ms,
while 6 Mbps data rate typically needs less than 4 ms of delay. Thus, the trade-offs are complicated between coding gain, delay, impulse protection, and application. This is why the ADSL
and LITE standards allow these many variables to be negotiated between modems upon initialization rather than imposed through mandated standardized values.

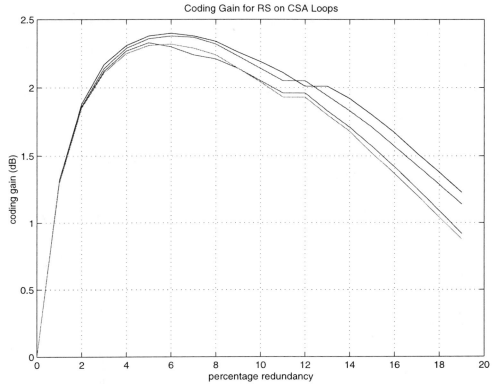

Figure 7.43 Coding gain for Reed-Solomon at 6.4 Mbps (no trellis) versus parity overhead percentage.

7.4.6   CRC Checks

DSLs use **cyclic redundancy code (CRC)** checks for detection of errors in a block of bits/bytes transmitted through the DSL link. One to two parity bytes are appended to a sequence of message bytes and used to corroborate the correct receipt of the message bytes in the receiver. Usually CRC violations, which correspond to errors detected, are used by higher-level maintenance functions to diagnose or reset the DSL link. These CRC checks are outside of RS coding, and thus detect errors not corrected by the RS code.

For bit-level CRC checking based on binary cyclic codes, a sequence of message bytes is treated as a sequence of bits, and a binary polynomial is formed representing the message as (see Section 7.4.2)

$$m(Z) = m_0 \cdot Z^{k-1} + m_1 \cdot Z^{k-2} + \ldots + m_{k-1} \tag{7.152}$$

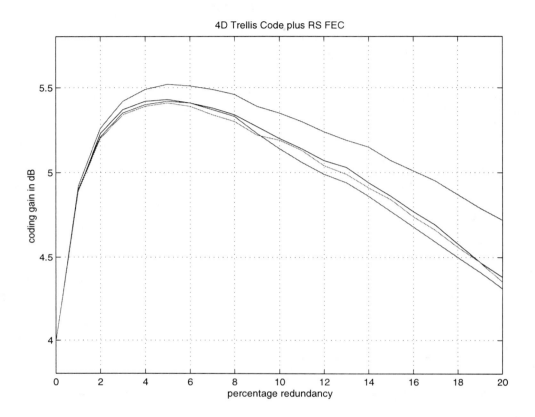

Figure 7.44   Coding gain for Reed-Solomon at 1.6 Mbps (with 4D 16-state trellis coding) versus parity overhead percentage.

Usually, $r = 8, 12,$ or 16 bits (1 to 2 bytes) are appended to the message in the form of a binary check polynomial $c(Z) = c_0 \cdot Z^{r-1} + c_1 \cdot Z^{r-2} + ... + c_{r-1}$ through the binary addition

$$m(Z) \cdot Z^r + c(Z)$$

(7.153)

The check polynomial is determined as the remainder of the division of $m(Z)$ by $g(Z)$, a binary polynomial that characterizes the CRC code:

$$c(Z) = m(Z) \,\text{modulo}\, g(Z)$$

(7.154)

At the receive side, the division is repeated and the remainders compared, which is equivalent to forming the syndrome as in Section 7.4.2. If they do not match (i.e., the syndrome is

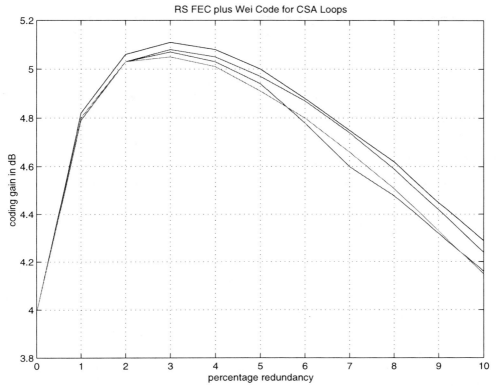

Figure 7.45   Coding gain for Reed-Solomon at 6.4 Mbps (with 16-state 4D trellis code) versus parity overhead percentage.

nonzero), a CRC violation is declared, and it is certain that an error has occurred in transmission. Only very special error patterns in the channel will lead to a division that erroneously produces the same remainder, thus causing errors to go undetected. These special patterns correspond to codewords in the binary cyclic code.

The performance of the CRC check is determined by the code polynomial $g(Z)$. The choice almost universally used in CRC checks is for $g(Z)$ to be the product of a primitive polynomial of degree $r - 2$ and the factor $(1 + Z)$. Table 7.2 summarizes the four types of CRC checks most commonly found in digital communications.

The first bit transmitted is m_0 and message bits follow up to m_{k-1} followed by the check bits. The circuit to implement the CRC division is shown in Figure 7.38. The delay elements are zeroed at the beginning of a message clock and contain the remainder/check bits after k message bits have been clocked into the circuit. An identical circuit in the receiver checks this remainder against the check bits out of the DSL detector.

Table 7.2   CRC Polynomials for DSL

Number of bits r	$g(Z)$
8 (ADSL/LITE steady state)	$Z^8 + Z^7 + Z^2 + 1 = (Z^7 + Z + 1) \cdot (Z + 1)$ or
	$Z^8 + Z^4 + Z^3 + Z^2 + 1$ (ADSL and LITE standards)
CRC-12	$Z^{12} + Z^{11} + Z^3 + Z^2 + Z + 1 = (Z^{11} + Z^2 + 1) \cdot (Z + 1)$
CRC-16 (USA)	$Z^{16} + Z^{15} + Z^2 + 1 = (Z^{15} + Z + 1) \cdot (Z + 1)$
CRC-16 (ITU, Euro) ADSL and LITE initialization	$Z^{16} + Z^{12} + Z^5 + 1 =$ $(Z^{15} + Z^{14} + Z^{13} + Z^{12} + Z^4 + Z^3 + Z^2 + Z + 1) \cdot (Z + 1)$
6 HDSL	$Z^6 + Z + 1$

To determine the probability of errors being missed by the CRC detector, one notes [86, 95, 96] that:

1. All single-bit errors are detected (because an error polynomial with a single 1 digital anywhere could not be divisible by $g(Z)$.

2. All double-bit errors are detected (because $g(Z)$ has a primitive polynomial as a factor, which cannot be divided into any two-term error polynomial unless the errors are spaced by $2^{2^{r-1}} - 1$ bits, which places an upper limit on the message polynomial length that is never exceeded in practice).

3. Any odd number of bit errors are detected (because of the factor $1 + Z$ in $g(D)$, $g(Z)$ will not divide any error polynomial with an odd number of 1s in it). In the case of ADSL and LITE steady-state, this is not guaranteed

4. Any burst error with length less than or equal to r (because the remainder after division could be a multiple of $g(Z)$ unless it is of the same degree.)

5. Most longer burst errors.

The last longer-burst observation follows from the fact that only $1/2^r$ of the error patterns on average could be multiples of $g(Z)$ because it contains a primitive. Of these, only $1/2^{r-1}$ of the length $r + 1$ bursts go undetected, because the first and last elements in such a length error polynomial must be 1s.

Thus, the probability of missing errors with the CRC detector will essentially be

$$P_{missed-error} \cong \frac{1}{2^r} \cdot P_{4\,bit\,errors}$$

(7.155)

which has a very low value for random errors. Therefore, the dominant issue will be probability of 4 or more errors in a burst of length greater than r due to nonrandom errors like impulses. Thus, only 0.01 to 1% or less of the impulses that can break a system's forward error correction would go undetected.

An alternative viewpoint of the CRC detector is that it works on blocks of up to $2^r - 1$ message-plus-check bits. There are thus $2^{2^r-1} - 1$ possible error bursts over that block length. Of these, only 2^{2^r-1-r} can avoid detection, none of which can be less than r bits as a burst (length between first one and last one).

Clearly CRCs will rapidly ascertain if a DSL is not operating properly, but they cannot be entirely relied upon to assure that all data is always transferred correctly. Thus CRC failures are usually reported to maintenance facilities for the DSL, typically resulting in corrective action for the DSL equipment (replacement or repair) if several are reported.

ADSL and ADSL-LITE use the 8-bit CRC in Table 7.2 for steady-state transmission[22] and the ITU 16-bit CRC for critical initialization data like bits and gain information. HDSL apparently uses the unusual $r = 7$ primitive polynomial $g(Z) = Z^6 + Z + 1$ on the 4682 bits of an HDSL frame.[23] This polynomial does not contain the factor $1 + D$ and so will not detect any pattern of an odd errors, but does detect $2^{127} - 2^{120}$ of $2^{127} - 1$ possible error burst patterns for a block of 127 bits.

7.4.7   Scramblers

Generally, scramblers in data transmission are used with the hope of "randomizing" an input bit stream. Because strings of zeros or ones may occur in realistic data transfer more often than other strings (thus violating the independent message assumptions made throughout this chapter and DSLs in general), it is desirable to make such events less probable. Scramblers convert the input bits into an equivalent set of bits that appears truly as independent bits in practice. This process helps DSL digital signal processing algorithms to see more robust data and usually prevents DSL subsystems from locking into undesirable settings that might be caused by highly redundant data. Equalizers, echo cancelers, bit-swapping methods, channel-identification methods, and timing-recovery loops all often benefit from the use of a scrambler.

The basic idea is to map the input bit sequence into a random set of bits through a **scrambler** as in Figure 7.46 and then **unscramble** the bits at the receiver. The mapping algorithm is always one-to-one and known at both ends of the transmission link. There are two types of scramblers and both are based upon the same binary primitive polynomials seen previously in this section. The **synchronous scrambler** in Figure 7.46 adds a pseudorandom sequence generated by the primitive polynomial in binary (exclusive or) effectively to prevent long strings of 1s or 0s from occurring. The corresponding descrambler adds the sequence again at the exact corresponding instants in time. Thus, the synchronous scrambler must establish a common time mark

---

22.   The factor of $1 + Z$ is not included in ADSL, but the polynomial itself is primitive, resulting in the same total fraction of all bit error patterns being identified, however, this apparent oversight in ADSL and LITE leads to the possibility of only three channel bit errors (if in a burst longer than r) possibly leading to no detection. This means the likelihood of a undetected error is about 10 million times higher than it needed to be, although still very small and unlikely.

23.   This HDSL polynomial appears to assume tacit inclusion the factor $1 + D$ because spots for 6 bits of parity are left in the frame.

between scrambler and descrambler or face very poor performance. This extra complexity of synchronization is usually avoided when possible. A synchronous descrambler will produce a bit error at its output if and only if there is a bit error on its input at the same time instant. The scrambling or "spreading" sequence is a pseudorandom binary sequence generated through the primitive polynomial $g(D)$ through a nonzero initial condition in the linear shift-register circuit described by $\frac{1}{g(D)}$ (or the circuit in Figure 7.47 with zero input and nonzero initial condition). The sequence of bits produced will be periodic with period $2^m - 1$ where m is the degree of the primitive polynomial, will run through all possible non–all-zero patterns of length m once per period and has a nearly flat ("white") power spectrum when viewed as a bipolar sequence. Such a sequence thus "scrambles" the input bits so that patterns like all zeros, all ones, or other highly correlated bit sequences that occur in computer data or quantized speech are mapped to more spectrally rich sequences. Most of the theory (and the best operation) of most of the signal-processing technologies used in DSL depend on this "whiteness" of the input for best performance.

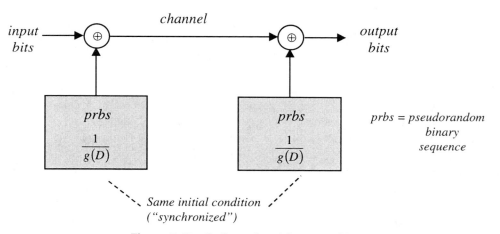

Figure 7.46   Self-synchronizing scrambler.

Self-synchronizing scramblers are prevalent in DSLs and do not require synchronization. Instead, the input bit stream is processed by a binary filter that is essentially the inverse of a primitive polynomial. The descrambler then filters by the primitive polynomial. The cascade of the two filters is the exact reproduction of the inputs, no matter what the relative delay between transmitter and receiver. However, a single input bit error into the descrambler can lead to a finite number of descrambler output bit errors. The multiplication number of bit errors is equal to the Hamming weight of the primitive polynomial, typically about 3 to 5 in practice. This slight loss in bit error rate is more than offset by the aversion of having to synchronize in most applications. Thus all DSLs to date use self-synchronizing scramblers.

It is important to note that some higher-level packet-application functions (like ATM) use synchronous scramblers simply because cell boundaries must be maintained anyway and a tacit drawback of the self-synchronizing scrambler is that it needs a nonzero initial state each time a new packet/cell is initiated.

Downstream and upstream ADSL and ADSL-LITE, as well as upstream HDSL and upstream ISDN, all use the same 23-bit self-synchronizing scrambler shown in Figure 7.47. Downstream HDSL and downstream ISDN use an alternate 23-bit scrambler, as shown in Figure 7.47.

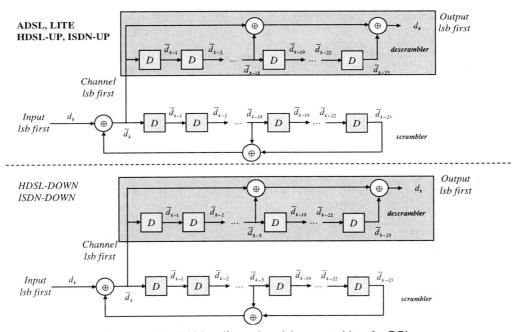

Figure 7.47 24 bit self-synchronizing scramblers for DSL.

When mapped into a two-level PAM sequence, the output of the scrambler has an autocorrelation function approximated by

$$E[x_k \cdot x_{k-n}] \approx \frac{1}{2^m - 1} \cdot \sum_{i=0}^{2^m-1} x_{k+i} \cdot x_{k+i-n} = \begin{cases} 1 & n = 0, \pm(2^m - 1), \pm 2(2^m - 1),... \\ \dfrac{-1}{2^m - 1} & \text{all other } n \end{cases}$$

$$(7.156)$$

which clearly for even small values of m very closely approximates the impulse autocorrelation function of a "white" signal. Even when not mapped to binary PAM, the highly uncorrelated nature of adjacent bits on the average is a desirable function for almost any transmission technique.

References

[1]   Forney, G. D., Jr., "Maximum Likelihood Sequence Detection in the Presence of Intersymbol Interference," *IEEE Transactions on Information Theory,* Vol. 18, No. 3, May 1972, pp. 363–378.

[2]   Porat, B., *A Course in Digital Signal Processing.* Wiley, New York, 1997.

[3]   Cioffi, J. M., Dudevoir, G. P., Eyuboglu, M. V., and Forney, G. D., Jr., "MMSE Decision Feedback Equalizers and Coding: Parts I and II," *IEEE Transactions on Communications,* Vol. 43, No. 10, October 1995, pp. 2582–2604.

[4]   Duttweiler, D. L., Mazo, J. E., and Messerschmitt, D. G., "An Upper Bound on the Error Probability in Decision-Feedback Equalization," *IEEE Transactions on Information Theory,* Vol. 20, No. 4, July 1974, pp. 490–497.

[5]   Tomlinson M., "New Automatic Equalizer Employing Modulo Arithmetic," *Electronics Letters,* Vol. 7, March 1971, pp. 138–139.

[6]   Hirashima, H., and Miyakawa, H., "A Method of Code Conversion for Digital Communication Channel with Intersymbol Interference," *IEEE Transactions on Communications,* Vol. 20, August 1972, pp. 774–780.

[7]   Laroia, R., Tretter, S., and Fervardin, N., "A Simple and Effective Precoding Scheme for Noise Whitening on Intersymbol Interference Channels," *IEEE Transactions on Communications*, Vol. 41, No. 10, October 1993, pp. 1460–1463.

[8]   Lucky, R. W., Salz, J., and Weldon, E. J. Jr., *Principles of Data Communication,* McGraw-Hill, New York, 1968.

[9]   Kuo, F. F., *Network Analysis and Synthesis,* 2nd ed. Wiley, New York, 1966, pp. 291–294.

[10]   Al-Dhahir, N., and Cioffi, J. M., "MMSE-DFE's: Finite-Length Results," *IEEE Transactions on Information Theory,* Vol. 41, July 1995, pp. 961–976.

[11]   Henkel, W., "An Algorithm for Determining the DMT Time-Domain Equalizer Coefficients," *ETSI Contribution TD26,* Lannion, France, September 1997.

[12]   Cioffi, J. M., and Forney, G. D., Jr., "Generalized Decision-Feedback Equalization for Packet Transmission with ISI and Gaussian Noise," *Communication, Computation, Control, and Signal Processing* (a tribute to Thomas Kailath), Kluwer, Boston, 1997.

[13] Forney, G. D., Jr,. and Calderbank, A. R., "Coset Codes for the Partial Response Channel," *IEEE Transactions on Information Theory,* Vol. 34, September 1989.

[14] Viterbi, A. J., "Error Bounds for Convolutional Codes and an Asymptotically Optimum Decoding Algorithm," *IEEE Transactions on Information Theory,* Vol. 13, April 1967, pp. 260–269.

[15] Anderson, R. R., and Foschini, G. J., "The Minimum Distance for MLSE Digital Data Systems of Limited Complexity," *IEEE Transactions on Information Theory,* Vol. 21, September 1975, pp. 544–551.

[16] Ferro, M. , and Kliger, A., *ANSI Contribution T1E1.4/95-069,* "Report on Measurements of 2.048 Mbps Single Pair HDSL" (CSELT and Metalink), June 5, 1995, Eatontown, NJ.

[17] Wilson, S. G., *Digital Modulation and Coding.* Prentice-Hall, Upper Saddle River, NJ, 1996.

[18] Eyuboglu, V. M., and Qureshi, S. U. H., "Reduced-State Sequence Estimation with Set Partitioning and Decision Feedback," *IEEE Transactions on Communications,* Vol. 36, January 1988, pp. 13–20.

[19] Duel, A., and Heegard, C., "Delayed Decision-Feedback Sequence Estimation," *Allerton Conference on Communications and Control,* October 1985.

[20] Moon, J., and Carley, R., "Performance Comparison of Detection Methods in Magnetic Recording," *IEEE Transactions on Magnetics,* Vol. 26, November 1990, pp. 3155–3172.

[21] Cover, T., and Thomas, J., *Elements of Information Theory.* Wiley, New York, 1991.

[22] Sklar, B., "A Primer on Turbo Code Concepts," *IEEE Communications Magazine*, December, 1997, pages 94–102.

[23] Forney, G. D., Jr., "Approaching Channel Capacity with Coset Codes and Multilevel Coset Codes," *1997 International Symposium on Information Theory.*

[24] Shannon, C. E., "A Mathematical Theory of Communication," *Bell Systems Technical Journal,* Vol. 27, 1948, pp. 379–423 (Part I), pp. 623–656 (Part II).

[25] Chow, P. S., "Bandwidth Optimized Digital Transmission Techniques for Spectrally Shaped Channels with Impulse Noise," Ph.D. thesis, Stanford University, May 1993.

[26] Chow, P., and Cioffi, J., "Method and Apparatus for Adaptive, Variable-Bandwidth, High-Speed Data Transmission of a Multicarrier Signal over Digital Subscriber Lines," U.S. patent 5,479,447, December 29, 1995.

[27]   Campello, J., "Optimal Discrete Bit Loading for Multicarrier Modulation Systems," *1998 IEEE International Symposium on Information Theory,* August 1998, Cambridge, MA.

[28]   Hoo, L., Tellado, J., and Cioffi, J., "Dual QoS Loading Algorithms for Multicarrier Systems Offering CBR + VBR Services," *Globecom '98,* November 1998, Sydney.

[29]   Hughes-Hartog, D., "Ensemble Modem Structure for Imperfect Transmission Media," U.S. patent 4,679,227 (July 1987) , U.S. patent 4,731,816 (March 1988), and U.S. patent 4,833,706 (May 1989).

[30]   Jacobsen, K. et al., "VDSL Alliance SDMT VDSL Draft Standard Proposal," *ANSI Contribution T1E1.4/98-036,* Austin, TX.

[31]   ITU-T g.992.2 Standard, "Digital Transmission System for ADSL Transmission on Metallic Local Loops with Provisioning to Facilitate Installation and fr Operation in Conjunction with Other Services."

[32]   Fan, J., and Cioffi, J. M., "A Real-Valued Discrete Multichannel System based on the Discrete Hartley Transform, and its relation to Vector Coding and DMT," *IEEE International Conference on Communications,* Atlanta, GA, June 1998.

[33]   Vaidyanathan, P. P., *Multirate Systems and Filter Banks.* Prentice Hall, Englewood Cliffs, NJ, 1993.

[34]   Vetterli, M., and Le Gall, D., "Perfect Reconstruction FIR Filter Banks: Some Properties and Factorizations," *IEEE Transactions on Acoustics, Speech, and Signal Processing,* Vol. 37, July 1989, pp. 1057–1071.

[35]   Sandberg, S. D., and Tzannes, M. A., "Overlapped Discrete Multitone Modulation for High-speed Copper-Wire Communications," *IEEE Journal on Selected Areas in Communication,* Vol. 13, December 1995, pp. 1571–1585.

[36]   Isaksson, M., et al., "Pulse Shaping with Zipper – Spectral Compatibility and Asynchrony," *ANSI Contribution T1E1.4/98-041,* (Telia), March 2, 1998, Austin, TX.

[37]   Spruyt, P., Reusens, P., and Braet, S., "Performance of Improved DMT Transceiver for VDSL," *ANSI Contribution T1E1.4/96-104,* April 22, 1996, Colorado

[38]   Bingham, J., "Digital RF Cancelation with SDMT," *ANSI Contribution T1E1.4/96-083,* April 1996.

[39]   Al-Dhahir, N., and Cioffi, J., "Optimum Finite-Length Equalization for Multicarrier Transceivers," *IEEE Transactions on Communications,* Vol. 44, No. 1, January 1996, pp. 56–64.

[40] Zervos, N., Pasupathy, S., and Venetsanopoulos, A., "The Unified Decision Theory of Non-Linear Equalization," 1984 *IEEE Global Telecommunications Conference Proceedings*, Atlanta, December 1984, pp. 683–687.

[41] Chow, J. S., and Cioffi, J. M., "Method for Equalizing a Multicarrier Signal in a Multicarrier Communication System," *U.S. patent No. 5,285,474*, issued to Stanford University, February 8, 1994.

[42] Wiese, B., and Bingham, J., "Digital Radio Frequency Cancelation for DMT VDSL," *ANSI Contribution T1E1.4/97-460*, December 1997, Sacramento, CA.

[43] Cioffi, J.M., *Digital Data Transmisson, Vol. 3*. EE379C Course Textbook, Stanford University; see Chapter 11 of world-wide-web site http://www-leland.stanford.edu/class/ee379c/reader.html.

[44] Kasturia, S., and Cioffi, J. M., "Vector Coding with Decision Feedback Equalization for Partial Response Channels," *Globecom'88*, December 1988, Ft. Lauderdale, FL, pp. 0853–0857.

[45] Kasturia, S., and Cioffi, J. M., "Precoding for Block Signaling and Shaped Signal Sets," *ICC'89*, June 1989, Boston, pp. 1086–1090.

[46] Cheong, K. W., and Cioffi, J. M., "Precoder for DMT with Insufficient Cyclic Prefix," *Globecom'98*, Sydney, Australia, November 1998.

[47] Van Kerckhove, J. F., and Spruyt, P., "Adapted optimization criterion for FDM-based DMT-ADSL Equalization," *Proceedings ICC*, New Orleans, June, 1996.

[48] ITU-U g.922.1 Standard, "Digital Transmission System for ADSL on Metallic Local Lines, with Provisions for Operation in Conjunction with Other Services."

[49] Wulich, D., "Reduction of Peak-to-Mean Ratio of Multicarrier Modulation Using Cyclic Coding," *Electronics Letters*, Vol. 32, No. 5, February 1996, pp. 432-433.

[50] Ochai, H., and Imai, H., "Block Coding Scheme Based on Complimentary Sequences for Multicarrier Signals," *IEICE Transactions on Fundamentals*, Vol. ESO-A, No. 11, November 1997, pp. 2136-2143, see also "Block Codes for Frequency Diversity and Peak Power Reduction in Multicarrier Systems," *1998 IEEE ISIT*, Cambridge, MA August 1998.

[51] Davis, J. A., and Jedwab, J., "Peak-to-Mean Power Control and Error Correction Using Golay Complementary Sequences and Reed-Muller Codes," *Electronics Letters*, Vol. 33, No. 4, February 1997, pp. 267-???, see also *1996 IEEE ISIT*, Cambridge, MA, August 1998.

[52] Patterson, K. G., "Generalized Reed-Muller Codes and Power Control in OFDM Modulation," *HP Technical Report HP198-57*, Bristol, England, March 1998, see also *1998 IEEE ISIT*, Cambridge, MA, August 1998.

[53]   Tellado, J., and Cioffi, J., "PAR Reduction in Multicarrier Trnasmission Systems," *ANSI Contribution T1E1.4/97-367,* Sacramento, CA, December 1997.

[54]   Tellado, J., and Cioffi, J., "Revisiting DMT's Peak-to-Average Ratio," *ETSI Contribution TM6/98-TD08,* Antwerp, Belgium, April 22, 1998.

[55]   Tellado, J., and Cioffi, J., "PAR Reduction with Minimum or Zero Bandwidth Loss and Low Complexity," *ANSI Contribution T1E1.4/98-173,* Huntsville, AL, June 1, 1998.

[56]   Gatherer, A., and Polley, M., "Controlling Clipping Probability in DMT Transmission," 1997 *Asilomar Conference,* November 1997.

[57]   Kschischang, F., Narula, A., and Eyuboglu, V., "A New Approach to PAR Control in DMT Systems," *ITU-T Study Group 15/Q4 Contribution NF-083,* May 11, 1998, Nice, France.

[58]   Friese, M., "Multicarrier Modulation with Low Peak-to-Average Ratio," *Electronics Letters*, Vol. 32, No. 8, April 1996, pp. 713-714.

[59]   Bingham, J. A. C., and Cioffi, J. M., "Dynamic Scaling for Clip Mitigation in the ADSL Standard Issue 2," *ANSI T1E1.4/ 96-019,* Los Angeles, January 1996.

[60]   Mesdagh, D., and Spruyt, P., "A Method to Reduce the Probability of Clipping in DMT-based transceivers," *IEEE Transactions on Communications,* Vol. 44, No. 10, October 1996, pp. 1234–1238.

[61]   Eetvelt, P., Wade, M., and Tomlinson, M., "Peak to Average Power Reduction for OFDM Schemes by Selective Scrambling," *Electronic Letters,* Vol. 32, No. 21, October 1996, pp. 1963–1964.

[62]   Baum, R. W., Fischer, R. F. H., and Huber, J. B., "Reducing the Peak-to-Average Ratio of Multicarrier Moudlation by Selected Mapping," *Electronics Letters,* October 1996, pp. 2056–2057.

[63]   *Asymmetric Digital Subscriber Line (ADSL) Metallic Interface,* ANSI Standard T.413-1995, ANSI, New York.

[64]   Muller, S. H., and Huber, J. B., "A Comparison of Peak Power Reduction Schemes for OFDM," *Globecom'97,* Vol. 1, November 1997, pp. 1–5, Phoenix, AZ. (See also same authors, *Electronics Letters,* Vol 33, No. 5, Feb. 1997, pp. 3680-369.)

[65]   Friese, M., "Multitone Signals with Low Crest Factor," *IEEE Transactions on Communications,* Vol. 45, No. 10, October 1997.

[66]   Li, G., and Cimini, L., (eds.), *Orthogonal Frequency Division Multiplexing for Wireless Communication.* Prentice Hall, Upper Saddle River, NJ, 1999.

[67]   Verbin, R., "Efficient Algorithm for Clip Probability Reduction," *ANSI Contribution T1E1.4/97-323,* Minneapolis, MN, September 1997.

[68]   Cooley, J. W., and Tukey J. W., "An Algorithm for the Machine Calculation of Complex Fourier Series," *Mathematical Computation,* Vol. 19 (90), 1965, pp. 197–301.

[69]   Jacobsen, K., "Spectral Compatibility of ADSL and VDSL, Part I: The Impact of VDSL on ADSL Performance," *ANSI Contribution T1E1.4/97-404R1,* Sacramento, CA, December 1997.

[70]   Jacobsen, K., "Spectral Compatibility of ADSL and VDSL, Part 2: The Impact of ADSL on VDSL Performance," *ANSI Contribution T1E1.4/98-035,* Austin, TX, March 1998.

[71]   Ho, M., Cioffi, J., and Bingham, J., "High-speed Full-duplex Echo Cancellation for DMT Modulation," *ICC'93,* Geneva, May 1993, pp. 772-776.

[72]   Younce, R. C., Melsa, P. J. W., and Kapoor, S., "Echo Cancellation for ADSL," *Proceedings 1994 ICC,* May 1994, New Orleans, pp. 301–305.

[73]   Isaksson M., et al., "Zipper – A Flexible Duplex Scheme for VDSL," *ANSI Contribution T1E1.4/97-016,* February 3, 1997, Austin, TX.

[74]   Marple, S. L., *Digital Spectral Analysis with Applications.* Prentice-Hall, Englewood Cliffs: NJ, 1987.

[75]   Kay, S., *Modern Spectral Estimation, Theory and Applications.* Prentice-Hall, Englewood Cliffs: NJ, 1988.

[76]   Cioffi, J., Mallory, M., and Bingham, J. A. C., "Analog RFI Cancellation with SDMT," *ANSI T1E.4/96-084,* April 1996, Colorado Springs.

[77]   Forney, G. D., Jr., "Coset Codes – Part I: Introduction and Geometrical Classification," *IEEE Transactions on Information Theory,* Vol. 34, No. 5, October 1988, pp. 1123–1151.

[78]   Forney, G. D., Jr., "Coset Codes – Part II: Binary Lattices and Related Codes," *IEEE Transactions on Information Theory,* Vol. 34, No. 5, October 1988, pp. 1152–11871.

[79]   Ungerboeck, G., "Trellis Coded Modulation with Redundant Signal Sets: Parts I and II," *IEEE Communications Magazine,* Vol. 25, February 1987, pp. 12–21.

[81]   Forney, G. D., Jr., "Trellis Shaping," *IEEE Transactions on Information Theory,* Vol. 38, March 1992, pp. 281–300.

[82]   Fortier, P., Ruiz, A., and Cioffi J. M., "Multidimensional Signal Sets Through the Shell Construction for Parallel Channels," *IEEE Transactions on Communications,* Vol. 40, March 1992, pp. 500–512.

[83]   Calderbank, A. R., and Ozarow, L., "Nonequiprobable Signaling on the Gaussian Channel," *IEEE Transactions on Information Theory,* Vol. 36, July 1990.

[84]   Zogakis, T. N., Aslanis, J. T., Jr., and Cioffi, J. M., "A Coded and Shaped Discrete Multitone System," *IEEE Transactions on Communications*, Vol. 43, No. 12, December 1995, pp. 2941–2949.

[85]   Laver, J., and Ciofi, J., "Turbo-DMT," *Globecom'98*, Sydney, November 1998.

[86]   Peterson, W., and Brown, D., "Cyclic Codes for Error Detection," *Proceedings of the IRE,* January 1961.

[87]   Wicker, S., *Error Control Systems for Digital Communication and Storage*. Prentice-Hall, Upper Saddle River, NJ, 1995.

[88]   Lin, S., and Costello, D. J., Jr., *Error Control Coding: Fundamentals and Applications*. Prentice-Hall, Upper Saddle River, NJ, 1983.

[89]   Berlekamp, E., "The Technology of Error Correcting Codes," *Proceedings of the IEEE,* May 1980.

[90]   Zogakis, T. N., Tong, P. T., and Cioffi, J. M., "Performance Comparison of FEC/Interleave Choices with DMT for ADSL," *ANSI Contribution T1E1.4/93-091,* April 14, 1993, Chicago.

[91] Blahut, R., *Theory and Practice of Error Control Codes*. Addison-Wesley, NY, 1983.

[92]   Tong, P., "Efficient Address Generation for Convolutional Interleaving Using Minimal Amount of Memory," *U.S. patent Number 5,764,649,* June 9, 1998.

[93]   Zogakis, T. N., Aslanis, J. T., and Cioffi, J. M., "A Coded and Shaped Discrete Multitone System," *IEEE Transactions on Communications,* Vol. 43, No. 12, December 1996, pp. 2941–2949.

[94]   Zogakis, T. N., Tong, P. T., and Cioffi, J. M., "Performance Comparison of FEC/Interleave Choices with DMT for ADSL," *ANSI COntribution T1E1.4/93-091*, April 14, 1993, Chicago.

[95]   Stallings, W., *Data and Computer Communications*. Macmillan, New York, 1988.

[96]   Tannenbaum, A. S., *Computer Networks,* 2nd ed. Prentice Hall, Upper Saddle River, NJ, 1988.

[97] Cioffi, J. M., "A Multicarrier Primer," *ANSI Contribution T1E1.4/91-157,* November 1991, Clearfield, FL.

[98]   Ruiz, A., Cioffi, J. M., and Kasturia, S., "Discrete Multiple Tone Modulation with Coset Coding for the Spectrally Shaped Channel," *IEEE Transactions on Communications,* Vol 40, No. 6, June 1992, pp. 1012–1029.

[99]   Pal, D., Cioffi, J., and Iyengar, G., "A New Method of Channel Shortening with Applications to Discrete Multitone (DMT) Systems," *Proceedings 1998 IEEE International Conference on Communications*, Atlanta, June 1998, paper no. S22.1.

[100] Parlett, B. N., *The Symmetric Eigenvalue Problem*. Prentice-Hall, Englewood Cliffs, NJ, 1980, Chapter 13.

[101] Jones, D. C., "Reducing the Complexity of a Cyclic Echo Synthesizer for a DMT ADSL Frequency Domain Echo Canceller," (Bellcore), *ANSI Contribution T1E1.4/93-255*, October 1993, Alexandria, VA.

CHAPTER 8

Initialization, Timing, and Performance

8.1 Initialization Methods

Initialization of DSL modems is crucial to proper performance during steady-state transmission of data. Initialization is usually accomplished through the use of handshaking start-up sequences that typically execute the following steps in order: activation, gain setting/control, synchronization, echo cancellation (if any), channel identification, equalization, and secondary channel identification/exchange. The more sophisticated recent transmission methods like ADSL and VDSL execute all steps, while earlier lower-performing DSLs may execute only a subset.

The high-level timing diagram of initialization appearing in Figure 8.1 does not in detail correspond exactly to any particular DSL, but generally all DSLs can be found to conform to it.

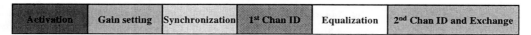

| Activation | Gain setting | Synchronization | 1st Chan ID | Equalization | 2nd Chan ID and Exchange |

Figure 8.1   General timeline for initialization of a DSL.

Echo cancellation, while not shown, tacitly can occur during any of the segments of transmission following gain setting.

8.1.1 Activation

Activation is the process of the DSL modems at each end of the link notifying each other that service needs to be established.

Activation usually occurs through a modem at one end transmitting a request to initialize, and the other modem recognizing that request and responding. When both modems have acknowledged each other, initialization can continue as described in subsequent subsections. At the time of final writing, the ITU Study Group 15 is trying to develop a universal DSL start-up mechanism or "g.hs" (hs abbreviates "handshake," now called g.994.1 [1]) so that any two xDSL modems may negotiate to a common initialization. That specific initialization, as well as DSL initialization in general, usually conforms to what is described in this section. Of course there will always be deviations at a detailed level, but this section tries to capture the fundamentals.

8.1.1.1   ADSL T1.413 Activation

T1.413 ADSL modems [2] can activate by the ATU-C transmitting one of four possible tones for 32 ms, 207 kHz, 189.75 kHz, 224.25 kHz, and 258.75 kHz for each of C-ACT1, C-ACT2, C-ACT3, and C-ACT4 that specify different options used by the modem to the ATU-R. These signals are transmitted for 16 ms at –4 dBm and then for an additional 16 ms at –28 dBm. The reduction in power level allows the receiver to avoid saturation on a very short loop. The ATU-R responds by sending back tones of frequencies 43.125, 34.5, or 60.375 kHz to represent one of three possible acknowledgments for 16 ms at –2 dBm followed by another 16 ms at –22 dBm. The ATU-R can also initiate connection (unless the ATU-C has previously disabled it by ordering it to remain quiet with a tone of 310.5 kHz for 16 ms each at –4 dBm and –28 dBm) by transmitting the upstream 34.5 kHz tone. The options are specified in the T1.413 standard [2].

Each tone is followed by 32 ms of silence to avoid overlap and echo problems. In ADSL, after activation is established, different signals are used for gain, synchronization, and the remainder of training.

8.1.1.2   HDSL Activation

HDSL activation begins with the HTU-C transmitting only ±3 level 2B1Q signals according to the output of a scrambler initialized with all ones, which is known as S0. The HDSL scrambler is defined in Ref. [3]. The HTU-R must return a similar signal called R0 within 1.9 to 2 seconds. Between 5 to 10 seconds after receipt of R0, the HTU-C transitions to sending scrambled data at all four levels (CS1) until it believes it is ready and then switches to use of actual DS1 bits (within 30 seconds from starting). The HTU-R can switch to four-level scrambled data (RS1) within 4 seconds after receipt of four-level data from the HTU-C and also switches when ready. The external DS1 link is presumed capable of establishing connection after actual data is valid.

All remaining segments of training are executed upon the S0/R0 and S1/R1 signals. An interesting drawback of the HDSL modem activation is that the modem temporarily increases its power by 7 dB, thus increasing crosstalk into its neighbors. The two HDSL loops are distinguished in the receiver by different synch symbols each 6 ms frame. Loop 1 uses + 3 + 3 + 3 – 3 – 3 + 3 – 3, while loop 2 uses – 3 + 3 – 3 – 3 + 3 + 3 + 3. For more information, see Ref. [3].

8.1.1.3   ISDN Activation

ISDN activation [4] is achieved by both ends sending 10 kHz (+ 3 + 3 + 3 + 3 – 3 – 3 – 3 – 3 repeating pattern) and then progressing through a phase similar to HDSL (see Ref. [4] for more details).

8.1.2   Gain Initialization

The settings of gain throughout a DSL transmission system can affect the performance, cost, and radiated line-emissions for the system. This section focuses entirely on analog gain settings for both the transmitter and the receiver. Section 8.1.2.1 begins with a discussion of the data-converter precision requirements that are important for proper setting of gain. Subsequent sections then discuss transmit power control and automatic gain setting. While the principles here are described in terms of analog signals and power levels, they are also useful inside the digital signal processing at interfaces between various processing stages where gain levels may need to be automatically set as a function of line rather than fixed, so that maximum use of internal precision can be maintained. However, a discussion of the internal precision requirements of various signal processing algorithms is beyond the scope of this text.

8.1.2.1   Data Conversion

Modulated signals in DSL are today usually generated/detected by digital signal processing and converted to/from analog line signals by digital-to-analog converters (DACs) and analog-to-digital converters (ADCs). These devices are linear converters (unlike the μ-law or A-law converters used for voice-signal digital conversion). Nonetheless, they will exhibit nonlinearity at a level of approximately half the magnitude of the least significant bit. Thus, the precision of the DAC and ADC for a DSL is important and is a significant contributor to system gain. For systems without echo cancellation, a simple rule-of-thumb can be used independent of line-code for DSL transmission systems that operate at maximum margin. The authors have never seen this rule in error, despite experience with a significant number of DSL designs.

To ascertain correct precision, the designer must estimate (or measure) the output SNR for the DSLs of interest as a function of used frequency, $SNR(f)$, across the transmission band used by the DSL line code. The best frequency SNR is

$$S_{max} = \max_f \{SNR(f)\}$$

$$(8.1)$$

For a good design, the number of bits in both the DAC and the ADC is

$$b = \frac{1}{2}\log_2\left(1 + \frac{S_{max}}{\Gamma}\right) + 3 + \log_4(PAR) + b_\Delta$$

$$(8.2)$$

where Γ is the system-design SNR-gap described previously in Chapters 6 and 7, and PAR is the ratio of the largest acceptable one-dimensional signal peak to the average squared one-dimensional signal value. The first term represents essentially the number of bits/dimension in a constellation at the frequency of the largest used SNR, the 3 bits represent a good level of quantization between signal levels (some designs use as low as 2 bits, but 3 bits usually leads to less than 0.25 dB degradation), and the third term represents the extra bits to capture analog signal peaks into the line. The last term b_Δ is usually zero in good DSL designs. It is the log base 4 of the ratio of signal levels on two disparate frequencies that have the same best signal-to-noise ratio. In DSLs, such signal differences can often be eliminated through analog compromise equalization.

As examples of use of the formula, let us suppose ISDN is used on a long line where the maximum SNR versus frequency occurs at 30 kHz and is 30 dB. Then since the gap is about 9.8 dB at 1e-7 error rate with no code and 2B1Q, $b = \frac{1}{2}\log_2\left(1 + 10^{2.02}\right) + 3 + \log_4\left(10^{.9}\right) = 3.5 + 3 + 1.5 = 8$ bits (without echo cancellation, 3.5 bits corresponds to the 30 dB of SNR and consequent 3.5 bits/dimension, 3 bits is for 0.25 dB or less quantization error, and 1.5 bits are for 9 dB of peak-to-average ratio in one dimension). The PAR of 9 dB takes into account some mild filtering that enlarges the nominal 6 dB PAR of the constellation in actual designs. For an HDSL system, that same frequency might have an SNR maximum of 42 to 54 dB at the same frequency, thus requiring about 10 to 12 bits of precision (without echo cancellation). Since ISDN and HDSL both use echo cancellation, an extra 1 to 2 bits over the PAR need be allocated for the ADC where the echo exceeds signal peaks by 6 to 12 dB.

For ADSL designs, the maximum used SNR may be as high as 60 dB, and the PAR can be as high as 15 dB (true for both CAP and DMT designs), so 12 bit DAC/ADC (or more) precision is often used. An echo-canceled ADSL system thus usually requires about 14 bits precision.

Gain levels of signals entering DACs or ADCs need to be set so that the full range of precision is used when best performance is desired. Some caution (common sense) needs to be applied for very short loops where S_{max} can be very high. Then, the SNR(f) can and should be limited by the ADC/DAC quantization noise at a level sufficient to ensure performance as good or better than on long loops.

8.1.2.2   Transmit Power Control

DSL twisted-pair channel responses can span a very wide range of signal attenuation, with long loops typically exhibiting very small signals at their output and short loops very large signals. This dynamic range may span as much as 9 orders of magnitude for the most sensitive DSLs. The level of the transmit signal thus cannot be constant in all DSLs. Fortunately, for setting transmit power level, only the average received signal power is of interest if ADCs and DACs have been designed as in Section 8.1.2.1. Average signal attenuation[1] is usually no worse than –40 to –50 dB. Since receiver gain control circuits can usually accommodate (see Section 8.1.2.3) up to 30 dB of range, DSL transmitters may need to reduce power by as much as 20 dB on short loops to prevent saturation of receiver electronics. Furthermore, in systems for which crosstalk into other services is important (i.e., ADSL, HDSL2, VDSL), transmitting at an excessive level leads to greater crosstalk noise on other loops, and a good network design would prevent one system from transmitting more power than it needs for reliable operation. Thus, a DSL transmitter must decide at what level to transmit.

---

1.   Average signal attenuation is the ratio of launched to received total power.

This is usually accomplished by initiating transmission at a very low level and increasing transmit power steadily while the receiving modem monitors and responds with a control signal that indicates when sufficient signal level is being received. After this indication, the level of transmit power is maintained. This operation is typically referred to as **power control.**

ADSL Systems   To date, ADSL is the only standardized transmission system that uses power reduction. The control algorithm is simplified with respect to a full adaptive power reduction. Basically, the power on upstream tones 7 to 18 is measured during early initialization (it is assumed that since the upstream signal occupies one-eighth the bandwidth that dynamic range associated with loop-length variation does not require upstream power reduction). The following table states the reduction in transmit power spectral density mask that must then be used:

measured up power	<3	4	5	6	7	8	9	dBm
downstream PSD	−40	−42	−44	−46	−48	−50	−52	dBm/Hz

The Good Neighbor Algorithm[2]   For VDSL systems, it is likely that more sophisticated power reduction will be used because of the larger band of crosstalk interference generated. The signal processing to control power cutback is at the receiver, which provides the control signal. It is desirable for systems like VDSL to operate at a fixed performance level (10^{-7} error rate with 6 dB of margin or perhaps an extra factor to account for the introduction of other crosstalk-recipient DSLs in the same cable later) so that they thus generate the minimum amount of crosstalk for reliable performance. In this mode of operation, the receiver AGC is set at maximum gain level. A control signal (low power sinusoid) is returned to the VDSL transmission unit at the ONU (VTU-O) while the received signal level is less than the ADC peak voltage by more than $log_4(PAR)$ bits. The transmitter responds by increasing transmit power by some increment $\delta\mathcal{E}_x$ from a very low initial level every τ seconds. Typically, $\delta\mathcal{E}_x = 0.5$ dB and $\tau = 500\,\mu s$. When the transmit power has been increased so that the received signal power is such that the PAR objective has been attained, this first segment of power reduction is complete. Noise power estimation then occurs with the AGC setting fixed and the transmitter silenced. The receiver can then estimate its worst-case margin. If this margin is significantly in excess of that specified, then a second control signal is returned requesting the transmitter to reduce power further by the amount of excess measured margin at the receiver. When such request for further power reduction has been made, the transmission system may want to allow for real-time power increase during later operation (sometimes called "gain-swapping," see Section 8.2.2.4).

Recently, emerging VDSL standards mandate this second margin-based power control based on an "average" loop so that all systems perform like this average loop (if possible).

---

2.   There is also a corresponding "nasty neighbor" algorithm that may be used by some vendors when they detect or suspect their competitor's product is operational in the same binder. This practice should not be used in DSLs (nor in public/marketing presentations or standards meetings!).

Power control may become increasingly important as DSLs become deployed in large volumes. The authors advise that standards efforts following the date of publication may proceed with the introduction of power control methods for later revisions of ADSL and HDSL standards, as well as its expected use in some form in VDSL.

8.1.2.3   Receiver Gain Control

Automatic gain control (AGC) is an operation that scales received signals so that precision of an ensuing analog-to-digital converter is best used (bits are not wasted). An AGC precedes the ADC and drives its output average power (measured over long periods in time) to a constant level. The output of an AGC is typically set PAR dB below the maximum voltage that can be tolerated by an ADC. The algorithm for gain control is a first-order integration that is typically implemented in discrete time.

$$Gain(k+1) = Gain(k) + \mu \cdot error(k) \tag{8.3}$$

where the error is the difference between the desired gain level and the most recently measured gain level. The loop constant μ determines the trade-off between adjustment speed and unwanted jitter of the loop gain. Often, the AGC gain is held fixed after initialization ($\mu = 0$).

8.1.3   Synchronization (Clock, Frame)

Acquisition of synchronization is crucial to adequate performance in DSLs. Even small errors in clock frequency can lead to significant transmission performance degradation. The accuracy of clock recovery can be fairly quickly estimated for a DSL if the channel transfer characteristic and the desired SNR are known. A nominal transmit signal $x(t)$ is essentially offset to $x(t+e)$ when a receiver does not know the transmit clock exactly. The phase error e should be small and can be time-varying. A Taylor series expansion of the signal about time t is

$$x(t+e) \approx x(t) + \dot{x}(t) \cdot e \tag{8.4}$$

Clearly the second term on the right is distortion proportional to e. Thus to make phase-error distortion negligible,

$$\frac{\varepsilon_X}{\varepsilon_X \cdot \sigma_e^2} << SNR \tag{8.5}$$

For a sinusoid of frequency f_m, or for a signal with frequencies no greater than f_m, the bound above can be rewritten

$$\sigma_e^2 << \frac{1}{4\pi^2 f_m^2 \cdot SNR}$$

(8.6)

which shows that for an SNR = 20 dB (100) and $f_m = 100$ kHz, then the standard deviation of the timing error should be 160 ns or less. For an SNR = 40 dB and a maximum frequency of 1 MHz, this bound suggests that timing error standard deviation be less than 1.6 ns. Required clock accuracy can thus vary from about 1 to 2% error for ISDN to levels as low as 0.1% for ADSL or VDSL.

Initial synchronization is almost always achieved by transmission of a mid-band sinusoid from the Central Office modem to the remote modem. The remote modem samples at what it expects to be the zero-crossings of the sinusoid. The signal level at the sampling instants forms a phase error that is used by a phase-lock loop in the remote modem to recover clock using a PLL as described in Section 8.4. In the future, one could expect broadband (multiple) pilots to be sent to speed timing acquisition.

Most DSLs use **loop timing**, which means that the clock recovered by the remote modem is reused for upstream transmission. Thus, only one phase-lock loop is necessary because the other non–PLL-end modem may use its transmit sample clock to derive its receive sample clock.

8.1.4   First Channel Identification

Channel identification is necessary for transmission designs that use an equalizer, particularly a DFE or TEQ, and also for DMT. Channel identification may execute twice in the most sophisticated designs for best practical optimization of the transmitter of a DSL (see Section 8.1.6). Some transceiver designs, for instance, HDSL, combine channel identification and channel equalization in a single step. This is certainly convenient but does not easily allow for adjustment of a transmit filter or spectrum, which can improve performance. In any case, this step of channel identification can be performed indirectly in modems that only equalize at the receiver (see Section 7.2.5).

Generally, channel identification methods measure the channel pulse response and noise power spectral density or their equivalents (i.e., direct measurement of SNRs without measuring pulse response and noise separately). Multitone channel identification directly estimates signal and noise parameters for each of the subchannels, but these estimates can always be converted to an aggregate channel and noise estimate for single-channel transmission. This section will focus on a method of channel identification specific to DMT, but for which the channel impulse response and noise correlation function can be computed for other transmission methods. This method is not necessarily efficient in its use of data, but is matched well to that often used in T1.413 ADSL modems. More sophisticated and efficient methods exist but are not presented here, although some issues and directions for improvement are outlined at the end of the section.

The method of this text separately identifies first the pulse-response DFT values at the subchannel center frequencies used by a DMT system and second the noise-sample variances at

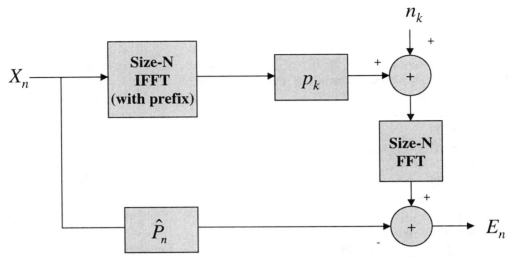

Figure 8.2   Channel identification, gain estimation, and noise estimation.

these same frequencies. Channel identification then finishes by computing the SNR_n $n = 0,...,\frac{N}{2}$ from the results of gain and noise estimation. The measurement of the pulse-response DFT values is called **gain estimation**,[3] while the measurement of the noise variances is called **noise estimation**.

The channel noise is an undesired disturbance in gain estimation, and this noise needs to be averaged in determining subchannel gains for the channel. When gain estimation is complete, the estimated (noiseless) channel output can be subtracted from the actual channel output to form an estimate of the channel noise. The spectrum of this channel noise can then be estimated as in Section 8.1.4.2. In gain estimation, care is taken to ensure that any residual gain estimation error is sufficiently small that it does not significantly corrupt the subsequent noise spectral estimation process. This can require large dynamic range in estimating the subchannel gains.

8.1.4.1   Gain Estimation

Figure 8.2 illustrates gain estimation. The known training sequence, x_k, is periodic with period N equal to or greater than the number of coefficients in the unknown channel pulse response p_k ("periodic" can mean use of the DMT cyclic prefix of Section 7.2 on otherwise nonperiodic training data, as is the case in the Medley sequence of ADSL's T1.413 [2]). The channel output sequence is

$$y_k = x_k * p_k + u_k \tag{8.7}$$

---

3.   This gain is usually a complex gain, meaning that both subchannel gain and phase information are estimated.

where u_k is an additive noise signal assumed to be uncorrelated with x_k. The channel distortion sequence is u_k here, instead of n_k, to avoid confusion of the frequency index n with the noise sequence n_k and because practical multichannel implementations may have some residual signal elements in the distortion sequence u_k (even though independence from x_k is assumed for simplification of mathematics).

Gain estimation constructs an estimate of the channel pulse response, $\hat{p}_k$, by minimizing the mean square of the error

$$e_k = y_k - \hat{p}_k * x_k \tag{8.8}$$

Ideally, $\hat{p}_k \neq p_k$. The unitary-normalized DFT of x_k is

$$X_n = \frac{1}{\sqrt{N}} \cdot \sum_{k=0}^{N-1} x_k \cdot e^{-j\frac{2\pi}{N}kn}. \tag{8.9}$$

When x_k is periodic with period N samples, the DFT samples X_n are also periodic with period N and constitute a complete representation for the periodic time sequence x_k over all time. The N-point DFTs of $y_k, u_k, \hat{p}_k, p_k$, and e_k are $Y_n, U_n, \hat{P}_n, P_n$, and E_n. Since x_k is periodic, then

$$Y_n = P_n \cdot X_n + U_n \tag{8.10}$$

The corresponding frequency-domain error is then

$$E_n = Y_n - \hat{P}_n \cdot X_n \quad n = 0, \ldots, N \tag{8.11}$$

(Recall that subchannels 0 and N are one-dimensional in this development.) A vector denotes one cycle of any of the periodic sequences in question, for instance

$$x = \begin{bmatrix} x_{N-1} \\ \vdots \\ x_1 \\ x_0 \end{bmatrix} \tag{8.12}$$

Then, the relations

$$X = Q^{*}x \quad ; \quad x = QX \tag{8.13}$$

follow easily, where Q^{*} is the matrix describing the DFT ($q_{N-1-i,N-1-k} = e^{-j\frac{2\pi}{N}ki}$) and $QQ^{*} = Q^{*}Q = I$. Similarly, $y, Y, e, E, p, P, \hat{p}, \hat{P}, u, U$ as corresponding time- and frequency-domain vectors. For the case of the nonperiodic sequences u_k and thus y_k and e_k, a block time index l differentiates between different channel-output blocks (symbols) of N samples, i.e. y_l.

The MSE for any estimate p is then

$$
\begin{aligned}
MSE \quad &= \quad E\left\{\left|e_k\right|^2\right\} \\
&= \quad \frac{1}{N} \cdot E\left\{\|e\|^2\right\} = \frac{1}{N} \cdot \sum_{k=0}^{N-1} E\left\{\left|e_k\right|^2\right\} \\
&= \quad \frac{1}{N} \cdot E\left\{\|E\|^2\right\} = \frac{1}{N} \cdot \sum_{k=0}^{N-1} E\left\{\left|E_k\right|^2\right\}
\end{aligned}
\tag{8.14}
$$

The tap error is $\delta_k = p_k - \hat{p}_k$ in the time domain and $\Delta_n = P_n - \hat{P}_n$ in the frequency domain. The autocorrelation sequence for any discrete-time sequence x_k is defined by

$$r_{xx,k} = E\left\{x_m x_{m-k}^{*}\right\} \tag{8.15}$$

with corresponding discrete power spectrum

$$R_{xx,n} = Q r_{xx,k} \tag{8.16}$$

The norm tap error (NTE) is defined as

$$
\begin{aligned}
NTE \quad &= \quad E\left\{\|p - \hat{p}\|\right\} = E\left\{\|\delta\|\right\} \\
&= \quad E\left\{\|P - \hat{P}\|\right\} = E\left\{\|\Delta\|\right\} \\
&= \quad E\left\{\sum_{n=0}^{N-1}\left|P_n - \hat{P}_n\right|\right\}
\end{aligned}
\tag{8.17}
$$

When the estimate $\hat{p}$ equals the true channel pulse response p, then $e_k = u_k$ and

$$r_{ee,k} = E\left[e_m \cdot e_{m-k}^{*}\right] = r_{uu,k} \quad \forall \ k = 0, \ldots, N-1 \tag{8.18}$$

where $r_{ee,k}$ is the autocorrelation function of e_k, and $r_{uu,k}$ is the autocorrelation function of u_k. The MMSE is then, ideally,

$$MMSE = r_{ee,0} \tag{8.19}$$

the mean-square value of the channel noise samples. $\sigma_u^2 = \dfrac{N_0}{2}$ for the AWGN channel.
In practice, $\hat{p}_k \neq p_k$ and the excess MMSE is

$$\sigma_{emse}^2 = MMSE - \sigma_u^2 \tag{8.20}$$

The SNR of the nth channel is

$$SNR_n = \frac{E_n |P_n|^2}{R_{uu,n}} \tag{8.21}$$

The gain-estimation SNR is a measure of the channel-output signal power on any sub-channel to the excess MSE in that same channel (which is caused by $\hat{p} \neq p$), and is

$$\gamma_n = \frac{E_n \cdot |P_n|^2}{R_{ee,n} - R_{uu,n}} \tag{8.22}$$

γ_n should be 30 to 60 dB for good gain estimation. To understand this requirement on γ_n consider the case on a subchannel in DMT with $SNR_n = 44$ dB, permitting 2048 QAM to be transmitted on this subchannel. The noise is then 44 dB below the signal on this subchannel, but to estimate this noise power to within 0.1 dB of its true value, any residual gain estimation error on this subchannel, $R_{ee,n} - R_{uu,n}$, should be well below this mean-square noise level. For 0.1 dB degradation in SNR (0.1 dB = (1+1/40)), $R_{ee,n} - R_{uu,n} = \frac{1}{40} \cdot R_{uu,n}$ or 16 dB below $R_{uu,n}$. Then, we can tolerate only (1 + 1/40) = 0.1 dB excess error in gain estimation. Then, γ_n should be at least 60 dB (= 44 + 16 dB). Similarly, on a comparatively weak subchannel for which we have $SNR_n = 14$ dB for 4-point QAM, one would need γ_n of at least 30 dB (= 14 + 16).

8.1.4.2   An Accurate Gain Estimation Method

A particularly accurate method for gain estimation uses a cyclically prefixed training sequence x_k with period, N, equal to or slightly longer than ν, the length of the equalized channel pulse response. The receiver measures (and possibly averages over the last L of $L + 1$ cycles) the corresponding channel output and then divides the DFT of the channel output by the DFT of the known training sequence.

The channel estimate in the frequency domain is

$$\hat{P}_n = \frac{1}{L}\sum_{l=1}^{L}\frac{Y_{l,n}}{X_{l,n}}$$

(8.23)

The output of the nth subchannel on the lth symbol produced by the receiver DFT is $Y_{l,n} = P_n \cdot X_{l,n} + U_{l,n}$. The estimated value for $\hat{P}_n$ is

$$\hat{P}_n = P_n + \sum_{l=1}^{L}\frac{U_{l,n}}{L\cdot X_{l,n}}$$

(8.24)

so

$$\Delta_n = -\frac{1}{L}\sum_{l=1}^{L}\frac{U_{l,n}}{X_{l,n}}$$

(8.25)

By selecting the training sequence, $X_{l,n}$, with constant magnitude, the training sequence becomes $X_{l,n} = |X| \cdot e^{j\theta_{l,n}}$.

The error signal on this subchannel after $\hat{P}_n$ has been computed is

$$\begin{aligned}
E_n &= Y_n - \hat{P}_n \cdot X_n = \Delta_n \cdot X_n + U_n \\
&= U_n + \frac{1}{L}\sum_{l=1}^{L}U_{l,n}\cdot e^{j(\theta_n - \theta_{l,n})}.
\end{aligned}$$

(8.26)

The phase term on the noise will not contribute to its mean-square value. U_n corresponds to a block that is not used in training, so that U_n and each of $U_{l,n}$ are independent, and

$$E\left\{\left|E_n\right|^2\right\} = R_{ee,n} = R_{uu,n} + \frac{1}{L}R_{uu,n} = \left(1+\frac{1}{L}\right)\cdot R_{uu,n}$$

(8.27)

The excess MSE on the ith subchannel then has variance $\left(\frac{1}{L}\right)\cdot R_{uu,n}$ that is reduced by a factor equal to the number of symbol lengths that fit into the entire training interval length, with respect to the mean-square noise in that subchannel. Thus the gain-estimation SNR is

$$\gamma_n = L\cdot\frac{R_{xx,n}\cdot\left|P_n\right|^2}{R_{uu,n}} = L\cdot SNR_n$$

(8.28)

which means that the performance of the recommended gain estimation method improves linearly with L on all subchannels. Thus, while noise may be relatively higher (low SNR_n) on some subchannels than on others, the extra noise caused by misestimation is a constant factor below the relative signal power on that same subchannel. For an improvement of 16 dB (suggested earlier to be good no matter what the subchannel SNR_n) gain estimation requires $L = 40$.

8.1.4.3 Windowing

When the FFT size is much larger than the number of significant channel taps, as is usually the case, it is possible to significantly shorten the training period. This is accomplished by transforming (DFT) the final channel estimate to the time domain, windowing to $v + 1$ samples, and then returning to the frequency domain. Essentially, all the noise on sample times $v + 2,..., N$ is thus removed without disturbing the channel estimate. The noise on all estimates, or equivalently the residual error in the SNR, ideally improves by the factor N / v with windowing.

8.1.4.4 Suggested Training Sequence

One training sequence for the gain estimation phase is a PRBS (e.g., with polynomial $1 + D + D^{16}$) with $b_n = 2$ on channels $n = 1,..., N - 1$ and $b_n = 1$ on the DC and Nyquist subchannels. One needs to ensure that the initial condition for the PRBS circuit is such that no peaks occur that would exceed the dynamic range of channel elements during gain estimation. Such peaks are usually compensated by a variety of other mechanisms when they occur in normal transmission (rarely) and would cause unnecessarily conservative loading if their effects were allowed to dominate measured SNRs. L blocks are used to create $\hat{P}_n$, which is computed recursively for each of the subchannels, one symbol at a time, by accumulation, and then divided by L. The L blocks can be periodic repeats or a cyclic prefix can be used to make the training sequence appear periodic. For reasons of small unidentified channel aberrations, like nonlinearities, it is better to use a cyclic prefix because any such aberrations will appear in the noise estimation to be subsequently described.

In complex baseband channels, a "chirp" training sequence for a periodic block of N samples sometimes appears convenient:

$$x_k = e^{j\frac{2\pi}{N}k^2}$$

(8.29)

This chirp has a the theoretical minimum peak-to-average (one-dimensional) power ratio of 2 and is a "white sequence," with $r_{xx,k} = \delta_k$, with δ_k being the Kronecker delta (not the norm tap error). While "white" is good for identifying an unknown channel, a chirp's low PAR can leave certain nonlinear "noise" distortion unidentified, which ultimately would cause the identified SNR_n s to be too optimistic.

8.1.4.5 Noise Spectral Estimation

It is also necessary to estimate the discrete noise power spectrum (clearly the noise can not have been whitened if it is unknown, and so this step precedes any noise whitening that might occur with later data transmission). The noise estimates can be mean-square distortion at each of the demodulator DFT outputs. Then the receiver can compute the SNRs necessary for the bit distribution using the channel gains and noise estimates,

$$NSR(i) = \frac{\hat{\sigma}_i^2}{\left|\hat{P}_i\right|^2}$$

(8.30)

Noise estimation computes a spectral estimate of the error sequence, E_n, that remains after the channel response has been removed. The training sequence used for gain estimation is usually continued for noise estimation.

The variance of L samples of noise on the nth subchannel is:

$$\hat{\sigma}_n^2 = \frac{1}{L} \cdot \sum_{l=1}^{L} \left|E_{l,n}\right|^2$$

(8.31)

where $E_{l,n}$ is computed as defined in Section 8.1.4.2. For stationary noise on any tone, the mean of this expression is the variance of the (zero-mean) noise. The "variance of this variance estimate" is computed as

$$\text{var}\left(\hat{\sigma}_n^2\right) = \frac{2}{L} \cdot \sigma_n^4$$

(8.32)

for Gaussian noise. Thus, the excess noise in the estimate of the noise is $\sqrt{2/L} \cdot \sigma_n^2$ and the estimate noise decreases with the square-root of the number of averaged noise terms. To make this extra noise 0.1 dB or less, we need $L = 3200$. T1.413 ADSL uses $L = 4000$. It is possible that the designer wants to ensure a 0.1 dB maximum error on 99 of 100 trains, which actually means that the extra noise should correspond to the 3-sigma point of its distribution, which for Gaussian excess noise would require $L = 32000$ for this kind of training.[4]

The signal-to-noise ratio is then estimated for each subchannel independently as

$$SNR_n = \frac{E_n \cdot \left|\hat{P}_n\right|^2}{\hat{\sigma}_n^2}$$

(8.33)

---

4.   For such "high-confidence" training, more sophisticated noise-elimination methods usually replace those listed here, rather than significant training-time increases.

For $L = 40$ in GE and $L = 3200$ in NE, SNR deviation should be less than 0.2 dB.

8.1.5   Channel Equalization

Channel equalization attempts to reduce the severe frequency variation in the channel on long twisted pairs. Usually channel equalization for DSLs is separated into two parts: analog and digital. The combination of the analog and digital filters attempts to realize the MMSE characteristics in Section 7.1.2.

Usually, the analog equalizer in a DSL is a fixed filter (sometimes one of several fixed filters is selected based on received signal level). Since DSLs all exhibit some attenuation with increasing frequency, an analog filter is used in the receiver to simplify some of the post-matched-filter equalization discussed earlier, prior to sampling and the matched filter (the MF is absorbed into the equalization). Such filtering can reduce the precision required by the ADC when the ADC quantization noise floor is the dominant source of noise at high frequencies. Of course, not too much high-frequency increase can occur in a fixed filter because short loops might then have increased distortion artificially created by the analog equalizer.

The remaining more exact channel-dependent filtering needs to be implemented adaptively and digitally. Simple modems will send an equalizer training sequence that is known to both transmitter and receiver. The receiver then adaptively adjusts its digital filter until the MSE error between the training sequence and the equalizer output is minimum. This adjustment can be executed immediately after synchronization and the channel ID of Section 8.1.4.

More sophisticated modems often have a periodic training sequence. A DFT of the periodically averaged channel output (with period equal to that of the training sequence) is divided by the known DFT of the training sequence to set the equalizer rapidly.

8.1.5.1   Decision Feedback

Decision feedback equalizers can also be trained through MMSE adjustment with the LMS adaptive algorithm simultaneously on both filters for a known training sequence, as in Section 8.2.1.1. Such training is often extremely slow and prone to numerical error in limited-precision implementation.

More reliably, a periodic sequence is sent and the channel directly identified (rather than equalized). Once the channel is identified, the formulas or equivalent algorithms in Section 7.1.2.6 can be used to calculate the proper equalizer coefficients. If Tomlinson or Laroia precoding is used, the feedback section needs to be returned to the transmitter through a reliable reverse low-speed transmission.

8.1.5.2   DMT's TEQ

The TEQ exhibits effects similar to those of the DFE. The only difference is that the feedback section need not be causal and the relative delay of feedforward and feedback sections can be adjusted to reduce substantially the number of coefficients required in the feedforward filter. Those coefficients are the TEQ.

TEQ training identifies the channel and then computes the coefficients according to the TEQ formulas of Section 7.2.5. Such computation may require the determination of an eigenvector corresponding to a minimum eigenvalue, often known as a **Rayleigh quotient** in mathematics. Algorithms to compute such are known as **Lanczos algorithms** [5]. Alternately, the simplified approximate Chow algorithms of the next subsection can be used.

8.1.5.3   Chow's Eigen-Updating and Related Methods

Direct implementation of the matrix inversions in mathematical descriptions of equalizers can sometimes be difficult. **Chow's eigen-updating methods** [6], [7] attempt to approximate iteratively such inversions using algorithms/hardware necessary for other essential functions in DMT transmission (namely, FFT and IFFT). This method is often effective, but not necessarily globally convergent, and is used in many early ADSL systems. An improvement occurs with limited precision in Ref. [7], but unfortunately neither the methods in Ref. [6] nor Ref. [2] universally converge to a global optimum.

To use **eigen-updating**, the input to the transmission system is periodic with period N samples greater than the length of the channel response to be equalized. Typically N is equal to the FFT size in a DMT system so that the FFT hardware or software in place for steady-state data transmission can be reused. The periodic data is typically robust in that all frequencies carry nonzero energy (subject to satisfaction of various spectral masks that may be imposed on the signal). Since the signals are cyclic, FFTs correspond to eigen-decompositions. Thus, a receiver FFT operation on the channel output decomposes the channel output into its independent eigenvalues/modes. The required equalization value for each such independent mode can be determined easily and then the time-domain equivalent computed via IFFT.

The basic algorithm makes use of the four basic steps iteratively until a threshold criterion is met and the algorithm terminates. These steps are:

1. Update b
2. Window b
3. Update w
4. Window w

It is desirable to perform the updating in the frequency/eigen domain. Thus, initially the equalized-by-w channel output corresponding to one period of the periodic training sequence is FFT'd and the resultant channel response identified by dividing by the known DFT values for the training sequence. This estimate may be averaged with previous estimates of the same quantity, thus implementing step 1. This channel response may have up to $N/2$ complex values (for a real input) in the frequency domain that correspond to N values in the time domain, exceeding the allowable length L_b of the vector b corresponding to one of the equalization filters. Time-domain windowing of b to the L_b consecutive taps with greatest energy then corresponds to the second step. For DFE systems, the positions of these taps are determined and fixed after the first iteration, while for the TEQ their position can be varied in subsequent iterations. The windowed response is then returned to the frequency domain via FFT and the corresponding "target" channel output is formed by multiplication with the known DFT of the training sequence. The difference between this response and the equalized channel output (in frequency domain) is used to update the frequency-domain equivalent of w, implementing step 3. Step 4 then is implemented by converting w to the time domain via IFFT and windowing for the L_w consecutive samples with largest energy. The FFT of this windowed equalizer is then computed for use in updating w in subsequent iterations.

The update of each filter executes independently on each tap in the frequency domain and can be an instantaneous **zero-forcing** update according to

$$
B_i = \frac{\overline{W}_{i,old} \cdot Y_i}{X_i} \quad \text{and} \quad W_i = \frac{\overline{B}_i \cdot X_i}{Y_i} \tag{8.34}
$$

where Y_i is the FFT of the channel output, X_i is the FFT of the known training sequence, and bars denote filters that corresponding to time-domain windowing. Several successive zero-forcing updates can be averaged if the designer desires. Alternately, averaged updating can be executed with the mean-square-error-minimizing LMS algorithm:

$$
B_i = \overline{B}_{old,i} + \mu_i \left(\overline{W}_{i,old} \cdot Y_i - \overline{B}_{old,i} \cdot X_i \right) X_i^*
$$

$$
W_i = \overline{W}_{old,i} + \mu_i \left(\overline{B}_i \cdot X_i - \overline{W}_{old,i} \cdot Y_i \right) Y_i^* \tag{8.35}
$$

Van Kerckhove has noted [7] that leakage factors on those frequency taps that correspond to regions of little or no channel output energy can dramatically reduce precision required to implement updating of the TEQ filter. Mixture of updates sometimes also occurs, with a zero-forcing update of b followed by an LMS update of w being popular because the division can be easily implemented for careful choices of the periodic training sequence (as in T1.413 ADSL).

Various modifications exist to the eigen-updating that use criteria other than zero-forcing and MMSE. Both Al-Dhahir [8] and Henkel [9] note that the geometric SNR of all the (used) subchannels should be maximized. Such a geometric SNR should include the "noise" distortion that occurs because the filter b has been windowed in the time domain. The squared-FFT values for the error between the windowed and unwindowed b (times the corresponding input squared energy $E\left\{ |X_i|^2 \right\}$) should be added to the measured noise on each of the corresponding subchannels. The product of (1 plus) the individual subchannel SNRs (each divided by the appropriate gap) should then be minimized. Various iterative algorithms that exist for minimizing such a geometric SNR would replace the LMS updates with updates based on incremental changes in the direction of the negative gradient of geometric SNR with respect to each of the time-domain filter coefficients.

8.1.6   Secondary Channel Identification and Exchange

A second instance of channel identification is often performed after equalizers have trained and set. This second step allows the measurement of SNRs with the equalizer in place rather than theoretical projections of the same. Thus, nonlinearity, finite length, and finite precision effects, as well as any training imperfections are all measured. The results of this step are usually used for loading in DMT methods, like ADSL, resulting in an exchange of channel characteristics from receiver back to transmitter for appropriate setting of any channel-dependent transmit parameters. The channel identification in this step can be the same as in Section 8.1.4. The exchanged characteristics need to be conveyed without error to the transmitter, so a low-speed highly reliable transmission format is used. For instance, in T1.413, two sets of four tones independently carry the same information at 2 bits/tone (or 1 byte, which is well below the maximum capable on nearly all DSLs) on tones 6 to 9 and again on 10 to 13. Additionally, a CRC check on the information allows detection of erroneous channel-characteristic information.

Generally, many different equivalent sets of channel characterization information can be conveyed back to the transmitter by a number of reliable methods as long as the channel allows bidirectional transmission. T1.413 ADSL was the first standardized DSL to use any exchange mechanism.

8.2 Adaptation of Receiver and Transmitter

Time variation of the channel response in a DSL is very slow with respect to most other transmission applications, notably with respect to wireless. Sources of time variation are typically temperature and/or other environmental changes, typically taking several seconds or minutes for appreciable change. Heavily loaded DSL cables, however, do have the potential for crosstalk noise to change suddenly as service is introduced or deleted on one or more of the pairs in the bundled cable. Such change is nearly instantaneous, although then fixed thereafter (after initial training transients have abated). Fortunately, these sudden crosstalk noise changes occur at absolute voltage levels well below the signal levels, so that the positive margins inherently provided for operation provide some immunity against the sudden small increase of noise. Nonetheless, the margin is reduced unless the transmission system takes corrective action. Again, this corrective action can be relatively slow ***if proper planning/provision of the services in the cable has occurred a priori.***[5] AM radio stations are also known to change their power levels during the

---

5.   While easily said, such proper planning and provisioning is the main focus of "spectrum compatibility" studies for DSL services that are active at time of writing in both the ANSI T1E1.4 and ETSI TM6 groups. The output of these studies is an expected set of guidelines for provisioning services so that sudden huge jumps in crosstalk noise would not disable service (exceed allocated margin) and thus allow moderately slow reaction of the DSLs to the change to be sufficient. As DSL service becomes widely deployed, more global "network-adaptive" methods will be required to accommodate the service needs and potential interference of many users. Crosstalk cancellation and coding methods are also progressing and may provide a greater degree of immunity to crosstalk noise changes in the future.

day (increasing power at night) and thus constitute another source of time-varying noise.

Given these sources of time variation, the DSL must have some degree of continuous adaptation of the receiver, and with ADSL/VDSL and other more advanced DSLs, updating at the transmitter also. This section describes some methods in use today and provides some guidelines for study in the future. The section opens with a review of heavily used adaptive algorithms for DSLs and then proceeds to more sophisticated, better-behaved methods for strained DSL transmission in Section 8.2.1. Section 8.2.2 covers transmitter adaptation, including the so-called "bit-swapping" methods (which actually are more sophisticated than the trading of bits implied by the name) that characterize DMT.

Recent interest in splitterless ADSL has identified an abrupt channel change when a telephone goes off-hook and the impedance changes. Such abrupt changes cause at least partial interruption in service for retraining or adjustment [10], [11].

8.2.1 Receiver Equalization Updating

Equalizers were discussed in Section 7.1 for mitigation of ISI and in Section 7.2 for channel shortening in multichannel transmission designs. Changes in the transmission line or noise characteristics of the DSL necessitate changes in the equalizer, requiring the equalizer to be adaptive for best performance.

Equalizers are usually updated during data transmission to maintain MMSE settings if any changes in the channel or noise occur. The updating is usually performed at regular intervals within the complexity constraints of the implementation (the slower the updating, the more negligible the cost). The most commonly encountered method for maintaining the MMSE setting is known as the least mean square or LMS algorithm. The algorithm requires a **desired response**, which for equalization is the channel input data symbol (delayed appropriately and possibly filtered in the case of PR equalizer, TEQ, or sometimes DFE updating). This input is assumed to be equal to the output of the decision device (i.e., correct decisions are assumed) for partial response or TEQ, the desired response is the convolution of the PR channel with the estimated channel input symbol sequence.

Adaptive filters form an error between the decision output and the decision device input,

$$e_k = \hat{x}_{k-\Delta} - \sum_{i=0}^{L_w-1} w_i \cdot y_{k-i} - \sum_{i=1}^{L_b} b_i \cdot \hat{x}_{k-i} \tag{8.36}$$

where $b_k = 0$ if the equalizer is linear and nonzero for the DFE, TEQ, or PR equalizers. The length, N_b, and any fixed or zeroed terms of b_i depend on whether DFE, TEQ, or PR is being updated. The error and filters will be complex with QAM or CAP.

8.2.1.1 The LMS Algorithm

The basic LMS algorithm [12] then computes new settings for the ith tap of these filters according to

$$w_{i,k+1} = w_{i,k} + \mu \cdot e_k \cdot y_{k-i}$$
$$b_{i,k+1} = b_{i,k} + \xi \cdot e_k \cdot \hat{x}_{k-i}$$

(8.37)

The parameters μ and ξ are **step-size** constants chosen to determine tracking speed and residual error from true MMSE setting. The LMS algorithm has long been a hallmark of voice-band modem equalizer designs (perhaps with some modifications going under a variety of names). Sometimes "synch symbols" or "training patterns" are regularly embedded in the transmit signals so that equalizers can be trained without concern that decisions, $\hat{x}_{k-i}$, are incorrect. For early DSLs, the LMS algorithm can be sufficient. However, because of the increasing use of longer, possibly oversampled, DFE filters and TEQs, the LMS algorithm may have severe limitations of slow convergence and near-explosive numerical precision problems. These problems occur particularly if the filters have significant singularity (i.e., the Fourier transform of the correct filter setting has large dynamic range — very typical in DSLs — and many orders of magnitude larger than evident in voice-band modems where LMS is usually sufficient). In this case, use of the LMS algorithm often becomes frustrated, and the equalizer will not converge or track changes and performance degrades. In this case, more complicated updating algorithms should be used — some are minor modifications and some are completely different. To understand the effects, a few basic principles help:

The **excess MSE** of the LMS algorithm is a level of fluctuation above the absolute minimum MSE setting that is caused by the adaptive updating, just as it was for channel identification. In adaptive filter design, best performance suggests small excess MSE, but increasing the speed of updating causes more excess MSE. The formula for the excess MSE of an LMS adaptive equalizer is

$$EMSE \leq \frac{max(\mu,\xi)}{2}\,\text{trace}\langle R_{YY} \rangle$$

(8.38)

where trace is the sum of the diagonal elements of the autocorrelation matrix of the channel output for a linear equalizer and is the sum of diagonal elements of the combined input/output autocorrelation matrix for the full DFE, denoted here by an uppercase Y. When the two step sizes are equal, the inequality becomes an equality. Seemingly small step sizes would reduce the EMSE to an acceptably small level.

However, it is also possible to determine a set of **time-constants** for the algorithm. Roughly speaking, these time constants are inversely proportionally to the step size and unfortunately each independently to the inverse of an eigenvalue of the matrix R_{YY}, where some of the latter eigenvalues can be very close to zero. The closer these eigenvalues to zero, the slower the convergence. Unfortunately in DSLs, the small values can result in time constants literally on the time frame of hours or days in systems operating near the limits of the channel. Since these small energy areas contribute little to the performance, one could legitimately ask "Why not just avoid them in training?" This is good insight and leads to the essence of a solution. Unfortunately, the simple LMS algorithm does not have this capability, thus forcing sophisticated solutions that do essentially ignore regions of zero energy.

Furthermore, it can be shown [13] that the numerical errors induced by updating in finite precision can accumulate in the LMS algorithm. Those errors accumulate to a level inversely proportional to the smallest eigenvalue of R_{YY}, again causing near-catastrophic levels of accumulation in a DSL operating near limits. Thus, the LMS algorithm may become impractical in DSLs using a wide bandwidth, especially with DFEs or TEQs.

One method for partially reducing the effect of finite-precision error accumulation (at the expense of a further increase of EMSE) is the so-called leaky LMS algorithm [13] in which the updates are modified to

$$w_{i,k+1} = \lambda \cdot w_{i,k} + \mu \cdot e_k \cdot y_{k-i}$$
$$b_{i,k+1} = \lambda \cdot b_{i,k} + \xi \cdot e_k \cdot \hat{x}_{k-i} \tag{8.39}$$

where the factor λ is slightly less than one and used to "leak" the coefficients of numerically introduced errors. For a more complete discussion, see the tutorial by Cioffi [13].

Leakage can also reduce the effect of error propagation in a DFE [14]. Leakage, however, causes a bias from the true solution, and the trade-off between the size of this bias and acceptable tracking speed is often unacceptable.

8.2.1.2   Chow's Eigen-Updating Used Continuously

During transmission of data, the updating of equalizers, particularly the TEQ, may be complicated by the lack of readily available known training data. One solution is to use the detected data, which of course may have errors. Another is to periodically insert a known training symbol (as for instance, the "synch" symbol of T1.413 ADSL).

Then, the updating algorithms (Chow's eigen-updating) of Section 8.1.3.3 can be applied on those symbols selected for updating — since the amount of computation is large, updating is typically executed only once every several symbols (100s to 1000s) so that computational increase is minimal.

8.2.1.3   Matrix-Based Equalizer Updating

The solution to the numerical problems and slow convergence can be to compute the equalizer coefficients via direct or indirect implementation of the matrix inversion described in Sections 7.1 and 7.2. The matrix-inversion problem is highly structured and a variety of "fast computation" methods have been studied [8], [9] for a treatment of this particularly for the DFE). Given that equalizer updating may be executed at initialization or at infrequent intervals in time, straightforward implementation of the matrix equations, based on identified values for the channel response, the known input autocorrelation matrix, and the noise autocorrelation matrix can be used to compute values for the equalizer. The identification of channel and noise can be executed as in Section 8.1.4 (again using detected channel inputs or periodically inserted training symbols).

8.2.1.4   RLS Methods

Recursive least squares (RLS) are high-performance adaptive algorithms that converge and track more rapidly than LMS methods (see Refs. [15] and [16]). These methods operate by efficiently minimizing a sum of exponentially weighted squared errors over the settings for the equalizer. Usually, RLS methods are not necessary when the training sequence or synch symbols can be carefully selected in DSLs.

8.2.2   Transmitter Adjustment

Continuous transmitter adjustment depends on the return of channel response and noise information to the transmitter regularly for optimization of the transmit filter, energy and bit distribution, and/or the symbol rates and carrier frequencies in a QAM/CAP implementation. This section starts with general discussion of the identification of information important to transmitter adjustment independent of line code and then proceeds to describe how such information is used in multichannel transmission only.

8.2.2.1   Channel Gain/Response Updating

The channel can be identified by an LMS adaptive filter, fortunately without the same degree of numerical and convergence problems as with the equalizer in Sections 8.1.3 and 8.1.1.1, as shown in Figure 8.2. The input to the channel must be known, either as past decisions or as known periodically inserted training patterns need to be used after training(e.g., known synchronization symbols are embedded periodically in HDSL and ADSL for potential use in updating as well as other purposes).

There are a variety of more efficient methods for channel identification known and used in wireless transmission where more rapid channel variation is common. For identification of the DSL channel response (not noise, just channel), such methods are not usually necessary and the simple frequency-domain LMS is usually sufficient. Some level of energy (below spectral mask levels) should be inserted in all frequency bands to prevent undesirable effects that encompass the use of identified channel characteristics in the computation of equalizers in a recursive manner.

8.2.2.2   Noise Monitoring

Noise can vary more rapidly in DSLs (e.g., crosstalkers turning on or off) and thus some level of efficiency in noise estimation is desirable. Noise estimation is performed on the error sequence in the channel identification of Section 8.2.2.1, just as in Section 8.1.4. However, since execution of channel identification updates may be infrequent for complexity reasons, it is desirable to make better use of the error sequence. Basically, noise estimation amounts to determining the power spectral density of the noise, a problem well known in signal processing as **spectral estimation** [17], [18]. The basic concept is that it is inefficient to identify 256 noise variances when the noise could have been modeled by a smaller number of parameters. In this case the autocorrelation function of the error sequence is computed and windowed before more powerful signal processing methods are used to identify the noise accurately for a limited number of observations. Such spectral estimation is an active area of research in DSLs and beyond the scope of this book. Nonetheless, the authors believe that the "ragged" nature of actual crosstalk interference on an individual transmission line (which is below, but much less continuous with frequency than the simple kf^x bounds used for crosstalk power spectral density) may lead to all 256 parameters being necessary. Should the situation of rapid change of such a truly complicated noise become evident in DSL deployment, then the frequency of update of channel identification (and thus noise monitoring) would need to be increased (with consequent complexity implications) or margin would be reduced.

The next subsection does, however, include some basic methods for improvement in the DMT case.

8.2.2.3   Combined Channel/Noise for DMT

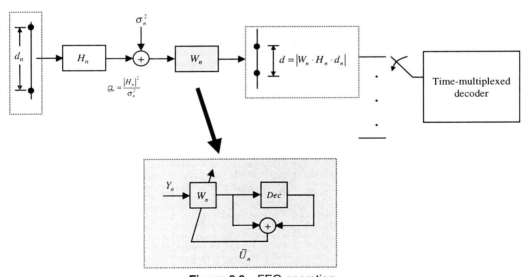

Figure 8.3   FEQ operation.

Figure 8.3 illustrates the operation of an FEQ. The FEQ has no direct impact on performance or SNR because both signal and noise are scaled by the same value W_n. When DMT transmission is used, W_n is a complex scalar. Nominally, the FEQ allows a common decision device to be time-shared for the decoding of all the subchannel outputs. Such normalization of distance satisfies:

$$d = \left| W_n \cdot H_n \cdot d_n \right|$$

(8.40)

where d_n is the minimum distance between points on the nth subchannel constellation on the transmitter side. The FEQ forces all subchannels to have output distance d, independent of the number of bits carried or signal to noise ratio. Clearly,

$$W_n = \frac{d}{H_n \cdot d_n}$$

(8.41)

However, since the subchannel may change its gain or SNR, the value for W_n is adaptively computed periodically (as often as every symbol is possible). The error, $\tilde{U}_n$ between the input and output of decision device shown in the enlarged normalizer box estimates the normalized noise (as long as the decision is correct). The updated value for W_n can be computed in a number of ways that average several successive normalized errors, of which the two most popular are the **zero-forcing algorithm:**

$$W_{n,k+1} = W_{n,k} + \mu \cdot \tilde{U}_n \cdot X_n^*$$

(8.42)

that produces an unbiased estimate of the signal at the normalizer output, equivalently $E\,[\tilde{U}_n/X_n]$ contains only noise, or the more common **MMSE algorithm:**

$$W_{n,k+1} = W_{n,k} + \mu \cdot \tilde{U}_n \cdot Y_n^*$$

(8.43)

which instead produces a biased estimate and an optimistic SNR estimate of X_n. The ZF algorithm is better because there is no issue of noise enhancement with the single tap of the FEQ. The parameter μ is a positive gain constant for the adaptive algorithm selected to balance time-variation versus noise averaging.

The channel gain is estimated by

$$\hat{H}_n = \frac{d}{d_n \cdot W_n}$$

(8.44)

The mean square noise, assuming "ergodicity" at least over a reasonable time period, is estimated by time-averaging the quantity $\left| \tilde{U}_n \right|^2$:

$$\tilde{\sigma}_{n,k+1}^2 = \left(1 - \mu'\right) \cdot \tilde{\sigma}_{n,k}^2 + \mu' \cdot \left|\tilde{U}_{n,k}\right|^2 \tag{8.45}$$

where μ' is another averaging constant. Since this noise was scaled by W_n also, the estimate of g_n is then

$$\hat{g}_n = \frac{\left|\hat{H}_n\right|^2}{\dfrac{\tilde{\sigma}_n^2}{\left|W_n\right|^2}} = \frac{d^2}{d_n^2} \cdot \frac{1}{\tilde{\sigma}_n^2} \tag{8.46}$$

Since the factor $d^2 \big/ d_n^2$ is constant, and the loading algorithm need only know the relative change in the loading-algorithm's[6] incremental energy, i.e., the incremental energy table is scaled by

$$\frac{g_{n,old}}{\hat{g}_{n,new}} \tag{8.47}$$

The new incremental-energy entries can be estimated by simply scaling up or down the old values by the relative increase in the normalizer noise.

When a swap is executed, the value for W_n and the corresponding noise estimate needs to also change. The new values for W_n and $\tilde{\sigma}_n^2$ are

$$W_{n,new} \rightarrow W_{n,old} \cdot \frac{d_{n,old}}{d_{n,new}} \tag{8.48}$$

and

$$\tilde{\sigma}_{n,new}^2 \rightarrow \tilde{\sigma}_{n,old}^2 \cdot \frac{d_{n,old}^2}{d_{n,new}^2} \tag{8.49}$$

---

6.   We presume here the use of the Greedy-type loading algorithms, which make use of the incremental energy tables.

Accurate noise estimation process is just as important as in channel ID. The value for the constant μ' needs to be carefully selected as a function of the expected time variation and the required accuracy.

A more general approach to noise estimation considers the squared error values for the FEQ outputs as inputs to a digital filter $F(D)$ that averages these squared values to produce the estimate. The noise power estimate will have D-transform $S_\sigma(D)$ and the input squared error has D-transform $\varepsilon(D)$, so that

$$S_\sigma(D) = F(D) \cdot \varepsilon(D)$$

(8.50)

The notation here drops the subscript n as it is understood to be the same applied to all subchannels, and the subscript then refers to symbol time index. The simple first-order updating loop for noise power would then be as above

$$\hat{\sigma}_{k+1}^2 = (1 - \mu') \cdot \hat{\sigma}_k^2 + \mu' \varepsilon_k$$

(8.51)

which corresponds to

$$F(D) = \mu' \cdot D \Big/ \left(1 - [1 - \mu'] \cdot D\right)$$

(8.52)

Note the restrictions on the real filter $F(D)$ in general that it be unbiased $F(1) = \sum_k f_k = 1$, and be causal and stable. In general, $F(D)$ is a lowpass filter designed to follow changes in the noise, but to also average fluctuations in the estimation of such noise. It is useful to define $G(D) = F(D) \cdot F(D^{-1})$. Of importance (as usual in filter design) is that the gain of the filter be low $g_0 = \sum_k f_k^2 < 1$ for good rejection of fluctuations, but that the time constants reflecting the tracking ability of the filter (i.e., the geometric series corresponding to the poles) be sufficiently large to track expected time variation of the noise.

From Section 8.1.4 on channel identification, the variance of estimate of the noise is important because the standard deviation from the noise estimate must be insignificant with respect to an SNR-estimation error that would cause incorrect loading (which is in turn a function of β). There, the variance for channel identification was found to be $var(\hat{\sigma}^2) = \frac{2}{L} \cdot \sigma^4$. The comparable quantity for time-varying loading with the above algorithm is

$$var(\hat{\sigma}^2) = g_0 \cdot \sigma^4$$

(8.53)

Then the standard deviation of such estimated noise $\sqrt{g_0} \cdot \sigma^2$ should be much less than the change in SNR that would produce a change in bit distribution, $\left| e_n\left(B_n\right) - e_n\left(B_{n\pm I}\right) \right|$. For the first-order filter, $g_0 = \dfrac{\mu'}{2 - \mu'}$, so that to keep the error in SNR estimate below, say, 0.25 dB, would lead to $1 - \mu' > .993$ and thus about 1000 to 2000 symbols are necessary for significant change in the noise estimate to occur. This number can be reduced with higher-order filtering.

8.2.2.4 Bit-Swapping

Any of the Greedy or Campello loading algorithms in Section 7.2.3 can be run periodically to assess changes in the bit distribution and/or gain settings for a multichannel transmission system. The results then can be conveyed to the transmitter (equivalently, the recomputed SNR distribution can be also be conveyed to the transmitter and loading executed on the transmit side). In either case, the system must have in place a mechanism for communicating the changes in bit distribution reliably. In T1.413 ADSL, this mechanism is called the **bit-swap.** Bit-swapping allows for one (or two in extended bit swapping) bit to be moved from one subchannel to another. Special cases allow a bit to be deleted from any subchannel (without movement to another subchannel, thus decreasing bit rate) or a bit to be added to any subchannel (thus increasing data rate). Issue 2 T1.413 ADSL also provides a mechanism for quick transmission of an entirely new bit distribution from receiver to transmitter with short interruption of service.

Upon execution of a swap, the bit tables on both ends need to be updated in synchronism (see Refs. [2] or [19]) for a protocol that ensures synchronization of the ends in updating. Contrary to some publicized but erroneous beliefs of a few misinformed engineers, this protocol is very robust against errors [20]. A receiving DMT modem can easily determine by inspection when and/or if a previous "swap" request was implemented by a transmitter, even when all commands are corrupt (see Ref. [20]).

Furthermore, swapping is essential (see Ref. [20]) to maintain margin and performance of a DMT modem. For this reason, T1.413 ADSL and g.dmt ADSL (see Refs. [2] and [21]) mandate bit-swapping. Unfortunately, because of misinformation (of dubious intent) spread by essentially a minority, g.lite only strongly recommends bit-swapping, rather than mandating it [11]. As Ref. [20] clearly shows, the simple situation of going from 0 to 1 crosstalker disables a nonswapping modem, forcing it to train again. Therefore, when a nonswapping LITE modem is on the line, g.lite modems have a fundamental flaw. Amazingly, the same fundamental patent that applies to bit-swapping [22] also still applies to the training and retaining of a LITE modem. Thus, while some may have thought they could avoid a patent license by creating a dysfunction in the g.lite specification, this is not true. It is recommended here that designers of LITE modems use the bit-swap to ensure proper operation of their products.

While MA bit-swapping will continue to move bits from channels of higher energy cost to those of lower energy cost, there is still a barrier to moving a bit that is the difference between the cost to add a bit in the lowest cost position of the incremental-energy table relative to the savings of deleting that same bit on the highest-cost position in that same table. In the worst case, this cost can be equal to the cost of a unit of information. On a two-dimensional subchannel with

$\beta = 1$, this cost could be as much as 3 dB of energy for a particular subchannel. The probability of error on that subchannel could then degrade by as much as roughly 3 orders of magnitude before the swap occurred. A particularly problematic situation would be a loss of, for instance, 2.9 dB that never grew larger and the swap never occurred. With a number of subchannels greater than 1000, the effect would be negligible, but as N decreases, the performance of a single subchannel can more dramatically affect the overall probability of error.

There are two solutions to this potential problem. The first is to reduce β, which may be difficult. The second is to reduce energy distributed flatly over all subchannels to increase margin and instead apply some or all of this energy to the weakest subchannels. This corrective action is known as a **gain swap**. With large numbers of subchannels, the likelihood that the sub-channel SNRs are such that a gap of 3 dB between swapping levels exists is small. Thus, gain-swapping is rarely used, especially when margin is sufficiently high. T1.413 ADSL and g.dmt have gain-swapping included in the bit-swap protocol.

8.2.2.5   Continuous Transmit Filter Adjustment

Transmit filter adjustment can be of great benefit to single-carrier transmission systems like QAM or CAP, but only of such great benefit when the transmit symbol rate and carrier/center frequency are optimized. Section 7.2.8 discusses transmit filter optimization for multichannel transmission with too-short guard period, and Section 7.1.5 discusses the more difficult transmitter optimization for single-carrier methods (when such optimization can be done, which may not the case in CAP/QAM-based DSLs). Recent CAP/QAM ADSL and VDSL designs appear to introduce this concept, but the degree of implementation is not presently known by the authors of this book. In both those sections, a variety of situations for transmit filter optimization are considered, both with and without receiver optimization. In all cases, the channel response and sometimes the channel noise power spectra must be known to compute the optimum transmit filter, symbol rate, and carrier/center frequency. The methods of this chapter can be used to identify the best bands and then determine best carrier/center frequencies and symbol rates, as well as the best filter settings.

The present state of the art is to compute the settings for these filters according to the formulas at regular intervals in time, communicate them to the transmitter through a "bit-swap-like" protocol and periodically insert. The author is aware of no current standards that allow for this. V.34 modems do allow for exchange of filter parameters, but only once upon start-up. Such once-optimized transmit filters are not appropriate in DSL, however, because of the dependency on changing noise in the channel.

Some level of adaptive transmitter adaption may be possible through the "duality" concepts of Cioffi and Forney in Ref. [23].

8.3 Measurement of Performance

The measurement of the performance of DSL systems has been very important. Typically, DSL modems are qualified by a service provider for deployment only after passing a series of tests designed to represent worst-case conditions in the loop plant.

8.3.1 Test Loops and Noise Generation

The process of DSL qualification and interoperability testing involves the measurement of transmission performance over a specified set of conditions. Both ANSI and ETSI have specified sets of loops (along with RLCG parameters characterizing the various loop segments — see Chapter 3) for DSLs. In addition, these groups have specified noise models for white background noise, crosstalk noise, radio noise, and impulse noise. In addition to the standards groups sets of loops, a number of service providers have specified their own models for testing. This section enumerates some of the standardized test loops and noises in Section 8.3.1.1 as well as suggesting a general approach to laboratory measurement in Section 8.3.1.2.

8.3.1.1 ANSI/ETSI Testing Specifications

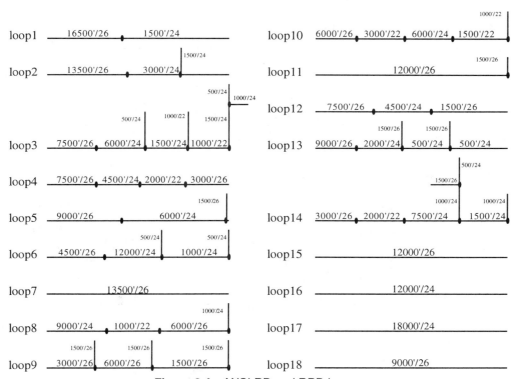

Figure 8.4 ANSI RD and RRD loops.
(The notation 16500'/26 means 16,500 feet of 26-gauge twisted pair.)

ANSI and ETSI have specified test loops and associated noises for the qualification of DSL modems. Figure 8.4 illustrates **resistance design** loops that generally represent the worst

CSA Loops

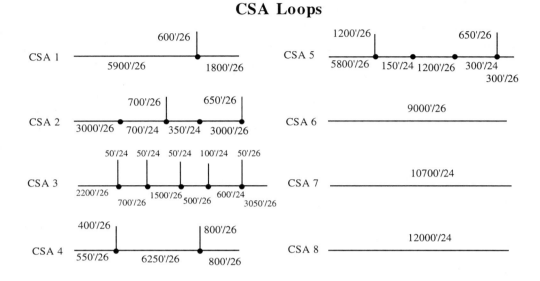

MID-CSA Loops

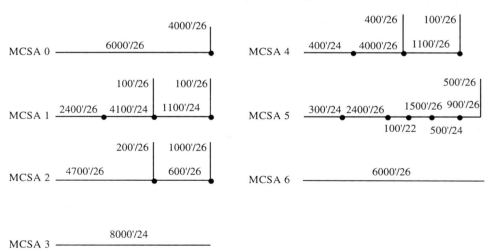

Figure 8.5 ANSI CSA and mid-CSA loops.

10% of nonloaded twisted-pair lines in North America. Loops 1 to 3 correspond to very old deployment practices and have DC resistances that exceed 1300 ohms. Loops 4 to 18 satisfy the so-called **revised resistance design (RRD)** line rules that mandate DC resistance of less than

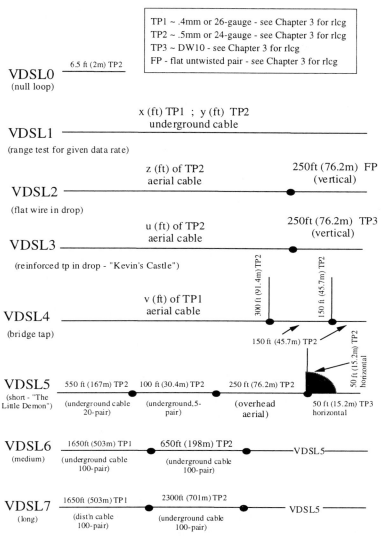

Figure 8.6   ANSI DSA (VDSL) loops.

1300 ohms. Nearly all loops on which DSL would be used are RRD loops or better. Loops that are not RRD require repeaters or special installation practices for DSL service installation or rate-adaptive operation.

RRD loops were originally specified for ISDN testing in the mid-1980s at 160 kbps symmetric data rate. Later, ADSL service at 1.6 Mbps downstream and 160 kbps upstream also used

RRD loops. Good design of ISDN and ADSL modems will ensure these data rates are achieved on nearly all RRD loops.

In the early 1970s, the Bell System began deploying new loops according to **carrier serving area (CSA)** requirements [3]. CSA loops are shorter than RRD loops, roughly two miles maximum. CSA loops are specified for 768 kbps transmission with HDSL and also for 6 Mbps/640 kbps transmission with ADSL. Figure 8.5 (top) lists eight worst-case CSA loops specified by ANSI for testing.

Mid-CSA loops were contrived by ANSI and do not conform to service provider deployment rules; however, these loops have been used for testing DSLs at data rates in excess of 6 Mbps for ADSL downstream. Figure 8.5 (bottom) also illustrates mid-CSA loops.

As fiber loop carrier systems are deployed in North America (and basically elsewhere), the loop lengths are shorter. About 15% of the North American loops use fiber loop carriers and correspond to the shorter loops in Figure 8.6. These loops are usually 3000 feet in length or less, with certain exceptions in remote regions potentially going to 1 mile. These loops are sometimes called **distribution service area (DSA)** loops. DSA loops are specified [73] by ANSI for VDSL service at 26 Mbps/3 Mbps asymmetric service and 13 Mbps symmetric service.

Table 8.1 specifies the crosstalk levels (see Chapter 3) and types of crosstalkers, along with an associated loop for various DSLs. Each of these noises is modeled as Gaussian, which according to Bellcore [24] is accurate and has a power spectral density determined by multiplication of the power spectral density of the corresponding type of crosstalker by the functions $10^{-13} f^{1.5}$ for NEXT and $8 \times 10^{-20} \cdot l \cdot |H|^2 \cdot f^2$ for FEXT where l is the length of the loop in feet and $|H|^2$ is the insertion-loss of the loop, as defined in Chapter 3. The NEXT for downstream receivers from T1 noise can be reduced by 5.5 dB because of the noncolocation of T1 repeaters with respect to other DSL receivers [2].

Actual Crosstalk Variations in Field Deployment   The crosstalk models are typically overly pessimistic for two reasons in practice: First, actual measurements [24], [25] reveal a ragged nature on any line to the crosstalk coupling, leaving the models in Chapter 3 and above as upper bounds on coupling that only apply at a select number of frequencies. Thus, actual crosstalk on any line exhibits significant spectral variation, sometimes showing rippling every few 100 kHz, and potentially spanning 10 to 20 dB in amplitude. Such rippled noise is equivalent to ISI as discussed earlier on equivalent channel analysis. Thus, transmission systems that have effective methods for countering ISI will also perform better with actual crosstalk than with the worst-case models. Second, crosstalk noise is typically synchronous with line signals and other crosstalkers because all DSLs typically loop-time the modem clock to network clocks. In this case, noise in images beyond the Nyquist frequency of the crosstalk is strongly correlated with image-frequency noise below the Nyquist frequency. Thus, a wideband system (like ADSL or VDSL) that spans several images of narrowband crosstalk (like ISDN or HDSL) may perform better than theoretical projections based on Gaussian noise (which is uncorrelated at all frequencies in theory) if the receiver digital signal processing includes a wideband equalizer (the TEQ that is common in ADSL designs).

Table 8.1 ANSI DSL Crosstalk Testing Summary

DSL Type	Loops	Noises (all have AWGN at −140 dBm/Hz included)	Performance $\bar{P}_b$
ISDN (T1.601)	ANSI 1-3	49 ISDN NEXT	1e-7 @ 0 dB
	ANSI 4-15	49 ISDN NEXT	1e-7 @ 6 dB
HDSL	CSA 1-8	49 HDSL	1e-7 @ 6 dB
T1	6000 ft.	None	Recorded
ADSL (T1.413)		(all tests also with imp 1 & 2)	
1.544 M / 176k	ANSI 7,9,13	24 ISDN	1e-7 @ 6 dB
6.144 M / 640k	CSA 4,6,8	10 ISDN + 10 HDSL +24 ADSL	1e-7 @ 6 dB
6.144 M / 640k	midCSA 6	10 ISDN + 24 adj T1	1e-7 @ 6 dB
6.144 M / 640k	CSA 6	4 adj T1	1e-7 @ 0 dB
VDSL – all tests, 5 rates	Record length of		
A:52 M / 6.4M, 26/3.2, 13/1.6	26-gauge 24-gauge	WGN+20 VDSL FEXT	1e-7 @ 6 dB
S: 26/26, 13/13		WGN+20 VDSL FEXT+RFI Noise Model A &B (see [27])	
	+50' bridge tap	WGN+20 VDSL FEXT	
HDSL2	CSA1-8	Reduced noise	Reduced spec

Crosstalk is often the source of noise specified for compliance testing of DSLs. Table 8.1 lists several crosstalk tests for DSL, each row of which corresponds to a DSL test, the loop type from the earlier figures, the type of crosstalker and number of lines in the binder creating the noise, and the performance in terms of $\bar{P}_b$ and specified margin on the crosstalker.

AM Radio Frequency Noise AM radio noise is specified in the informative annex for ADSL standard T1.413.[7] For this test , the AWGN noise level of –140 dBm/Hz for the ANSI loops above is altered to be –100 dBm/Hz from DC to 100 kHz, linear decay at –40 dB/decade to 1 MHz, and –140 dBm/Hz above 1 MHz. 10 sinusoids of power –70 dBm each are added at the frequencies of Table 8.2 in kHz:

Table 8.2 AM Radio Interference Frequencies for ADSL and VDSL Testing

99	207	33	387	531	603	711	801	909	981

No other crosstalk is added (so the higher PSD at –100 dBm/Hz represents crosstalk). While AM radio noise is typically 10 kHz in width, the sinusoids are simpler to generate and generally test the ability of the ADSL system to reject narrowband noise.

---

7. The informative annex was supplied by ETSI to ANSI for European loop testing, and a new set of loops for Europe can be found in that annex (Annex H in Issue 1 ADSL). Two noise masks are also specified, one with AM radio noise. This noise mask is often also used for testing with American loops.

For VDSL, the AM radio noise specification has been modified with respect to ADSL. The reader is referred to the evolving VDSL specifications in Refs. [26], [27], both of which have the same AM radio model. This model includes 10 AM radio noises from 660 kHz to 1600 kHz. The AM sources are modeled by a fixed frequency carrier 30% modulated with a flat (±3dB) Guassian noise source band limited to 0 to 5 kHz. There are three different tests or "threats" with signal levels from -30 dBm to -70 dBm on each of the 10 AM radio signals.

Amateur Radio Noise Amateur radio noise is only of concern for VDSL as the lowest amateur radio band is 1.8 to 2 MHz. Table 8.3 lists the amateur radio bands specified (in MHz):

Table 8.3   Amateur Radio Bands

1.8–2.0	3.5–4.0	7.0–7.3	10.1–10.15	14–14.35	18.068–18.168	21–21.45	28–29.7

An amateur radio signal is modeled by a SSB-modulated carrier that changes frequency every 2 minutes by at least 50 kHz and visits all amateur radio bands in each test. The baseband signal is speech weighted noise, interrupted on a 15-second period with 5 seconds on and 10 seconds off. The baseband signal is also interrupted on a period of 200 ms with 50 ms on and 150 ms off. The doubly interrupted signal is bandlimited to 4 kHz and subject to 6 dB per octave pre-emphasis.

Amateur Radio Emissions The area of radio emissions has been of extreme concern for VDSL. The power spectral density in the radio bands is 20 dB lower (-80 dBm/Hz versus -60 dBm/Hz) in all amateur radio bands.

Impulse Noise **Impulse noise** is specified only for ADSL testing at the time of writing, although it is anticipated for VDSL also. Two impulses are specified and unfortunately not modeled mathematically, but instead enumerated in terms of 140 and 480 successive sample values at 160 ns intervals. Amplitudes can be as high as 20 mV. These impulses are in Annex C of Ref. [2].

The probability of an error second that was discussed in Section 6.1 is used for impulse noise testing because it is most likely (with 6 dB of margin specified on other types of noises) that it is an impulse that actually leads to an error on a DSL when in operation. Bellcore studied P_{es} for ADSL with no interleaved forward-error correction and empirically has determined that the probability of an errored second can be related to the measured amplitude (in mV, a scale factor equal to the peak voltage) of the two types of impulses. Two levels of impulse voltage are identified according to the scaling that causes errors half the time with a 15-time-repeated impulse test. u_{e1} is the level at which half the impulses cause errors for impulse 1, and u_{e2} is the level at which half the impulses cause errors for impulse 2. Then,

$$P_{es} = 0.0037 \cdot P(u > u_{e1}) + 0.0208 \cdot P(u > u_{e2}) \tag{8.54}$$

where

$$P(u > u_e) = \begin{cases} \dfrac{25}{u_e^2} & u_e < 40 \, \text{mV} \\[2ex] \dfrac{0.625}{u_e^2} & u_e \geq 40 \, \text{mV} \end{cases}$$

(8.55)

is the probability that an impulse of that amplitude can occur in practice.[8] The impulses are applied 15 times at 1-second intervals. Error second tests based on empirical formula are provided in Ref. [2]. Impulse testing is an area for which considerable further study is prudent in DSLs.

8.3.1.2   General Measurements and Noise Injection

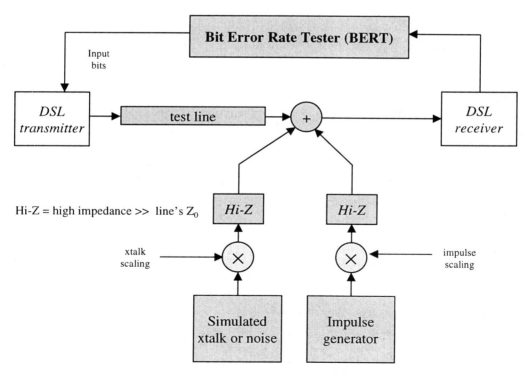

Figure 8.7   General performance measurement for a DSL.

---

8.   Clearly this is a test created, and the probabilities stated cannot be universally applicable.

A general diagram of the method for measuring DSL performance appears in Figure 8.7.

A **bit-error-rate tester (BERT)** supplies and retrieves a suitably random bit stream for measurement of $\overline{P}_b$. Transmission lines can either be actual twisted pairs or can be emulated by what are known as "line simulators," devices sold by a few manufacturers to simulate the behavior of the ANSI and other twisted pairs without taking as much lab space. Sometimes the line simulators also allow injection of noise, which also appears in Figure 8.7 (as part of a simulator or separately implemented). The noise sources can be generated as shown in Figure 8.8.

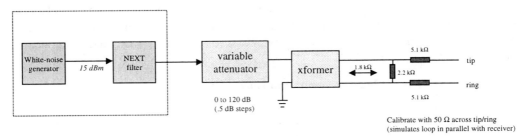

Figure 8.8   Injection of noise for DSL measurement.

The variable gain (or variable attenuator) allows margin to be evaluated at some fixed probability of error measured on the BERT. A NEXT filter imposes the spectral shaping discussed in Chapter 3 on each of the types of crosstalk (or other noise). The transformer allows an unbalanced noise to be differentially coupled into a balanced transmission line with high impedance. Noise is added at the output of the transmission line, where it appears like its own characteristic impedance in parallel with that of the receiver. Usually, this is 100 ohms in parallel with 100 ohms, but some noise PSDs (e.g., HDSL) have PSD listed for other impedances (like 135 ohms). Then a correction factor on the attenuator gain should be applied if the noise is calibrated for a 100 ohm line.

The noise generator itself should be white Gaussian noise, which is hard to generate, but a crest factor of at least 5 should be used (meaning that $P_e = 10^{-7}$ is the smallest probability of error that can be measured). Larger crest factors may be advisable. Care should also be taken to ensure that the noise source has a period significantly longer than that of the interval between bit errors.

Insertion loss can be verified by measuring the line response to any white noise source (the noise injector without the NEXT filter is acceptable with transformer matching to 100 ohms) before and after loop insertion.

Longitudinal balance, which is known by nontelecommunications and radio engineers as **common-mode rejection ration (CMRR),** measures the coupling between differential "longitudinal" (common-mode) signals on the transmission line and "metallic" (differential-mode) signals on the transmission line. The latter are referenced from either pair to ground, while the former are voltages measured across the tip and ring leads of the telephone wire. Longitudinal balance is defined as

$$L_B = 20 \log_{10} \left| \frac{e_l}{e_m} \right|$$

(8.56)

where e_l is the applied longitudinal signal and e_m is the measured corresponding metallic (differential) signal. Figure 8.9 shows a circuit for measuring longitudinal balance. The applied longitudinal signal causes some differential voltage to occur because of imperfect balancing of the signal caused by the DSL transceiver.

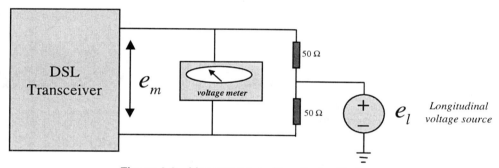

Figure 8.9 Measurement of longitudinal balance.

8.3.2 Measurement of Performance

As in all digital transmission, the most common measure of DSL performance is the probability of error. Usually, probability of bit error is desired, but sometimes probability of symbol error is also of interest. In either case, the engineer attempts to measure the probability of error by observation of a system's performance — this is usually achieved with transmission hardware through the use of a BERT. Typically, the BERT allows a test engineer to select one of a number of different bit streams (typically various lengths of pseudorandom patterns). The BERT essentially synchronizes to the receiver output bit pattern and compares it to the input pattern, while counting the number of error positions. The number of errors accumulated is periodically divided by the total number of bits measured to estimate the probability of bit error. As time increases, this average bit error rate should converge to the actual system value (when the system is not time-varying). A reset button allows bit error counts to be restarted at zero when necessary.

In general, bit error rate measurements become more reliable with time. The designer then needs to know how long the bit error rate needs to be observed before any derived bit error rate is sufficiently accurate. This is a basic statistical problem that involves measurement of a distribution. Let us suppose that bit (or symbol) errors are made with some unknown, but fixed, probability p. One measures p by counting errors in successive observations of the channel output. Let the kth experiment be denoted by p_k where

$$p_k = \begin{cases} 1 & \text{error measured } (p) \\ 0 & \text{no error measured } (1-p) \end{cases}$$

(8.57)

Then, an estimate of the probability of error, based on N independent measurements, is

$$\hat{p}(N) = \frac{1}{N} \cdot \sum_{k=0}^{N-1} p_k$$

(8.58)

This estimate has an average value

$$E\left[\hat{p}(N)\right] = p$$

(8.59)

and a variance about this average of

$$\hat{\sigma}_p^2 = \text{var}[\hat{p}(N)] = \frac{p}{N}(1-p) \approx \frac{p}{N}$$

(8.60)

Clearly, this estimate converges to the true probability of error as N gets large. However, N can be much larger than sometimes expected. For instance, the standard deviation is the square root of the variance. Thus, for a system where $p = 10^{-7}$, for the probability of error estimate deviation from to have a standard deviation of 10% of the value of p, then

$$10^{-8} = \sqrt{\frac{10^{-7}}{N}} \quad \text{or} \quad N = 10^9$$

(8.61)

In fact, a single standard deviation may not be sufficient to guarantee good accuracy of measured probability of error.

The distribution of the random variable $\hat{p}(N)$ has a binomial distribution given by

$$f_{\hat{p}}(k) = \Pr\{N \cdot \hat{p}(N) = k\} = \binom{N}{k}(1-p)^{N-k} p^k$$

(8.62)

The test engineer desires to ensure that the probability of error estimate deviates less than an amount $\varepsilon = p/L$ from the true value with a high degree of confidence. Let us say that we desire $(1-\delta)$ (90% is $\delta = 0.1$) confidence that the measurement deviates less than ε from the true value. Corresponding to this value of ε is a range of values for the index $k(\varepsilon)$ such that the estimate is close enough, mathematically stated precisely as

$$Pr\{|\hat{p}(N) - p| < \varepsilon\} > 1 - \delta \;=\; \sum_{k(\varepsilon)} \binom{N}{k} (1-p)^{N-k} p^k \tag{8.63}$$

Clearly, just to have a nontrivial set for $k(\varepsilon)$, then $N \geq L/p$. Evaluation of the sum can be excessively intensive and so a rough use of the central limit theorem is applied to the distribution to say that for large N, the distribution is approximately Gaussian and so the probability is then approximated by

$$Pr\{|\hat{p}(N) - p| < \varepsilon\} \approx \int_{-\varepsilon = -p/L}^{\varepsilon = p/L} \frac{1}{\sqrt{2\pi}} e^{-\frac{x^2}{2p/L}} dx = 1 - 2Q\left(\frac{\sqrt{Np}}{L}\right) = 1 - \delta \tag{8.64}$$

or then

$$Q\left(\frac{\sqrt{Np}}{L}\right) = \frac{\delta}{2} \tag{8.65}$$

For 90% confidence, $\delta = 0.1$ that the error is less than $p/L\%$, then the above equation produces

$$N \geq \frac{2.7 \cdot L^2}{p} \tag{8.66}$$

so, for instance, 10% accuracy at $p = 1e - 7$ requires that nearly 2.7 billion bits be tested. Thus, at a speed of 10 Mbps, this takes about 270 seconds, or approximately 4.5 minutes. For 1 Mbps transmission, the test would require 45 minutes. The measurement time can be reduced most easily by reducing L to 2, which corresponds to only about an 0.2 dB SNR difference. Even then, 1 Mbps DSL transmission at $1e - 7$ error rate may take 2 minutes for a measurement, while a lower speed of 100 kbps would take 20 minutes. Such measurement intervals are typical in performance comparison tests sponsored by standards groups like ANSI.

8.3.2.1   Effect of Input Bit Sequence

Clearly, the input bit sequence will need to be periodic for any practical implementation of a BERT. The period of this sequence should be such that it exceeds the memory of the transmission system significantly. Such sequence length is necessary to ensure that all possible channel output conditions are excited. Given that DSL transmission systems may have long memory, a 24-bit pseudorandom pattern is most likely used (with a period of $2^{24} - 1$ bits and running through all 24-bit sequences once and only once per period). Some sequences with lengths greater than 24 will not have equal likelihood of occurrence and can bias probability of error measurements, but this effect is usually presumed small by DSL engineers.

8.3.2.2   Period of Injected Gaussian Noise

Zimmerman [28] notes that most commercial line simulators make use of pseudorandom noise in generating Gaussian noise measuring DSL performance. An unfortunate consequence is that the peak noise samples generated do not accurately follow the Gaussian distribution tails, thus biasing probability of error measurements in an optimistic direction. Typically, line simulators generate noise by using some internal analog noise source and adding digitally generated noise to it. If the period of the latter digital "Gaussian" noise is M, then the peak value of the noise in a set of M samples is also Gaussian with mean

$$\mu = 1 - Q\left(\frac{1}{M\sigma}\right)$$

(8.67)

and variance

$$\sigma_{peak}^2 = \frac{(M-1) \cdot 2\pi\sigma^2 \cdot e^{\frac{\mu^2}{\sigma^2}}}{M^3}$$

(8.68)

To eliminate an optimistic bias, the tester would need $M > 10^7$, which complicates line simulator design. For the more typical value of $M = 8192$, the bias is optimistic by 2.4 dB (see also Ref. [28]), meaning that lab measurements for $M = 8192$ are then optimistic by 2.4 dB and should be reduced by such for field performance.

8.3.2.3   6 dB Margin and Importance Sampling

To avoid long measurement times, **importance sampling** is a method used by test engineers to test only the worst-case situations by increasing the occurrence of peak noise samples with respect to Gaussian noise. Such importance sampling must be very carefully applied for informative results. However, DSL engineers use a form of importance sampling in the concept of **margin**. Recalling that DSLs are specified to have a probability of bit error of 10^{-7} with a 6 dB margin. This means that the actual probability of error would be below 10^{-24}, requiring centuries of measurement time. Instead, testing is executed with noise increased by 6 dB so that reasonable measurement times can be used. The margin concept is one mechanism for importance sampling. DSL engineers, however, prefer the supposed practical interpretation that unforeseen noise disturbances of a temporary nature will not cause an error with such a large margin, although justification for such unforeseen noises at a level of 6 dB is difficult (either the noise change is much smaller for crosstalk changes or much larger for impulse or temporary RF disturbances).

8.4 Timing Recovery Methods

Timing recovery in DSLs is the extraction of the symbol rate frequency and phase by the receiver from the channel output. Such timing recovery often occurs at the remote or customer premises modem, and the recovered clock is used for both data detection as well as modulation in the reverse direction, a practice known as **loop timing.** For loop-timed DSLs, only one modem need recover symbol timing. The other modem can then use the same clock for transmission and reception without need of timing recovery (of frequency, although phase may still be important).

Timing recovery makes use of a **phase-lock loop (PLL)**, a basic operation in signal processing described in Section 8.4.1. Closed and open-loop timing recovery refer to the respective use, or disuse, of decisions in the operation of the PLL, as in Sections 8.4.2 and 8.4.3. Some add/delete and pointer methods appear in Section 8.4.4. These later methods find use in carriage of data bit clock information, like a network clock or timing reference, over modems with asynchronous symbol clocks.

It may be useful to note that DSLs do not use carrier heterodyning in that there is no intermediate carrier network in the twisted pair comprising the DSL. Thus, any carrier frequencies are always locked to the symbol clock (e.g., in QAM, the carrier frequency is locked to the symbol rate). Thus, carrier frequency and carrier phase jitter are not of concern in DSLs, even though carrier recovery may be very important in voice-band modems or wireless transmission methods.

Since modern DSL systems are largely implemented with digital signal processing in discrete time, this section focuses only on discrete-time PLLs.

8.4.1 Basic PLL Operation

The basic operation of a PLL is illustrated in Figure 8.10. Three components are essential in principal if not in implementation: a phase detector, a voltaged-controlled oscillator (VCO), and a loop filter. The phase detector computes the difference between the phase of a (usually noisy) reference signal, θ, and the phase of a local oscillator, $\hat{\theta}$, to get a phase error

$$\phi_k = \theta_k - \hat{\theta}_k \tag{8.69}$$

where k is the usual sampled-time index. There are many types of phase detectors. In DSLs, the

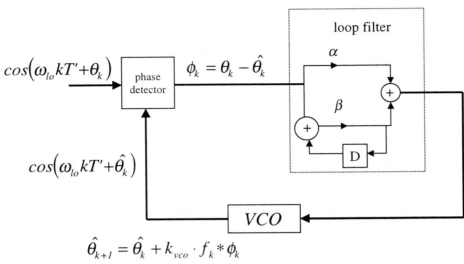

Figure 8.10   Basic PLL operation.

phase error is usually computed indirectly from modem transmission signals as discussed in Sections 8.4.2 and 8.4.3. The VCO in discrete time is any device or tacit mechanism that produces an output sequence whose frequency (or phase increment) is proportional to a control signal applied to the input of the VCO. Typically this control signal is a filtered version of the phase error. The filtering is performed by the loop filter, which usually is comprised of the sum of a fixed gain α on the phase error and a scaled (by β) accumulation of phase errors, as shown. The loop filtering corresponds to the following two equations for frequency offset Δ_k and phase increment $\hat{\theta}_k$:

$$\Delta_k = \Delta_{k-1} + \beta \cdot \phi_k$$
$$\hat{\theta}_{k+1} = \hat{\theta}_k + \alpha \cdot \phi_k + \Delta_k \tag{8.70}$$

The PLL is known as a **first-order PLL** if $\beta = 0$ and more generally a **second-order PLL** if $\beta \neq 0$. The gain of the VCO can be considered absorbed into the gain parameters α and β so that $k_{vco} = 1$. Of particular importance in analyzing and understanding the PLL is the phase error relationship to the input phase, namely, the transfer function

$$\frac{\Phi(D)}{\theta(D)} = \frac{(1-D)^2}{1 - (2 - \alpha - \beta) \cdot D + (1-\alpha) \cdot D^2} = H_{PLL}(D) \tag{8.71}$$

One can show that for any frequency and phase offset from a nominal $\hat{\theta}_0 = 0$, so that

$$\theta_k = \Delta \cdot k + \theta_0 \tag{8.72}$$

that the phase error of the second-order PLL will eventually go to zero as long as the roots of the denominator polynomial in the transfer function above have magnitude larger than one.[9] The design of PLLs requires selection of the values of α and β so the PLL tracks or converges with sufficient speed and that the amount of any noise in the phase-detection process is small enough to allow proper operation of the receiver.

A fundamental engineering trade-off occurs in the selection of these two PLL parameters or more generally for any choice of loop filter $H_{PLL}(D)$: The designer can reduce them to very small values close to zero and thus reduce the effects of phase noise from the phase detector to negligible values. However, small values for the parameters mean that the updating of the PLL is slow (small phase increments), and so it may take too long to converge or track any clock variation. If the input phase has noisy $n(t)$ in it, then the PLL processes this noise so that the variance of the phase error is computed by

$$\sigma_\phi^2 = \frac{1}{2\pi} \cdot \int_{-\pi}^{\pi} \left| H_{PLL}\left(e^{-j\omega}\right) \right|^2 \cdot S_n\left(e^{-j\omega}\right) d\omega \tag{8.73}$$

which depends on the PLL parameters and the power spectral density of the noise. Clearly, one desires the rms value of this **phase jitter** noise to be less than at least the largest tolerable phase error of the receiver (computed for instance as in Section 8.1.3 or again later in this section).

DSL systems may be cascaded with other transmission systems that require a stable embedded clock. Thus, in telecommunications, the **phase spectrum** is usually specified for both input clock accuracy and especially output clock accuracy. The input phase spectrum is that of the phase noise, $S_n(f)$. An example of such typical specification appears in Figure 8.11. Below some frequency, very low-frequency variation in the phase is often called **wander**. Wander is usually specified in unit intervals, where a unit interval is typically one sampling period or bit period for the data being transmitted. Wander determines the size of buffers that may be located between a DSL transceiver and other system elements. Above a certain frequency, the phase noise is **phase jitter**. Phase jitter is more important to ascertain how well a receiver works. Often lower-frequency phase noise is more tolerable in that other system mechanisms can adjust to it so the typical phase spectrum decreases with frequency. The cut-off between jitter and wander depends on the line code selected, network synchronization models, and the data rate, but typically it is at least 1000 times slower than the bit rate.

The DSL designer can then also specify a phase spectrum for any clocks that the DSL transceiver may deliver. This phase spectrum will depend heavily on the parameters chosen for the PLL for jitter and on the mechanism for generating the bit clock from the symbol clock.

---

9. This stability condition is satisfied when $0 < \alpha < 2$ and $0 < \beta < 1 - (\alpha/2) - \sqrt{(\alpha^2/2) - 1.5\alpha + 1}$.

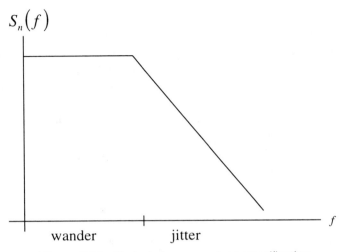

Figure 8.11 Typical phase spectrum specification.

The required accuracy of a phase lock loop can be related to the necessary SNR for satisfactory performance. Let us suppose that a signal $v(t)$ is sampled at an offset from a correct sampling phase at $t = kT' + \phi$. The signal with timing error has Fourier transform

$$v(t + \phi) \Rightarrow V(f) \cdot e^{-j2\pi f\phi}$$

(8.74)

With a reasonable PLL and system design the phase error may be time-varying but should be small and nearly constant over a few sampling periods, in which case the above equation can be approximated by

$$v(t + \phi) \Rightarrow V(f) \cdot \left[1 - j2\pi f\phi\right]$$

(8.75)

The distortion is clearly in the second term, which means that the distortion power caused by a phase error is roughly

$$(4\pi^2 f^2 \phi^2) \cdot \sigma_v^2$$

(8.76)

To make timing distortion negligible, this distortion should be much less than the noise in any frequency band SNR. Thus, to keep timing distortion to a minimum, the phase jitter should satisfy

$$jitter = \sigma_\phi^2 << \frac{2.5}{\pi^2 \cdot f^2 \cdot SNR(f)}$$

$$(8.77)$$

One can compute the jitter power from the PLL as in the earlier formula for the jitter and then compare it at worst case frequency value to that in this formula. Alternately, one can compute the jitter spectrum at each frequency and ensure that this formula is satisfied.

The PLL may generate a clock signal that is any rational fraction multiple, p/q, of the input frequency. The local oscillator frequency is then chosen close to the desired frequency, which is the rational fraction multiple times the input frequency. Then, the input frequency is divided by q (e.g., with a q-counter) and compared to every pth sample of the VCO output. Such rational fraction PLL circuitry is common in DSLs because bit clocks, sampling clocks, symbol clocks, and carrier-frequency clocks may all be derived as rational fractions of a single clock source. Typically, it is desirable to keep the values of p and q small (10 or less), so that a small amount of jitter is not excessively magnified as clocks are divided.

DMT systems sometimes take advantage of Equation (8.74) to avoid the use of an analog VCO. If the timing phase error of a PLL is computed, then the output of each FEQ is offset by

$$Z_n = Y_n \cdot e^{j((2\pi)/n)\, \phi \cdot n}$$

Thus, the linear phase offset can be implemented on each symbol for each tone, n, by a simple complex rotation of $((2\pi)/n)\, \phi \cdot n$ radians. When this offset is caused by a fixed-sampling-rate ADC,[10] the FEQ can be followed by a **rotor**, which computes the rotation by the upper equation in Equation (8.70). When Δ_k has grown in magnitude to the value corresponding to one sample, this sample is added or deleted from the cyclic prefix samples in the buffer of inputs to the receiver FFT, and Δ_k adjusted accordingly. Thus, **all-digital timing recovery** is facilitated for DMT, and expoited by several manufacturers.

8.4.2   Open-Loop Timing Recovery

Open-loop square-law timing recovery makes no use of the discrete nature of the transmit signal distribution, or equivalently of decisions, in the receiver PLL. The basic PLL needs a "true" reference phase. Thus, the phase detector often produces a noisy estimate of the phase error because any reference is also noisy. In open loop timing recovery, a variety of nonlinear processing of the receiver input waveform is used to generate such a noisy reference.

---

10.   Such fixed sampling leads to higher integration and, thus, less cost.

The classic open-loop timing recovery method is known as **square-law** timing. Figure 8.12 depicts the generation of the reference sinusoid for square-law timing.

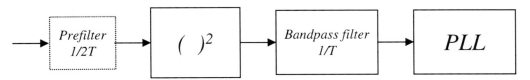

Figure 8.12   Open-loop timing recovery (generation of input reference to PLL).

Basically, the (prefiltered) channel output is squared and narrowband filtered around 1/T prior to any sampling (at a sampling frequency at least twice the symbol rate). As long as the original signal had energy at frequency *1/2T*, the squared signal will have energy at frequency 1/T. The prefilter and postfilter both act to reduce energy at frequencies distant from the desired rate *1/T*. The principle of operation follows from the equations:

$$y^2(t) = \left[\sum_k x_k\, p(t - kT) \right]^2$$

$$E\{y^2(t)\} = \sum_k \sum_l E_x \cdot \delta_{kl} \cdot p(t - kT) \cdot p(t - lT)$$

$$E\{y^2(t)\} = E_x \sum_k [\, p(t - nT)\,]^2$$

$$(8.78)$$

which is a periodic signal with period equal to the symbol period. This means the average squared value contains the timing information. Averaging is achieved by bandpass filtering at the symbol rate. Sometimes, if the signal *y(t)* has little energy near *1/2T*, meaning that $y^2(t)$ has small energy at *1/T*, then a different nonlinearity like absolute value (full-wave rectifier) replaces the squaring device. Such a nonlinearity will often translate more energy to the symbol rate. While the average value of the squared signal is periodic, there is nevertheless a significant variance about this average, which is known as **data-dependent jitter.** Such data-dependent jitter depends on the transmit signal statistics, but can be computed for various constellations. The more narrowband the filters, the less jitter, but any variation in the true symbol rate might not be tracked.

Open-loop timing recovery methods are common in simple transmission systems that do not operate at or near maximum limits where a larger phase error can be tolerated. High-performance transmission almost never uses an open-loop timing recovery.

8.4.2.1   Pilot and Regular-Insertion Timing Recovery

Open-loop timing-recovery methods can be made extraordinarily accurate through the use of a **pilot**. A pilot is usually a sinusoid inserted at a known frequency into the transmitted waveform $x(t)$. Since the pilot must pass through the channel also, it must be placed within the usable frequency range. In **pilot timing recovery**, the receiver bandpass filters the known pilot frequencies and pass the resultant filter output to a PLL. The PLL then causes the local frequency to match (some rational multiple) of the frequency of the pilot to generate the sampling and symbol rate clocks. A more general use of the term **pilot training** includes the embedding of regular repeated known patterns in the transmitted waveform. These patterns need not be simple sinusoids, but they carry sufficient information for the receiver to determine a phase error with respect to local estimate of the symbol rate.

Pilot timing is most often used in multicarrier transmission like DMT. ADSL's T1.413 standard places a pilot at the frequency Nyquist over 4 $(1/8T')$, which is usually in the better regions of the DSL transmission band and also one of the DMT frequency bins. The bandpass filtering is easily achieved by the receiver FFT, allowing a reference signal for input to a PLL with essentially zero data-dependent jitter. Such a clean input to the PLL allows highly robust timing recovery (T1.413 modems are known for their exceptionally good synchronization robustness). However, pilots can also be used with PAM and QAM transmission. A DFE can insert a very narrow receiver notch at certain (but not all) frequencies with little signal degradation if sufficient filter length is used in the feedforward filter. A good choice is perhaps the edge of the transmission band, the Nyquist frequency for PAM and the baseband equivalent of the Nyquist frequency or its image for QAM/CAP, where the DFE can more easily notch. Thus, the DFE feedforward section passes the signal, while one minus the feedback section is the bandpass filter that generates an input for the PLL.

Usually with pilot timing recovery, the phase detector is simplified because the phase error is easily computed as certain (filtered) sample values at the estimated zero crossing sample times of the pilot.

The drawback of pilot timing recovery (in exchange for its excellent accuracy) is the loss of some signal bandwidth for the pilot. In a T1.413 modem, for instance, one bin of 256 is used. Depending on how many bins are active on a DSL and how many bits would have been carried on the pilot bin, this could be as much as a 1% bandwidth loss, usually perceived as negligible.

8.4.2.2   Band-Edge Timing Component

The pilot and open-loop timing recovery methods only recover frequency. Because of the unknown channel phase-shift in general, the phase at which the open-loop PLL converges may not be the optimum sampling phase for best digital transmission. Sampling at optimum phase as well as at optimum frequency becomes important when equalizers use symbol-spacing of taps or are very short (even when sampled at higher rates that the symbol rate). Sampling phase is also important in multicarrier systems when they use nearly all the available tones (and are not oversampled).

The best timing phase for these situations maximizes the geometric SNR over the used bandwidth. This geometric SNR is the critical performance measure for both DFE and multicarrier transmission systems. When the entire bandwidth is used, as for instance with the DFE, the optimum timing phase that maximizes this geometric SNR also causes the aliased bandedge signal component to be largest [29]. While the mathematics is beyond this text, one can intuitively understand this **bandedge component maximization** principle by recalling that the choice of phase that maximizes the amount of energy clearly should increase the high frequencies. The result of a phase error is essentially to lowpass filter the higher frequencies increasingly in DSP — thus by increasing the worst-case frequency's amplitude until it is maximum over all sampling phases, the receiver will have maximized the net of all the others as well.

For QAM (and related systems like CAP), a (nonpilot) open-loop timing recovery system can thus prefilter near the Nyquist frequency as illustrated earlier. The timing-recovery PLL needs to sample 90 degrees out of phase with the zero-crossings of the generated reference to maximize the bandedge component. Godard studied one popular implementation of such a method for general transmission with symbol-spaced equalizers in [30].

When a pilot is used, it is best to insert the pilot at the bandedge 90 degrees out of phase (essentially requiring double-rate sampling or analog pilot insertion). Then when the receiver looks for zero crossings of this pilot (even at the symbol rate), the bandedge component is automatically maximized. If this offset insertion at the transmitter is not used, then the receiver must attempt to ascertain not only the correct PLL frequency, but also at what phase is this same frequency producing maximum energy. While this latter problem can be solved, the offset insertion of the pilot by 90 degrees at the transmitter eliminates the need for detecting signal gain.

For multicarrier methods, pilot position can be automatically determined by used bandwidth measurements or be fixed. Most multicarrier systems do not use the Nyquist bin because it has only a real component and thus complicates regular processing of the carriers. It is common to insert a second pilot 90 degrees out of phase (essentially on the imaginary part anyway) in analog at the transmitter output. The receiver Nyquist bin output will have least energy when the bandedge energy is maximized, making the Nyquist FFT bin a phase error. Unfortunately, the DSL may not transmit much energy at the Nyquist frequency, making this bin unreliable in general. It is then possible to insert a pilot at the next bin after the largest used frequency by a multicarrier system, again with a nonzero component only on the imaginary dimension. The receiver-output real component on this same pilot frequency is a phase error, which when driven to zero will maximize the geometric SNR over the used carriers of the multicarrier system.

8.4.3   Decision-Directed Timing Recovery

Decision-directed timing recovery does not generate a local reference sinusoid. Instead the channel outputs and previous/current symbol decision are used to generate directly the phase error. Decision-directed timing recovery wastes no bandwidth (unlike pilot recovery), and is usually more accurate than (nonpilot) open-loop methods (but of course less accurate than a pilot).

Decision-directed methods most often chose MMSE criteria over the sampling phase, which is

$$
\min_{\tau} J(\tau) = E\left[\left|\hat{x}_k - z(kT + \tau)\right|^2\right]
$$

(8.79)

where $z(kT+\tau)$ is the decision-device input, and $\hat{x}_k$ is the decision. When this MSE is minimized, the probability of error should be minimized (in the absence of biases). The sampling phase is updated in the negative direction of the derivative of the MSE with respect to phase. That derivative is

$$
\frac{dJ}{d\tau} = \Re\left\{E\left[2e_k^* \cdot \left(-\frac{dz}{d\tau}\right)\right]\right\}
$$

(8.80)

The symbol phase offset updates are

$$
T_k = T_{k-1} + \beta \cdot \Re\{e_k^* \cdot \dot{z}_k\}
$$
$$
\tau_{k+1} = \tau_k + \alpha \cdot \Re\{e_k^* \cdot \dot{z}_k\} + T_k
$$

(8.81)

The instantaneous error is $e_k = \hat{x}_k - z_k$ where $\hat{x}_k$ is the decision and z_k is the decision-device input. The derivative $\dot{z}_k$ of the input to the decision device (usually the equalizer output) must be approximated.

If the device preceding the decision element is an equalizer, and if that equalizer is oversampled (i.e., fractionally spaced) by a factor of 2, then the derivative can be approximated accurately by (with an abuse of notation),

$$
\dot{z}_k = \frac{z_{k+1/2} - z_{k-1/2}}{T}
$$

(8.82)

and timing recovery will be very good if decisions are correct. In practice computing twice the number of outputs of the equalizer (even when fractionally spaced the output is computed usually only at the symbol rate) just for timing recovery is expensive. Instead, the derivative can be poorly approximated by

$$
\dot{z}_k = \frac{z_{k+1} - z_{k-1}}{2T}
$$

(8.83)

In either case, the advance signal may not be known. Thus, most decision directed systems try to produce a stochastic approximation, $\phi(z_k, \hat{x}_k)$ to the derivative that has the same average value. One such signal that depends only on current and past samples for symbol-spaced outputs is

$$\phi_k = \Re\left\{\hat{x}_k^* \cdot z_{k-1} - \hat{x}_{k-1}^* \cdot z_k\right\}$$

(8.84)

and the symbol phase update becomes:

$$T_k = T_{k-1} + \beta \cdot \phi_k$$
$$\tau_{k+1} = \tau_k + \alpha \cdot \phi_k + T_k$$

(8.85)

8.4.4   Pointers and Add/Delete Mechanisms

Data bit streams carried by DSLs can often be synchronized to network clocks, while signal processing clocks may be locally generated for a variety of reasons (cost, simplicity, etc.).[11] In these cases, the modem is capable of transmitting a higher data rate than the highest possible data rate of the supplied bit stream. In such systems, the phase error is used to alter the number of symbol values in a frame of symbols, usually by adding/deleting dummy symbol positions.

8.4.4.1   HDSL Stuff Quats

An example of such add/delete positions occur in HDSL. The nominal symbol rate is 392 kHz. With 2B1Q's 2 bits/symbol, this corresponds to a raw bit rate of 784 kbps on each of two coordinated transmission lines (the two together carry a 1.544 Mbps DS1 bit stream plus overhead). In HDSL, a symbol is often called a "quat" because it represents 2 bits of information, corresponding to four levels. HDSL superframes are organized as a nominal 2352 consecutive quats, corresponding to exactly 6 ms in time. These 2352 quats include HDSL overhead along with the DS1 bits.

To avoid a need to lock to the DS1 clock or network clock at the physical layer, HDSL actually transmits either 2351 quats or 2353 quats in a 6 ms frame. In the later 2353 case, the last two quats carry dummy or meaningless data. An HDSL phase error with respect to the incoming DS1 clock is generated by a phase detector between the 1.544 MHz incoming clock and the 392 kHz HDSL system symbol clock. When the phase error is positive, meaning HDSL is running too slow, then the two dummy "stuff quats" are not transmitted. When the phase error is negative, meaning HDSL is running too fast, the two stuff quats are inserted. The receiving modem

---

11.   Such synchronization is inherent both in asynchronous transfer mode (ATM) systems and the current synchronous switching hierarchy. It is not present in ethernet or data-communications routers/bridges where the data clock is presumed to be supplied by the modem to the data-supplying entity, because those systems operate in burst mode.

discerns the presence/absence of the quats by investigating the first 7 quats of each frame, which always contain a special sequence known as a double-Barker code. The receiving modem can also use the presence/absence of the stuff quats as a phase error for PLL that generates the output DS1 clock. In order for the HDSL with plus/minus two quats to have sufficient pull-in ability to track the DS1 clock, both DS1 clocks and the HDSL master system clock need to have 32 ppm accuracy, leading to an overall difference between clocks of up to 64 ppm (HDSL actually could track up to 130 ppm difference according to its specification and nominally the $\pm 1/2352$ ratio allows an even greater range).

8.4.4.2   ADSL Add/Delete Mechanism

ADSL uses a more complex scheme of SYNCH and FAST bytes to control stuffing of additional bytes in dummy byte positions known as the "AEX" (asymmetric extra) and "LEX" (low-speed extra) bytes, or deletion of the last byte in a symbol. ADSL does this because up to 10 asynchronous signals may be multiplexed in a T1.413-compliant modem. The principal for each of the 10 is the same as the HDSL add/delete mechanism, except the AEX/LEX extra bytes actually carry real data or dummy data depending on the indication of the control bytes. Further these extra bytes are shared over all 10 signals, depending on the control in the SYNCH and FAST bytes. At the time of the standards development for Issue 1 ADSL, ATM standardization was not sufficiently progressed to allow ATM to instead do such multiplexing. More recent Issue 2 ADSL sytems reduce the specification for the add/delete mechanism when ATM is also used. For more up-to-date information, see Ref. [21].

8.4.4.3   Pointer Methods

Pointer methods avoid the addition/deletion of bytes or symbols and instead rely on a synchronous network to carry asynchronous data, much as is done in ATM networks. In such networks, there is a master timing reference (typically 8 kHz) whose nominal period beginning is marked by pointers that are embedded in normal overhead for the synchronous network. The pointers indicate just which bit/byte position corresponds to the beginning of the network timing reference. Data is simply inserted into network cells on an as-available basis and that data along with the network timing reference is used to reassemble the data at the receiving end.

In this way almost any rate can be carried over a network as long as system bandwidth resources are managed productively. When a DSL carries ATM data, it need not concern itself with the individual clock speeds of the application signals, thus simplifying the DSL so that addition/deletion is performed at higher levels. However, it should be mentioned that ADSL does allow different quality of service to be offered at the physical layer (and is unique among current DSLs in this capability). This facility however, does not necessitate reinsertion of add/delete timing, as each quality of service path through the modem still has synchronization executed at a higher level.

While pointer methods do alleviate the need for add/delete synchronization at the physical layer, they can also require greater amounts of buffering memory to be used at digital modem interfaces.

8.4.5 Frame Synchronization

DSL modems must sometimes discern frame boundaries among groups of symbols, even if the symbol clock has been accurately recovered by a PLL. Such **frame synchronization** is achieved by searching for embedded known symbols in the received signal. Such a symbol sequence can be called s_k, $k = 0,...,L-1$, where L is the length of the synchronization pattern in symbols.

The signal processing problem associated with frame synchronization is to detect over some interval in time, say $\pm M$ symbol periods, which of $s_{k \pm m}$ $m = 0,...,M$ was transmitted. The synchronization symbols s_k traverse the same channel as the information symbols x_k, and so are subject to the same noise and intersymbol-interference contamination.

The maximum-likelihood estimate of the synchronization sequence position on an AWGN channel forms the cross-correlation of the incoming sequence segments in a sliding window of symbols of length L and searches for a maximum. The time of the maximum corresponds to the time in which the synch pattern is most likely to be situated in the length-L window. Mathematically, the operation is

$$r_k = \sum_{l=0}^{L-1} s_l \cdot z_{k+l}$$

$$(8.86)$$

which is a sequence that should have low value until the synchronization sequence and the equalized channel output z_k are aligned. This value of k corresponds to the place at which r_k reaches a maximum in magnitude.

For multicarrier methods, it can sometimes be easier to implement the autocorrelation in the frequency domain by matrix multiplication of the channel output DFT by the known conjugate DFT of the sequences $s_{k=m}$ for each value of m. The sum of the magnitudes of the DFT bin outputs then becomes the indicator of best phase alignment for a given value of m.

An example of a good synchronization pattern is the 7-symbol Barker code used in HDSL. This code transmits at maximum amplitude the binary pattern

$$+ + + - - + - \tag{8.87}$$

(or its time reverse on loop 2 of HDSL). This pattern has a maximum r_k value of 7 units when aligned, and otherwise a maximum value of 0 units (the 13 place autocorrelation function is -1 $0-10-1070-10-10-1$). Because there are adjacent transmissions, the minimum distance of 7 units may be reduced to 6 or 5, as long as the searching receiver has some idea when to look. Initially, when searching for synch, the Barker code can be isolated (preceded and followed by a few zeros). When searching through random data, the probability of false detection depends on the symbol size and the nominal probability of error. For instance, in HDSL with nominal minimum distance equal to 2 (forcing the Barker to +/-3 pattern) and 2 bits/symbol, the likelihood of occurrence of the Barker pattern in random data is $4^{-7} = 2^{-14}$, or about 10^{-5}. Given a modem has lost frame synchronization when this very low probability of false reacquisition can occur, it was found acceptable for HDSL, knowing also that the next occurrence of the synch pattern (2352 symbols later) should also have a peak. Searching for two successive peaks, reduces the false synch acquisition to 10^{-10}, 3 successive peaks to 10^{-15}, and so on.

ADSL uses a frequency-domain synchronization symbol that is generated by pseudorandom PRBS binary signals mapped to a four-point constellation value on each of the 256 tones. Tones that are unused are also zeroed for the synchronization symbol, which occurs once in every 69 symbols at the ADSL superframe boundary. Correlation in the frequency domain is achieved by multiplication of the possible offset patterns (recall an offset in time is a linearly increasing phase ramp in frequency) with increased phase slopes corresponding to the offset positions and searching for the maximum. Again the likelihood of false acquisition is very small, especially if successive synch symbols are used to verify a previous declaration of phase boundary. For more details on the synch pattern, see Ref. [2].

8.4.6 Discrete-Time VCO Implementation

Implementation of a discrete-time VCO can take a number of forms that avert the use of a separate analog component. These implementations can allow digital integration of the PLL, but sometimes increase jitter. Section 8.4.6.1 presents two such methods for the all-digital implementation. Section 8.4.6.2 presents a method for avoiding an extra DAC for the generation of the control waveform for situations when an analog VCO or VCXO is controlled by a digital signal.

8.4.6.1 Numerically Controlled Oscillators and Digital Interpolation

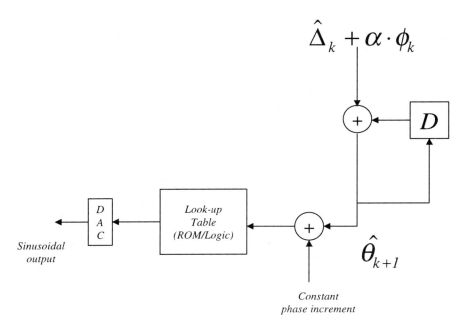

Figure 8.13   Numerically controlled oscillator.

Figure 8.13 shows a **numerically controlled oscillator (NCO).** The NCO computes the phase of a waveform digitally as in the equation:

$$\hat{\theta}_{k+1} = \hat{\theta}_k + control \tag{8.88}$$

The digitally accumulated phase is converted to a sinusoid through a table-look-up (or algorithmic sine implementation). For generation of a sampling clock, the most common situation in DSLs, the look-up table can simply be a threshold detect that produces a clock edge when the phase passes through a multiple of 2π (or any other desired values). More generally when a clock waveform of 50% duty cycle is desired, the look-up table essentially produces a 1 output when the phase is between 0 and π and a 0 output when the phase is between π and 2π. (For systems with carrier frequency recovery, NCOs actually produce a sinusoidal waveform that is converted with a DAC to analog, but there is no carrier in DSLs.) The effective sampling rate of the nominally added phase corresponding to the local frequency needs to be significantly higher than the desired sampling clock rate, nominally a factor of 100 or more in DSLs — however, for values of the computed phase $\hat{\theta}$ and the local phase, it is possible to interpolate so that the only element with high resolution is the actual device generating the sampling clock edge. Thus, the adds need not be implemented at full speed and instead an interpolator can be used, as in the next paragraph.

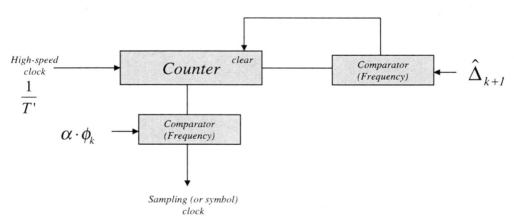

Figure 8.14   Digital counter implementation of VCO.

The phase of the NCO can be thought of as a counter value for a high-speed clock as in Figure 8.14. A high-speed counter runs at the high clock speed while its output is compared with the current phase value $\hat{\theta}$ of the NCO. When equal, the output edge is produced and the counter cleared to zero. The high-speed clock frequency, $1/T'$, determines the nominal value for the frequency comparator,

$$p = \frac{\frac{1}{T'}}{\frac{1}{T}} = \frac{T}{T'}$$

(8.89)

Any actual sampling frequency should be then in the range

$$\frac{1}{T'(p+1)} < \frac{1}{T} < \frac{1}{T'(p-1)}$$

(8.90)

to keep jitter and wander effects to a minimum. The worst-case timing (assuming converged) error is

$$T = \tfrac{1}{2} T'$$

(8.91)

An analysis of phase jitter follows the theoretical analysis in Section 8.4.1, with allowance for an additional "quantization step" input jitter of $T'/2$, which has mean zero and variance $(T')^2 / 12$.

8.4.6.2   Oversampled 1-bit Control of Analog VCO/VCXO

Digital implementation of the VCO function can lead to unacceptable level of jitter when the derived clock is used for sampling (as is often the case in loop-timed implementations of DSLs). Thus, analog VCOs are often used, in particular a specific form called the voltage-controlled crystal oscillator (VCXO) because its nominal frequency is based on a crystal (a device external to any chip). The crystal prevents significant deviation from the nominal frequency, thus average noise and limiting jitter in a nonlinear way. Such a device will require an analog control voltage. Because the PLL processing in DSLs is done precisely in digital signal processing, the control voltage (typically, the signal $\hat{\Delta}_{k+1} + \alpha \cdot \phi_k$ of the PLL) must be converted by a DAC. The use of an extra DAC is expensive. Instead typical implementation is by oversampling and the use of a serial stream of 1s and 0s that are fed through an analog RC circuit to the control voltage input of the VCXO in a manner similar to that used for compact disk players. This circuit is shown in Figure 8.15.

Because the error signal e_k is white, then differentiated noise will be highpass and not pass through the analog lowpass filter, leaving simply the control signal. The higher the oversampling (which means repeating the value for the input voltage shown as x_k), the more accurate the reconstruction of the control signal for the VCXO. Such a method is in common use in DSLs to save cost since the typical implementation of DSL sampling rates is well below the maximum speeds of digital VLSI.

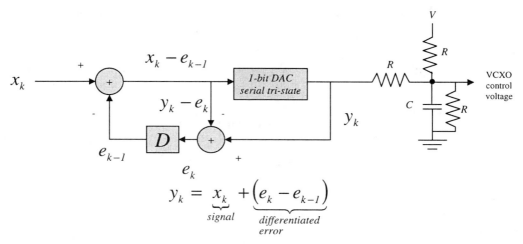

Figure 8.15   Oversampled 1/0 tri-state port generation of control voltage.

References

[1]   International Telecommunications Union (ITU) – Telecommunications Standard g.994.1, Geneva, "Handshake Procedures for Digital Subscriber Line (DSL) Transceivers."

[2]   *Asymmetric Digital Subscriber Line (ADSL) Metallic Interface,* ANSI Standard T.413-1995, ANSI, New York.

[3]   *HDSL.* ANSI Technical Report TR28. See Also Bellcore Technical Advisories T!—NWT-001210 and 001211.

[4]   *Integrated Services Digital Network (ISDN) — Basic Access Interface for Use on Metallic Loops for Application on the Network Side of the NT (Layer 1 Specification),* ANSI Standard T1.601-1992, New York.

[5]   Parlett, B. N., *The Symmetric Eigenvalue Problem.* Prentice-Hall, Englewood Cliffs, NJ, 1980, Chapter 13.

[6]   Chow, J. S., and Cioffi, J. M., "Method for Equalizing a Multicarrier Signal in a Multicarrier Communication System," *U.S. patent No. 5,285,474,* issued to Stanford University, February 8, 1994.

[7]   Van Kerckhove, J. F., and Spruyt, P., "Adapted optimization criterion for FDM-based DMT-ADSL Equalization," *Proceedings ICC,* New Orleans, June, 1996.

[8]   Al-Dhahir, N., and Cioffi, J., "Optimum Finite-Length Equalization for Multicarrier Transceivers," *IEEE Transactions on Communications,* Vol. 44, No. 1, January 1996, pp. 56–64.

[9]   Henkel, W., "An Algorithm for Determining the DMT Time-Domain Equalizer Coefficients," *ETSI Contribution TD26,* Lannion, France, September 1997.

[10]   L. Hoo, J. Tellado, and J. Cioffi, "Dual QoS Loading Algorithms for Multicarrier Systems Offering CBR and VBR Services," Globecom '98, Nov 1998, Sydney.

[11]   ITU-T g.992.2 Standard, "Digital Transmission System for ADSL Transmission on Metallic Local Loops with Provisioning to Facilitate Installation and fr Operation in Conjunction with Other Services."

[12]   Treichler, J. R., Johnson, C. R., and Larimore, M. G., *Theory and Design of Adaptive Filters.* John Wiley & Sons, New York, 1987.

[13]   Cioffi, J.M., "Limited-Precision Effects in Adaptive Filtering," *IEEE Transactions on Circuits and Systems,* special issue on adaptive filtering, Vol. 34, No. 7, July 1987, pp. 821–833.

[14]   Bednarz, P. S., "Decision Feedback Detection for the Digital Magnetic Storage Channel," Ph.D. thesis, Stanford University, March 1997.

[15]   Cioffi, J. M., and Kailath, T., "Fast, Recursive-Least-Squares, Transversal Filters for Adaptive Filtering," *IEEE Transactions on Acoustics, Speech, and Signal Processing,* Vol. 32, No. 2, April 1984, pp. 304–337.

[16]   Slock, D., and Kailath, T., "Numerically Stable Fast Transversal Filters for Recursive Least Squares Adaptive Filtering," *IEEE Transactions on Acoustics, Speech, and Signal Processing,* Vol. 29, No. 1, January 1991, pp. 92–114.

[17]   Marple, S. L., *Digital Spectral Analysis with Applications.* Prentice-Hall, Englewood Cliffs: NJ, 1987.

[18]   Kay, S., *Modern Spectral Estimation, Theory and Applications.* Prentice-Hall, Englewood Cliffs: NJ, 1988.

[19]   Hunt, R. R., and Chow, P. S., "Updating of Bit Allocations in a Multicarrier Modulation Transmission System," *U.S. patent No. 5,400,322,* March 21, 1995, issued to Amati Communications Corp.

[20]   J. Cioffi, "The Essential Merit of Bit-Swapping," ANSI Contribution T1E1.4/98-318, November 1988, Dallas, TX.

[21]   ITU-U g.922.1 Standard, "Digital Transmission System for ADSL on Metallic Local Lines, with Provisions for Operation in Conjunction with Other Services."

[22]   Chow, P., and Cioffi, J., "Method and Apparatus for Adaptive, Variable-Bandwidth, High-Speed Data Transmission of a Multicarrier Signal over Digital Subscriber Lines," U.S. patent 5,479,447, December 29, 1995.

[23]   Cioffi, J. M., and Forney, G. D., Jr., "Generalized Decision-Feedback Equalization for Packet Transmission with ISI and Gaussian Noise," *Communication, Computation, Control, and Signal Processing* (a tribute to Thomas Kailath), Kluwer, Boston, 1997.

[24]   Kerpez, K.J., "Near End Crosstalk is Almost Gaussian," *IEEE Transactions on Communications*, Vol. 41, May 1993, pp. 670–672.

[25]   Im, G. H., and Werner, L. H., "Bandwidth Efficient Digital Transmission over Unshielded Twisted-Pair Wiring," *IEEE Journal on Selected Areas in Communication,* Vol. 13, December 1995, pp. 1643–1655.

[26]   Cioffi, J. , "Very-High-Speed Digital Subscriber Lines System Requirements," Draft Technical Report, ANSI T1E1.4/98-043, March 1998.

[27]   "Transmission and Multiplexing (DTS/TM-06003-1(draft), V0.0.7 (1998-2)); Access Transmission Systems on Metallic Access Cables; Very High Speed Digital Subscriber Line (VDSL); Part I: Functional Requirements," approved by ETSI to become official VDSL TR in 1998.

[28]   Zimmerman, G., "On Testing Systems with Finite Length Noise Sequences," *ANSI Contribution T1E1.4/96-049,* January 22, 1996.

[29]   Mazo, J., "Optimum Sampling Phase of an Infinite Equalizer," *Bell Systems Technical Journal,* Vol. 54, No. 1, January 1975, pp. 189–201.

[30]   Godard, D., "Self-Recovering Equalization and Carrier Tracking in Two-Dimensional Data Communications Systems," *IEEE Transactions on Communications,* Vol. 28, November 1980, pp. 1867–1875.

Operations, Administration, Maintenance, and Provisioning

DSLs should provide the greatest loop coverage for the lowest cost. The cost of the transceiver ICs are a small fraction of the system life cost. The cost to turn up and maintain service is of great importance. To minimize costs, DSLs provide functions that assist with the installation and maintenance of service. DSLs are often integrated with transmission, switching, and operations systems to reduce cost and simplify operation. Network operators use operations systems (OS) to manage their platforms and services efficiently. The management functions fit into the categories of operations, administration, maintenance, and provisioning (OAM&P).

The ITU created the telecommunications management network (TMN) model to describe the management of telecommunications networks. TMN is a multivendor architecture using open systems interconnection (OSI) protocols, and consists of:

- Service management layer: manages services provided to customers.
- Network management layer: manages configuration and status of network.
- Element management layer: manages related network elements.
- Network element layer: units such as switches, multiplexers, and transmission terminals.

The telecommunications network management (TMN) model is an abstract model of the management of a telecommunication network, and has been developed by the ITU in the "M" series specifications.[1] The TMN model partitions the management of telecommunications func-

---

1.   ITU-T Recommendation M.3000: "Overview of TMN Recommendations."

tions into five areas: performance management, configuration management, fault management, accounting management, and security management.

In order to deal with the natural hierarchical nature of both telecommunications networks and the organizations that own and manage them, the TNM organizes these five functions into a hierarchy in which the management of elements at a lower level are seen as abstractions by the level above them. This hierarchy has four levels:

1. **The Element Management Level** — Is responsible for the management of actual physical elements (e.g., switches, cross connects, or multiplexors) that make up the network. The management models to implement the five functions at this layer represent the details of these devices. For example, the representation of an ADSL DSLAM would contain references to ADSL ports, line cards, and the media (such as a DS3 or SONET connection) that connect the DSLAM to the carrier network. Management functions at this layer control or survey the specific configurations of these elements alone, independent of specific knowledge of the network context the element is in. A management system that controls the elements is known as an *element manager*.

2. **The Network Management Level** — Views the assemblage of elements and their connections as a whole that make up a communications network. The network management level is responsible for management of constructs that occur across several network elements, such as a circuit or path. The management system that controls a network is known in the TMN as a *network manager*. A network manager communicates with other network managers (and the subtending element managers) to configure or control the network. For example, creating a circuit between two points may require configuration of several elements. A request to a network management system to configure the circuit from A to B may result in interactions with the individual element managers that know the details of the individual components that need to be configured to build the circuit. Similarly, a fault in the circuit would be interpreted by the network manager as a problem in a network construct (for example, if a particular circuit was down).

3. **The Service Management Level** — Allows the operator of a network to control their network in terms of specific services sold to specific customers. For example, a service configuration system for ADSL might allow an operator to give a particular customer ADSL access. The service management system would map the customers account to the parameters of the specific service they ordered. It would interact with the network management systems that actually control the networks that provide the service. Similarly, a test system at the service management level would allow a technician to run a test of a customer's service based on their account information. These service-level tests would use the functions of the network and, ultimately, element management levels. Information to bill for a customer's service would be a service management function based upon information gathered from the network and element management systems.

4. **The Business Management Level** — Is concerned with those management functions that oversee the entire structure of a carrier's networks, independent of the specific customer services. Examples of these functions include usage information required for long- and short-term planning of the network, or the company's network assets for accounting purposes.

The management legacy of most telephone companies, however, bares little relation to this abstract model. Instead, the phone companies have, over many years, automated the manual activities of their various operational departments. Because of this historical legacy, these operations support systems (OSS) do not contain a coherent model of the element, network, or services that they manage. Thus, interaction with each other is complex and often (years after the systems were developed) poorly understood. The evolution to management based upon coherent and extendable models such as the TNM is thus slow and expensive for telephone carriers.

An element management or network management system at a central site is connected to network elements located at many sites. This permits one technician to manage equipment dispersed over a large geographic area. The most common protocols used between the network element and the management system are TL1, CMIP, and SNMP.

Transaction language one (TL1) is the oldest, being introduced in the mid-1980s. TL1 is a man-machine language; it is intended to be directly usable by skilled technicians. Many persons feel that TL1 is old and poorly suited to the management of complex functions. TL1 does not include a standard for modeling systems, whereas SNMP and CMIP do. Nonetheless, TL1 is proving itself to be flexible and extensible.

Common management interface protocol (CMIP) was defined in the late 1980s as a powerful machine-machine management protocol. CMIP is used with common management information service element (CMISE) equipment. CMIP allows the definition of an object-oriented model of: elements, protocols, systems being modeled, and the management information base (MIB). The CMIP protocol enables manipulations of these models to manage the underlying systems. Vendors and carriers have been slow to adopt CMIP even though CMIP is endorsed by the ITU and other standards organizations as part of TMN.

Simple network management protocol (SNMP) is widely used to manage TCP/IP networks and private data networks, which often consist of LANs, bridges, and routers. SNMP-managed elements place configuration and status information in a management information base (MIB). However, unlike CMIP, SNMP uses a simple table structure to organize the information that is accessed. Although there are concerns about SNMP's suitability for management of large public telecommunications networks, its use is spreading into the telco sector.

9.1 OAM&P Features

DSLs may provide the following OAM&P features:

- **Loss of signal.** The DSL transceiver at each end of the loop may provide a local indication that it either is synchronized to the far-end signal or is not receiving a signal.

- **Performance monitoring.** A cyclic redundancy code (CRC) check is used to detect block errors in both directions. This error detection is performed continuously while the line is in service, and a record is maintained of the bit error rate (BER) during 15- or 60-minute periods. Monitoring BER only provides an indication after errors have occurred. Thus, BER provides little indication that transmission environment has almost reached the point of errors occurring. The error performance of a DSL using high-performance coding has a "cliff" characteristic. As the line conditions degrade, the DSL will perform nearly error-free up to the point where large numbers of errors result. Another drawback of BER as an indication of line transmission quality is that an accurate BER measurement requires error monitoring over a long period of time. Line quality can be more quickly determined by receiver's estimation of its signal-to-noise ratio (SNR). The receiver's SNR margin provides an accurate measure of how close the system is to the "cliff" where errors would occur. Thus, BER history is best for characterizing lines which are having errors, and SNR is best for measuring transmission quality on healthy systems.

- **Loop characterization.** DSL receivers adapt to the characteristics of each loop, and the DSP filter coefficients within the DSL transceiver contain much information about the loop's characteristics. However, thus far, little practical use has been made of this information. As a result, many systems rely on direct metallic measurements performed by loop test systems such as Lucent's mechanized loop test (MLT with LMOS, loop maintenance operations system) and Teradyne's 4TEL. These systems measure the resistance tip-to-ring, tip-to-ground, and ring-to-ground, leakage current, and capacitance tip-to-ring. Many loop faults may be detected and localized in this manner. The loop test system can determine if equipment is connected at the customer end of the loop. Furthermore, the loop length may be estimated from the loop capacitance. Advanced systems may measure loop attenuation at various frequencies if the far-end equipment generates a known wideband signal upon command. Background noise may be measured on lines where the far-end unit may be commanded to enter a quiet mode.

- **Loop-back (physical layer).** An in-band or out-of-band signal is sent that takes the line out of service and causes the equipment to send the same bit pattern that it is receiving. Addressable loop-backs select the location of the loop-back (near end, repeater, or far end), and its direction (reflecting data towards the customer, or the CO). The originating end sends a test pattern and looks for any mismatch in the received data. Loop-backs predate DSLs and, in the author's opinion, cause nearly as much trouble as they resolve. Loop-backs may be unintentionally triggered by certain user data patterns. To guard against this, loop-back control codes must be repeated for a certain duration. The length of this duration is a source of confusion. Once enabled for a test, the technician may forget to turn off the loop-back, leaving the line out of service. To guard against this, loop-backs are sometimes designed to automatically time out. This means that sometimes the loop-backs turn off unexpectedly. A variety of loop-back control codes are used for various systems and to permit selection of the point of loop-back (LT, repeater, NT, TE). As a result, loop-backs sometimes fail because the wrong control code is used. It may be difficult to send the loop-back control codes reliably on a line with a high BER. Yet, this is the very line that needs to be diagnosed. Most of these problems stem from the out of service nature of the loop-back test. CRC-based performance monitoring with far-end block error (FEBE) bits provide virtually the same information as a loop-back test without taking the line out of service. Since the performance monitoring is always functioning, there is no need for a complicated method to turn it on and off. Unlike many loop-back tests, performance monitoring will detect which direction of transmission was in error. The only advantage of loop-backs is more rapid measurement of a precise error rate; this is less of an issue for higher-speed lines. ATM layer loop-backs avoid many of the concerns previously discussed because the line is not taken out of service.

- **Self-test.** The near-end or far-end transceiver may be requested to perform an internal self-test, and report the results of this test.

- **Far-end status.** Since records may not be accurate, it is helpful to interrogate the far-end equipment to determine its type and option settings. The far-end equipment manufacturer, model, and version would be indicated. Furthermore, the far-end equipment may indicate when it is in a test mode and whether it is operating from its main or back-up power source

- **Provisioning of options in remote equipment.** Messages sent from a centralized network management center may set options in the customer-end equipment. This avoids the cost of sending a technician to the remote site.

- **Software/Firmware download.** To avoid the costs of visiting the customer site to update software or firmware, the system can download the new program via the DSL.

- **Metallic termination.** To facilitate network-based metallic-loop measurements, the far-end equipment provides a precise tip-ring resistance and capacitance. In the case of BRI, the terminating resistance has two values, controlled by the voltage placed on the loop. This permits the measurement of conductor resistance and tip-ring leakage resistance. In other cases, a separate metallic termination unit (MTU) or *half ringer* termination (15 kohm resistor in series with 0.47 µF capacitor and a back-to-back pair of zener diodes) may be placed at the NID.

- **Dying gasp.** The customer-end DSL transceiver may sense the loss of its local power. In the few milliseconds of power remaining in the power supply capacitor, the transceiver may quickly send a message that it is losing power. This may provide the network management system a helpful indication of why a line is dead.

9.2 Loop Qualification

Unfortunately, DSLs cannot operate on all loops. Thus, when a customer requests DSL-based service, it is first necessary to determine if the customer's loop will permit DSL operation. The only completely accurate way to determine the DSL suitability of a particular line is to connect a DSL transceiver to both ends of the loop and see if it operates with an acceptable error rate. However, this method is rarely used because of the costs involved in the numerous cases where service turn-up fails. To reduce provisioning costs, it is necessary to predict which lines are suitable before incurring the costs of connecting transceivers to the line.

All forms of predictive loop qualification involve a compromise between rejecting lines that would have actually worked versus accepting lines that actually would not work. The easiest and least accurate method of predictive loop qualification is to accept all loops running to sites within certain straight-line radius from the CO. One simply draws a circle on the map with the CO at the center. This method makes no allowance for many factors, including indirect cable routing, bridged taps, crosstalk, etc.

A more accurate method is to use loop make-up data (found in loop inventory OS systems, such as the LFACS, loop facilities assignment center system, database) to determine the length, gauge, and bridged taps for the particular loop. Identification of the types of transmission systems within the binder group is also useful. For a given DSL technology, a dB-per-meter loss can be ascribed for each wire gauge, and nominal dB loss may be ascribed for bridged taps of certain lengths. Furthermore, the loop records should indicate if the loop has loading coils (disqualifying for DSL operation) or digital loop carrier (DLC). If the loop is served via DLC, it is then necessary to determine if it is possible to provision the necessary DSL transceiver within the DLC remote terminal. The accuracy of loop records varies by area; in some areas, the use of this method may result in frequently incorrect predictive loop qualification.

Direct measurement of the loop characteristics via a metallic loop testing system can avoid problems resulting from inaccurate loop records. The two-ended loop measurement method is highly accurate but very costly since test equipment must be simultaneously connected at both ends of the line. Single-ended loop testing requires loop measurement equipment only at the CO, and thus is much less expensive. Systems such as MLT and 4TEL have been used for many years to perform single-end measurements of line characteristics for POTS transmission. It is expected that new single-ended loop test systems will provide useful wideband measurements to characterize DSL suitability. In addition to the conventional voice-band characterization, it is hoped that single-ended measurements may provide a good estimate of wideband loop loss, loading coils, bridged taps, and the degree of crosstalk noise. It should be noted that the CO-based single-ended measurement would be able to only test a portion of the loop in the preservice cases where no serving area interface (SAI) jumper has been placed to connect the distribution segment of the loop to the feeder section. In addition to the automated test systems, a technician may attach a hand-held loop tester at the CO (MDF), field (SAI), or customer site (NID).

Regardless of the whether loop records or direct line measurements are used, the loops may either be bulk prequalified or individually qualified as needed. With bulk prequalification, all loops in a CO are analyzed prior to service orders, and a database is maintained that the customer service representative may consult when a customer asks for service. This permits an instant answer on loop suitability. Alternatively, the loop may be analyzed in response to a service request. Given the necessary system automation, it may be possible to analyze the line and determine its suitability in a few seconds while the customer service representative is talking to the customer.

DSL in the Context of the ISO Reference Model

10.1   The ISO Model

In the late 1970s, the International Organization for Standards put forth both a model for the description of data communications and a series of protocols that met this model. This Open Systems Interconnection, or OSI, model defined a layered division of communications between applications on two computers or hosts. The layered approach hides details of information from the invoker of the service of a particular layer. Thus an application on a host requesting communication with its peer application does not need to know the details of its physical connection to a data network, communications used between itself and the network, or even the details of the protocols exchanged between the hosts supporting the application. The layered approach also simplifies the analysis of complex communication systems by segmenting the systems into several well-defined portions, facilitates reuse of portions of communications systems, and allows upgrades of portions of a communications system independently of each other.

For example, the functions of data delivery and connection management can be implemented in separate layers and therefore separate protocols. Thus, one protocol is designed to perform data delivery, and another protocol, layered above the first, performs connection management. The data-delivery protocol is fairly simple and knows nothing of connection management. The connection-management protocol is also fairly simple, since it doesn't need to concern itself with data delivery. Protocol layering produces simple protocols, each with a few well-defined tasks. These protocols can then be assembled into a useful whole. Individual protocols can also be removed or replaced as needed for particular applications.

In the OSI layered model, a *protocol* defines the actual communications between two entities over the network. A *service* defines the primitives that can be used by a higher layer to request the services of the layer immediately below it. The layered model services thus enforce a "need to know" environment. The requestor of a function need know nothing about implementation of the service it is requesting. It need only know the parameters to request the service of the layer immediately below itself. The requestor does not even need to know of the existence of any layers except that immediately below itself.

Figure 10.1 illustrates the OSI model and services. The application, such as spreadsheet or mail program, calls on the services of the application layer, the presentation layer, and so on until the message is sent on the physical medium. At any layer, communication occurs between peers on hosts at each end of the link.

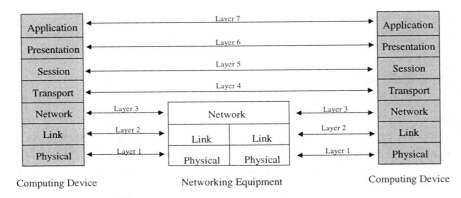

Figure 10.1   ISO model.

Layer 7 — the application layer, the highest layer of the model, defines the way applications interact with the network.

Layer 6 — the presentation layer, includes protocols that are part of the operating system. It defines how information is formatted for display or printing, how data are encrypted, and translation of other character sets.

Layer 5 — the session layer, coordinates communication between systems, maintains sessions for as long as needed, and performs security, logging, and administrative functions.

Layer 4 — the transport layer, controls the movement of data between systems, defines protocols for structuring messages, and supervises the validity of transmissions by performing error checking.

Layer 3 — the network layer, defines protocols for routing data by opening and maintaining a path on the network between systems to ensure that data arrive at the correct destination node.

Layer 2 — the data link layer, defines the rules for sending and receiving information from one node to another between systems.

Layer 1 — the physical layer, governs hardware connections and byte-stream encoding for transmission. It is the only layer that involves a physical transfer of information between network nodes.

10.2 Theory and Reality

Originally, ISO proposed a complete set of protocols encompassing all seven layers of the "stack." Although these protocols were fully defined in a series of standards, there has been relatively little actual implementation of the proposals. However the seven-layer model provides a departure point for discussions on data protocols even if the reality is far more complex then the original model. When discussing DSL technologies and network architectures that implement them, our concern is primarily with the first three layers of the stack as the functions represented by the transport, session, presentation, and application layers are almost entirely independent of the network that carries the communications.

10.3 The Internet Protocol Suite

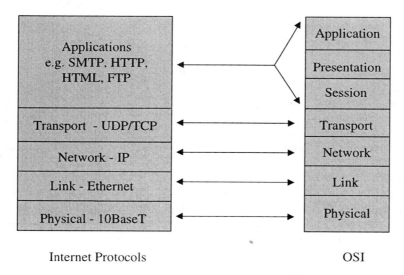

Figure 10.2 Relation between IP and OSI.

The Internet suit of protocols is also based upon a layered model. Starting at the physical layer, there is a parallel with the OSI model up through the transport layer. Thus, TCP and UDP provide transport layer functions, IP is a network layer protocol, and ethernet, provides a data link layer. 10baseT is an example of a physical layer. In discussion of IP, the three higher layers of the stack are lumped into an "application" layer. Figure 10.2 indicates the relationship between the ISO seven-layer model and the Internet IP stack.

10.4   ATM in the Seven-Layer Model

ATM can be considered a protocol at the data link layer (layer 2). However, a discussion of ATM requires that the details of its interaction with the physical layer also be specified. In the ITU's definition of ATM (or broadband ISDN), they define the following reference model for the two lowest layers of the stack.

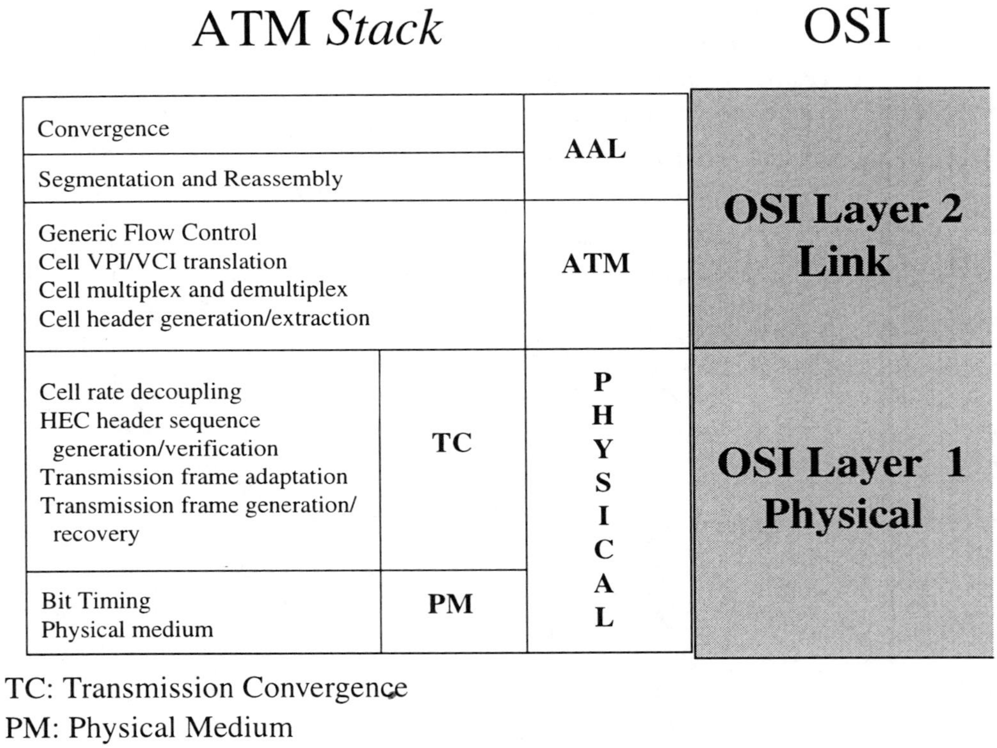

ATM *Stack*   OSI

Convergence		**AAL**	**OSI Layer 2 Link**
Segmentation and Reassembly			
Generic Flow Control Cell VPI/VCI translation Cell multiplex and demultiplex Cell header generation/extraction		**ATM**	

Cell rate decoupling
HEC header sequence
  generation/verification
Transmission frame adaptation
Transmission frame generation/
  recovery
	TC	**P H Y S I C A L**	**OSI Layer 1 Physical**

Bit Timing Physical medium	**PM**		

TC: Transmission Convergence
PM: Physical Medium

Figure 10.3   ATM reference model.

10.4.1 Physical Layer Functions

As shown in Figure 10.3, the physical layer is divided into two sublayers.

1. *Physical medium dependent (PMD) sublayer*
   The PMD sublayer contains only the physical medium–dependent functions. It provides bit transmission capability including bit alignment. It performs line coding. For DSL, these functions are provided by the modem. Chapter 11 will discuss the PMD sublayer functions for ADSL in detail.

2. *Transmission convergence (TC) sublayer*
   The TC sublayer handles five functions:
   1. Generation and recovery of the transmission frame, which contains the cells.
   2. Transmission frame adaptation adapts the cell flow according to the physical transmission.
   3. Cell delineation function enables the receiver to recover the cell boundaries.
   4. The HEC (header error correction) allows errors in the cell header to be detected.
   5. Cell rate decoupling inserts the idle cells in the transmitting direction in order to adapt the rate of the ATM cells to the payload capacity of the transmission system. It suppresses all idle cells in the receiving direction.

Chapter 12 discusses the functions of the transmission convergence sublayer for ATM in an ADSL environment. Similar functions must be performed when a packet-based link layer is used. Chapter 13 discuses the requirements for the transmission convergence function when a packet-mode data link layer is supported.

10.4.2 Link and Higher-Layer Functions

Chapter 14 will discuss the issues and requirements for end-to-end transport of data in an ADSL-based access network. It will touch on the requirements of the data link layer and network layer for both ATM and packet-based architectures.

ADSL: The Bit Pump

The standard physical medium–dependent sublayer for ADSL is defined in the ANSI T1.413 standard [1]. This chapter will provide an overview of how an ADSL modem based upon this standard prepares a user's data stream for modulation into the signal that is carried over the copper pair.[1]

11.1 ADSL System Reference Model

A pair of ADSL modems connects a user's computing environment via the ADSL link to a communications network. In order to place this in context, both T1E1.4 in ANSI T1.413 and the ADSL Forum in its TR-001 have defined reference models for the ADSL link. Since the ADSL Forum model is an extension of the ANSI T1.413 reference, we will use the ANSI T1.413 model for our discussion of the ADSL PMD (physical-medium dependent) sublayer of the physical layer.

---

[1]   All figures used in this chapter are reproduced with the permission of ATIS.

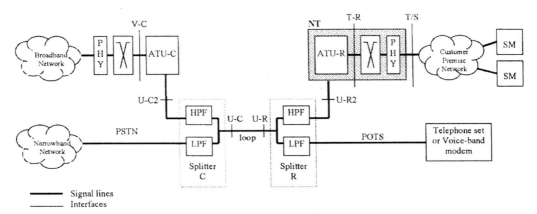

Figure 11.1   ADSL Forum reference model.

Source: ANSI draft T1E1.4/98-007R5.

Going from left to right:

V-C	Interface between the access node and the carriers data network
U-C2	The ADSL interface to the ATU-C without the 0 to 4 kHz POTS (voice band)
U-C	The ADSL interface to the ATU-C including the POTS band
U-R	The ADSL interface to the ATU-R including the POTS band
U-R2	The ADSL interface to the ATU-R without the 0 to 4 kHz POTS (voice band)
T-R	The interface between the ATU-R and a premises network[*]
T-S	The interface between the Premises network and the customers hosts

[*]The premises distribution network could be a local network such as a LAN or could be null as in the case of a direct connection between a modem and PC or an internal ADSL modem card and the computers bus.

For the purposes of this chapter, the U-C and U-R and the T-R and T-S interfaces can be merged. They will be referred to as the U and T interfaces, respectively.

The remainder of this chapter discusses the transformations that allow a pair of ADSL modems (an ATU-C and ATU-R) to transport data over their U interface between the V and T interfaces of this reference.

11.2   ATU-C Reference Model

In the ATU-C, up to seven "bearer channels" are defined at the V interface between the ATU-C and a carrier's transport network. These are labeled AS0 through AS3 and LS0 through LS2.

The AS-x channels are unidirectional simplex channels, while the LS-x channels are duplex. The standard allows combinations of the channels to be configured. A particular implementation of the V interface may enable from one to all seven of these channels. The simplex channels are used to support downstream communications. Typically, the duplex channels are used to support upstream communications (where only the upstream "half" of that channel is used).

Flowing from the V interface through the ATU-C to its output on the line at the U interface, the following activities occur:

1. Pacing the channel on one of the two "latency paths" supported on the ADSL interface.
2. Creation of cyclic redundancy codes (CRC) and error correction codes for the data.
3. Assembly of the data into an ADSL physical layer frame and superframe structures.
4. Encoding into the multiple tones of the DMT signal.
5. Analog output on the copper twisted pair.

Within the standard ATU-C, two "latency" paths are supported for the data: a fast and an interleaved path. In a particular configuration, a channel may be assigned to either one of the two paths. The interleaved path supports a convolutionally interleaved Reed-Solomon error correction, while the fast path does not support interleaving. The greater resistance to error in the interleave path is meant to support applications that are sensitive to degradation due to errors induced by line noise but are relatively tolerant to increased latency. Transmission of MPEG II video data is an example of such an application. The fast path provides less protection against transmission errors but less delay. It is meant to transport delay-sensitive applications such as interactive data.

A pair of modems may configure their communication to use either one or both of these paths. For example, a system implementing interactive video programming might place the "AS0" downstream bearer path in the interleaved path, while placing the LS0 duplex channel in the fast path. The AS0 channel would carry the MPEG date stream, while the other path would be used for the user interactive control data for the video system.

This can be contrasted with an implementation optimized for data communications. The AS0 channel carries the downstream ATM traffic in the fast path. The upstream traffic is carried in the LS0 "duplex" channel, which is also placed on the fast path (and carries traffic only in the upstream direction).

Although four downstream simplex and three upstream duplex bearer channels are defined in the standard, in any particular configuration of ATU-C and ATU-R only some of them may be used. For example, in the two examples above, only the AS0 and LS0 channels are configured for the connection. The available bandwidth for the ADSL link is allocated among the channels configured. Each channel can be allocated bandwidth in units of 32 Kbps up to the maximum bandwidth available for upstream or downstream communication on the ADSL link.[2]

---

2   The maximum bit rate for communication for a T1,413 compliant ADSL modem is 6144 Kbps downstream and 640 Kbps upstream. Due to loop conditions and length, the available bandwidth could be less.

In the case of support for ATM, the ATU-C and ATU-R are responsible for the ATM transmission convergence function (ATM-TC). A network timing reference is supported on the ATU-C. This allows the carrier's network to optionally provide an 8 kHz timing signal to the ATU-C and ATU-R for synchronization to the carriers network.

The channels supported over the two paths are combined into a physical layer frame and superframe structure. A frame is generated every 250 microseconds (there are 4000 in each second) and contains data bytes for all channels that are implemented over a connection. Channels carried over the fast latency path are placed in the frame before those sent over the interleaved path. Each frame is encoded in DMT as a single DMT symbol; that is, the entire frame is encoded at one time across all the tones supported. Thus, the size of a frame in bytes is a function of the line rate that is currently supported by connection between the two modems. A faster adapted rate will support larger frames than a slower one. Sixty-seven frames make up a superframe, which is terminated by a synchronization symbol. This flow is illustrated in Figure 11.2.

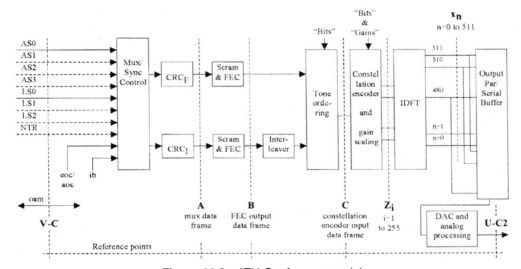

Figure 11.2   ATU-C reference model.

Source: ANSI draft T1E1.4/98-007R5.

11.3   ATU-R Reference Model

The ATU-R is similar to the ATU-C, however at the T interface the AS-x simplex channels are all receive only (at the ATU-C they are send only). See Figure 11.3. They constitute the downstream channels while the duplex LS-x interfaces can be configured to define the upstream only channels. Since the upstream bandwidth has a lower maximum (640 Kbps) than the downstream channel (6144 Kbps) only the three duplex channels are multiplexed in the frame for transmission on the ATU-R U interface. Like the ATU-C, both an interleaved and fast buffer are supported.

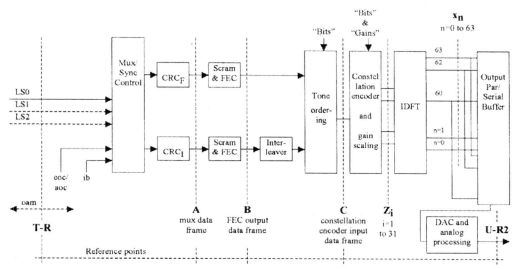

Figure 11.3   ATU-R reference model.

Source: ANSI draft T1E1.4/98-007R5.

11.4   Specific Configurations to Support ATM

When ATM is supported over the ASDL interface, the V interface supports two ATM channels: ATM-0 and ATM-1. ATM-0 is always supported, while ATM-1 is an optional second channel to allow support of dual latency (simultaneous use of both the fast and interleaved path). The ATM compliant ADLS V interface supports flow control between the ATM network and the ATU-C to ensure that the ATU-C's buffers neither underflow or overflow. If only one latency path is supported, "ATM-0" is configured with one simplex (downstream) bearer channel, AS0, and one duplex (upstream) channel, LS0. If dual latency is supported, a second ATM interface, ATM-1, is supported, and both the AS1 and LS1 bearer channels are used to bear the second latency path. Figures 11.4 and 11.5 illustrate these reference models for the ATU-C and ATU-R respectively.

11.5   Framing

An ADSL frame's length is determined by the adapted bit rate of the interface. DMT symbols are encoded at the rate of 4000 baud, that is, one complete transition every 250 microseconds. The actual amount of data that can be encoded in one transition is a function of line condition. Depending on the line conditions found at initialization, the number of tones supported and the

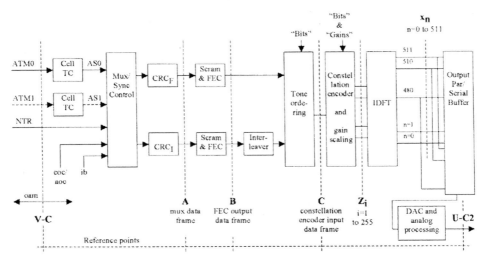

Figure 11.4   ATU-C reference model.

Source: ANSI draft T1E1.4/98-007R5.

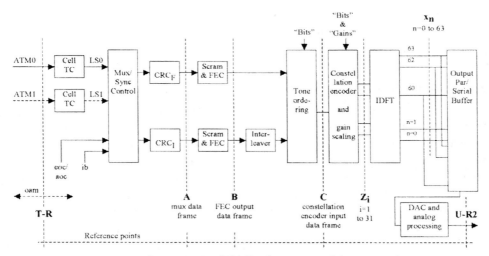

Figure 11.5   ATU-R reference model.

Source: ANSI draft T1E1.4/98-007R5.

amount of data encoded in each tone will vary. Tones that are in noisy areas of the spectrum are suppressed, and the complexity of the constellation of symbols encoded in each tone will be set to optimize transmission to the particular line conditions. One frame contains the data that can be transmitted over the DMT interface as one symbol, that is, at one time. Each frame contains data from both the fast data buffer and the interleaved data buffer as well as overhead bits used for error correction and administration and management of the ADSL link. See Figure 11.6. Since the frame length is related directly to how data is encoded using DMT, there is no need for a delimiter indicating the beginning or end of a frame.

11.5.1 Superframe Structure

A superframe is defined. Each ADSL superframe is made up of 68 ADSL frames followed by a synchronization symbol, which delimits the superframe. See Figure 11.6. Since the length of the frame is a function of the particular configuration of the ADSL U interface, the length of the superframe is also variable based upon configuration of the link.

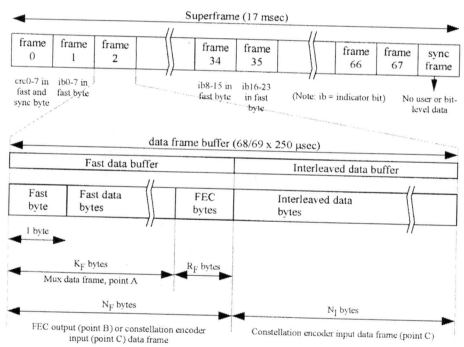

Figure 11.6 ADSL superframe structure.
Source: ANSI draft T1E1.4/98-007R5.

11.5.2 Fast Data Buffer Frame Structure

The data in the fast buffer is inserted in the first part of the frame. The first byte is the "fast byte" and contains either overhead or synchronization information. The data bytes from the fast buffer are inserted following the fast byte. Bytes for each bearer channel follow in order, as indicated by Figure 11.7.

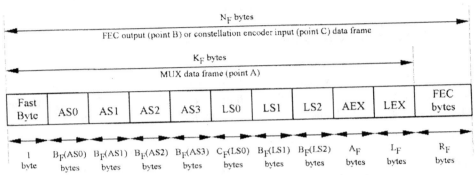

Figure 11.7 Fast latency path frame structure.

Source: ANSI draft T1E1.4/98-007R5.

If a bearer channel is not implemented, no data are inserted for that bearer channel. The data for the individual bearer channels are allocated proportionally within the frame based on the proportion of the ADSL link's total bandwidth allocated to each channel. If no data were sent over the fast path, then the fast portion of the frame would be made up solely of the fast byte. The fast buffer section of the frame is terminated by bytes holding synchronization information (the "AEX" and "LEX" bytes) and the forward error correction codes calculated for the fast path data in the frame.

11.5.3 Interleaved Data Buffer Frame Structure

The interleaved buffer is inserted in the frame after the fast buffer data. It is first assembled in a format identical to the fast frame. Like the fast frame, the amount of data for each bearer channel allocated to the interleave path is spread proportionally to the proportion of the bandwidth of the bearer channel to the total bandwidth of the ADSL link. Once the multiplex is assembled for the interleave path, frames are held in a buffer up to the defined interleave depth (N frames) and are convolved with each other. The result is an output that is the same length as the original input frame but contains bits from the previous N input interleaved frames. Thus, the data sent to the DMT encoder contains data from all N frames in the buffer. Thus data from any interleave frame is delayed up to N frame periods ($N \times 250$ microseconds) before it can be decoded at the far end. The output of the interleaver is appended to the output from the fast buffer to construct the frame. This is illustrated in Figure 11.8.

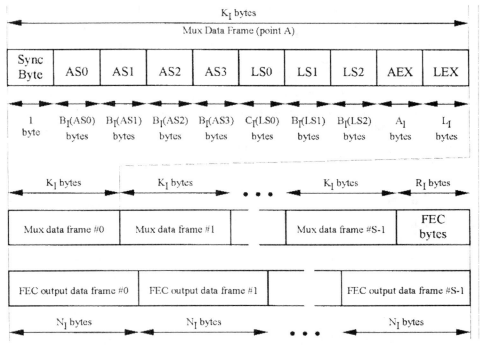

Figure 11.8   Interleave frame creation.

Source: ANSI draft T1E1.4/98-007R5.

11.6   Operations and Maintenance

The ADSL interface supports three methods for exchanging physical layer operational information between the ATU-C and the ATU-R: (1) the embedded operations channel (EOC), (2) the ADSL overhead control (AOC), and (3) the indicator bits.

The embedded operations channel supports the reading and writing of registers containing operational information on the ATU-R from the ATU-C. Registers have been defined in the standard to allow access to the fields containing the identification of the ATU-R including vendor, model, and serial number, current configuration of the ATU-R, self-test results, current line attenuation, and SNR (signal-to-noise ratio) margin. Several registers are devoted to vendor-defined fields, which can either be read or written via the EOC. The ATU-R can send the ATU-C a dying gasp message over the EOC when the ATU-R is powered down.

The EOC is implemented by using bits in the "fast byte" of frames 2 through 32 and 36 through 67 of a superframe. The EOC "frame" is made up of 13 bits, a 5-bit header, and an 8-bit payload. One EOC frame can be sent in every two ADSL frames. The payload can either contain a byte of data or a command (op-code) to the other end. In the T1.413 standard, except for the dying gasp message sent by the ATU-R, all commands are sent by the ATU-C to the ATU-R. The ATU-R may respond to the command with a reply containing data from a register.[3]

The AOC has a similar structure to the EOC. It is used to carry real-time information used to reconfigure the ADSL link should line conditions change. It is carried in bits in the synchronization bytes of the interleave section of the ADSL frame. An AOC "frame" is 13 bits long and contains a 5-bit op-code and 8 bits of data.

The 23 indicator bits are carried in the fast bytes of the fast portion of the ADSL frame. Each bit serves as an indicator to the receiving modem of some aspect of the state its peer at the other end of the link. Indicator bits are set when conditions such as transmission errors or loss of signal are detected at the far end of the ADSL link.

11.7   Initialization

The initialization process allows the ATU-C and ATU-R to establish their communications. The process allows the two modems to identify themselves to each other, determine line conditions available to support communications, exchange parameters that define the request connection, allocate resources, and begin normal communications. The process is divided into four segments:

1. **Activation and Acknowledgment:** The ATU-R begins the initialization process by transmitting the appropriate tones to the ATU-C. When this segment of initialization is complete, the ATU-R and ATU-C have negotiated the timing method used between them and have determined which device is the master. At the end of the procedure, the ATU-R and ATU-C are in a state capable of analyzing the line conditions. The following signal tones are used during the initialization phase: R/C-ACT 1,2,3,4, and REVEILLE.

2. **Transceiver Training:** During this process, the ATU-R and the ATU-C send signals that allow their partner to determine line conditions and adjust the equalization of their transceivers. Transceiver training also determines if ADSL is operating in FDM (frequency division multiplexing) or ECH (echo cancellation) mode. The following signal tones are used during the transceiver training phase: R/C-REVERB, and R/C-SEGUE.

3. **Channel Analysis:** The modems exchange information on the upstream and downstream bearer channels required for the connection, the latency paths they will be placed in, and the bandwidths for each channel requested. Information about specific features supported or requested is also exchanged. The modems then perform tests that

---

3   There is an informative annex in the proposed text for T1.413 issue 2 which enhances the EOC to allow it to be used as a full duplex data channel as well as supporting the command/reply syntax described above.

determine the loop quality and SNR for each specific 4 kHz DMT tone. The following signal tone is used during the channel analysis phase of initialization: <u>MEDLY</u>.

4. **Exchange:** Having gathered the information about the quality of the connection and the requested configuration, the modems configure themselves and exchange information about their configuration. The specific bandwidth allocated to the requested bearer channels is assigned, the specific DMT tones and the amount of data encoded in each tone are determined and assigned. The connection is tested in both directions after which each modem notifies its peer that it is ready to enter normal communications, known in the standard as "showtime." The following signal tones are used during the exchange phase of initialization: <u>R/C-REVERB</u>, and <u>R/C-SEGUE</u>.

Reference

[1]   American National Standards Institute. *Standards Project Relating to Carrier to Customer Connection of Asymmetric Digital Subscriber Line (ADSL) Equipment* T1.413 Issue 2, 1998.

ATM Transmission Convergence on ADSL

12.1   Functions of ATM Transmission Convergence

Chapter 11 described how the physical layer channels defined in the ANSI T1.413 standard for ADSL can be used to transport ATM cells. However, in order to transport ATM cells, a transmission convergence (TC) sublayer must be defined. In the TC sublayer, a mapping is provided between the definition of ATM cells and the ADSL physical layer constructs. The issues that must be addressed include:

1. Generation and recovery of the transmission frame that contains the cells.
2. Transmission frame adaptation adapts the cell flow to the physical transmission medium.
3. Cell delineation function enables the receiver to recover the cell boundaries.
4. The HEC (header error correction) allows errors in the cell header to be detected.
5. Cell rate decoupling inserts the idle cells in the transmitting direction in order to adapt the rate of the ATM cells to the payload capacity of the transmission system. It suppresses all idle cells in the receiving direction.

All ATM data is transferred in fixed-length "ATM cells." Each 53-byte (a byte is termed an "octet" in the standards that define ATM) cell has a 5-byte header and a 48-byte payload section. Figure 12.1 illustrates the layout of an ATM cell.

The header contains fields identifying the ATM virtual circuit that carries the cell and header error correction field that is used to identify those cells that have had their header information corrupted.

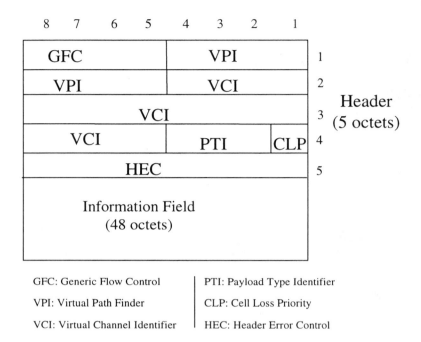

Figure 12.1 ATM cell format.

12.2 Transmission Convergence in an ADSL Environment

Since the ADSL frame, as described in Chapter 11, can be of differing lengths depending upon the particular conditions on the link at the time, there can be no definite relationship between the beginning of an ATM cell and the structure of an ADSL frame. The ADSL TC layer allows the recovery of the ATM cells at the far end of the ADSL connection.

The cell delineation function permits the identification of cell boundaries. The ADSL ATM transmission convergence sublayer is based on ITU-T recommendation I.432 [1]. The transmitter in the ATU-R or ATU-C transmits the cell bytes aligned to any bit in the ADSL frame. When first receiving ATM cells, the receiver starts at an arbitrary bit and assumes that the next 53 bytes are an ATM cell. If the HEC field in the assumed cell does not indicate an error-free header, the receiver assumes that cell alignment has not been found and shifts over one bit and performs the same test again. Ultimately the HEC field in the assumed cell indicates a correctly transmitted header. This is likely a valid cell. The test can now be performed on the next assumed cell header (the next 5 bytes) and so on. After a sufficient number of assumed cells have been found which have valid header HEC fields, the ATM transport is assumed to be in sync and cells can be detected continuously. Figures 12.2 and 12.3 illustrate this process of determining the cell alignment in the ADSL channel.

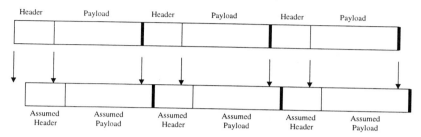

Transmitted Cell Stream
The Cell Headers Contain Valid HEC Fields

Trial Cell Stream at Receiver
Note the misalignment of the Cell Headers. When the HEC code is calculated for the assumed header it will not match what is found in the fifth byte (octet) received in the assumed header. The TC layer is not yet in Sync.

Figure 12.2 Recovery of cell delineation (receiver not in sync).

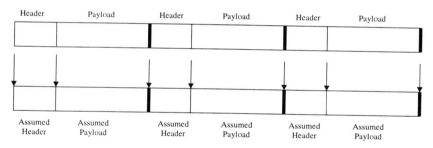

Transmitted Cell Stream
The Cell Headers contain valid HEC fields

Next Trial Cell Alignment at Receiver
The Receiver has shifted the received stream over by one bit.
This time the assumed cell boundaries match what is transmitted.
The calculated HEC matches what is received, the transmitter and receiver are now in alignment with respect to cell delineation.

Figure 12.3 Recovery of cell delineation (receiver now in sync).

ATM idle cells or ATM unassigned cells are inserted in the cell stream whenever the ATU-R or the ATU-C is not receiving cells with data in them from the networks connected to the ATU-C or ATU-R. This process, cell rate decoupling, ensures that the channel is filled with valid ATM cells regardless of the traffic being delivered from the hosts or network. By the use of the methods described in I.432, a continuous stream of ATM cells can be sent over the ADSL physical layer channels.

Reference

[1]   International Telecommunications Union, *ITU-T Recommendation I.432 - B-ISDN UNI Physical layer specification,* Geneva.

CHAPTER 13

Frame-Based Protocols over ADSL

As seen in the previous chapters, ADSL is a physical layer modulation scheme that provides an asymmetrical "bit pipe" between two end points, an ATU-C and an ATU-R. Transferring data over this "bit pipe" will require a data link layer. Although ATM is one logical option, frame-based protocols are more widely deployed in today's networks and should be considered as part of an ADSL offering. For certain environments they may be the preferred implantation. There are many different frame, or packet-based data link control (DLC) protocols, that could be used over ADSL. However, the ADSL Forum has issued recommendations [1] describing two possible frame-based solutions: the point-to-point protocol and FUNI (frame-based user network interface).

13.1 PPP over a Frame-Based ADSL

The Internet Engineering Task Force (IETF) originally drafted the point-to-point protocol (PPP) as a replacement for the serial line interface protocol (SLIP). In its current form, the PPP Internet standard (IETF RFC 1661) has defined a robust and adaptable protocol that can transport multiple packet datagram types over just about any kind of point-to-point link. This includes the ADSL link between an ATU-R and its adjacent ATU-C. Further, PPP has established a set of protocols that allow for the "online" negotiation of service parameters between adjacent point-to-point nodes. The link control protocol (LCP) supports the exchange of link layer information for the automatic establishment, configuration, and testing of the data link connection. The network control protocol (NCP) allows the information exchange of network layer information for the automatic establishment and configuration of different network-layer protocols.

As seen in the Figure 13.1, RFC 1661 defines the basic procedures, methodologies, and protocol formats of the PPP header, link control protocol, and network control protocols. Addi-

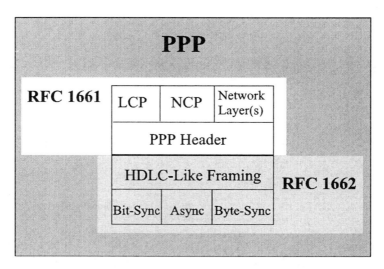

Figure 13.1   Functions of RFC 1662 and RFC 1661.

tional LCP and NCP protocols exist within the body of current and emerging IETF standard and draft RFCs. Table 13.1 lists some of the extensions that have been proposed to PPP (RFC 1661) along with a brief description of their intended purpose.

Table 13.1   Extensions and their purposes.

RFC	Purpose
RFC 1331	IP Control Protocol
RFC 2331	DHCP Protocol (Dynamic Address Assignment)
RFC 1877	Assignment of Location of DNS (Domain Name Server)
RFC 1334	PAP and CHAP — Authentication Protocols
RFC 1638	Transparent Ethernet Bridging
RFC 1664	IPX over PPP

13.1.1   RFC 1662 — PPP in HDLC-Like Framing

The ADSL Forum's Packet Mode subcommittee has specified that ADSL links supporting direct PPP mapping will operate under the RFC 1662, bit-synchronous mode. They require a physical link(s) that can provide a full-duplex, simultaneous bidirectional operation and ensure packet ordering.

As seen in Figure 13.2, RFC 1661 defines PPP's header, states, Link Control, and Network Control Protocols. RFC 1662 defines PPP's HDLC-like (high-level data link control) framing and operation in an asynchronous, bit-synchronous, or byte- (octet) synchronous environment. RFC 1662 refers to ISO 3309, where the HDLC protocol is defined.

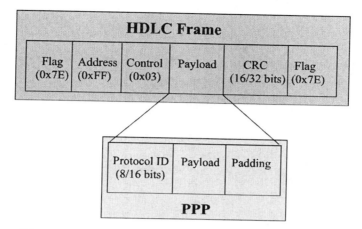

Figure 13.2   Format of HDLC and PPP protocol data units.

13.1.2   RFC 1661 — The Point-To-Point Protocol

The point-to-point protocol, as defined in IETF RFC 1661, performs three functions:

1. It provides a method for encapsulating multiprotocol datagrams.
2. It defines a link control protocol (LCP) for establishing, configuring, and testing the data link connection.
3. It allows support for network control protocols (NCPs) for establishing and configuring various network-layer protocols.

13.1.2.1 Encapsulation

PPP encapsulation of data allows the multiplexing of different network-layer protocols simultaneously over the same link. This encapsulation uses the HDLC framing defined in RFC 1662 to delimit the encapsulated data. The 16- or 32-bit protocol ID indicates which protocol is carried in the payload of the PPP message.

13.1.2.2 The Link Control Protocol

Link control protocol (LCP) is used to automatically negotiate the encapsulation format options, determine the packet size carried over the PPP link, detect common misconfiguration errors, and terminate the link when communications are complete. Optional facilities are provided to allow authentication of the identity of the peer on the link, and determination if the link is functioning properly.

To establish communications over a point-to-point link, each end of the PPP link sends LCP packets to configure and test the data link. After the link has been established, authentication of the connection can be provided using either the PAP or CHAP protocols defined in RFC

1877. Upon completion of link establishment, PPP uses one or more of the NCPs to choose and configure network-layer protocols. Once the chosen network-layer protocols are configured, datagrams from the chosen network-layer protocol can be sent using the encapsulation method. The link will remain configured until explicit LCP or NCP packets terminate communications on the link or the physical link fails and a time-out occurs.

13.1.2.3 The Network Control Protocols

The network control protocols allow the negotiation of the options for the particular protocols being carried over a PPP link. For example, the IP address assigned to a link must be assigned dynamically whenever PPP is used to transport IP protocol. NCP have been defined for a number of protocols including IP, IPX, and transparent Ethernet bridging. In an ADSL environment, support for IP using RFC 1332 will be most common.

13.2 FUNI over ADSL

ADSL Forum TR-003 also defines support for a packet-mode link layer based upon FUNI (frame-based user network interface) defined by the ATM Forum in ATM Forum document *str-saa-funi_01.01*. FUNI is a frame-based version of ATM in which variable-length frames are used instead of fixed-length 53-byte cells. Multiple virtual circuits can be simultaneously defined over the same physical link. Since the ATM UNI (user network interface) and the FUNI interface use the same method of addressing virtual circuits and the same basic constructs, FUNI-based access interfaces are commonly implemented on ATM network equipment.

13.2.1 FUNI Frame Structure

This interface allows FUNI data frames to be transported over an ADSL link. FUNI is similar to the PPP interface described earlier in this chapter as it uses an HDLC-type framing to define the link level frame.

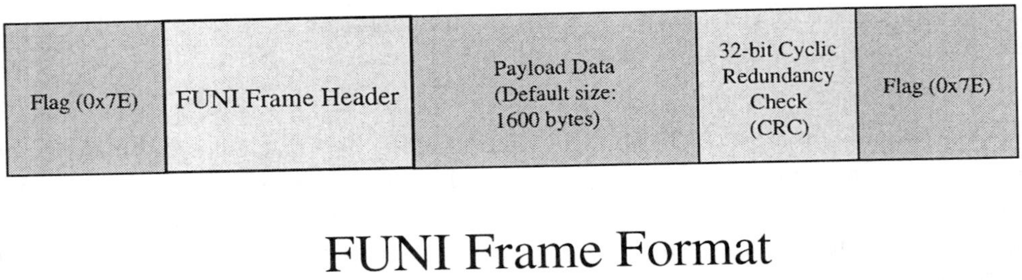

Flag (0x7E)	FUNI Frame Header	Payload Data (Default size: 1600 bytes)	32-bit Cyclic Redundancy Check (CRC)	Flag (0x7E)

FUNI Frame Format

Figure 13.3 Structure of a FUNI frame.

The FUNI frame uses HDLC start and stop flag bytes to define the beginning and end of each frame. Bit stuffing is used to prevent confusion between data and the frame byte. The ATM

FUNI frame header contains address and control fields. The address field encodes the virtual path identifier (VPI) and virtual connection identifier (VCI) of the virtual circuit that will transmit the frame. The use of a default data size of 1600 bytes facilitates encapsulation of 1500-byte Ethernet frames.

13.2.2 Encapsulation

The ADSL Forum technical requirement for FUNI-based packet mode encapsulates the data within the FUNI frames using the procedures defined in IETF RFC 1483 — *Multi-Protocol Encapsulation Over ATM AAL5*. RFC 1483 describes two different methods for carrying connectionless network interconnect traffic over an ATM network.

The first method, logical link control (LLC) encapsulation, allows multiplexing of multiple protocols over a single ATM virtual circuit. Prefixing the encapsulated protocol by an IEEE 802.2 logical link control (LLC) header identifies the protocol carried in a particular frame.

The second method, VC multiplexing, does higher-layer protocol multiplexing implicitly by ATM virtual circuits (VCs). In VC multiplexing, any payload may be transported in a FUNI frame without any additional encapsulation. When VC encapsulation is used, the end systems create a distinct virtual circuit connection for each payload protocol type. Since there is no protocol type discriminator in VC encapsulated RFC 1483 frames, there must be some out-of-band method for the end points to agree on the interpretation of the payload that follows the FUNI header.

According to ADSL Forum TR-003, when PPP is carried over an ADSL link supporting FUNI, the PPP protocol data units (PDUs) are always carried using VC multiplexing.

Reference

[1]   ADSL Forum *Technical Report TR-003 — Framing and Encapsulation Standards for ADSL: Packet Mode*, June 1997.

ADSL in the Context of End-to-End Systems

DSL technologies, especially ADSL, occur in an architectural context that has fundamental differences from that found when analog modems are used to connect a user's site to a remote site. In the case of these earlier technologies, communications occur over a path established by a telephone carrier which has been optimized for carrying voice traffic. This has the advantage of allowing for worldwide switched connectivity of data over the voice network. However, it also limits the bandwidth of a switched connection to, at most, 64 Kbps. This is the rate supported by the digital voice channels within and between the switches of the worldwide voice transmission network. DSL technologies are able to achieve their high bandwidths over the copper loop because they avoid the existing telephony network of switches and transmission channels optimized for voice traffic.

In a sense, the data connection "creatively abuses" the physical path that is established point-to-point for carrying voice traffic. This is true whether the user is using analog technologies over the public switched telephone network (PSTN) or leases private facilities from the carrier to achieve high-bandwidth digital communications between their sites. In either case the existing data streams are transported over the physical layer connection of the existing voice network. It is the function of equipment at the remote site and users site to create this data stream and manage the protocol stack. However, in the case of DSL technologies, one of the modems will reside at the user's site while the other will reside at the office of the carrier. The carrier must then provide a data network, which connects the user to the remote site.

In the case of a conventional modem, the network operator and, in certain senses, the modem itself are not concerned with the manipulation of the data except the physical layer encoding to place it on the network. Actual manipulation of the data is the responsibility of the hosts on each end. However, when ADSL is implemented, manipulation of the protocol stack will occur in the modem and also in the carrier's network.

Figure 14.1 schematically illustrates today's situation. The carrier provides physical layer connectivity. For analog modems this is through the PSTN, and for higher-bandwidth connections over private lines using the same transmission facilities as used for voice. For example, phone companies connect the routers that make up the backbone of the public Internet to each other using physical layer connections provided using the same transmission equipment that supports the voice network.

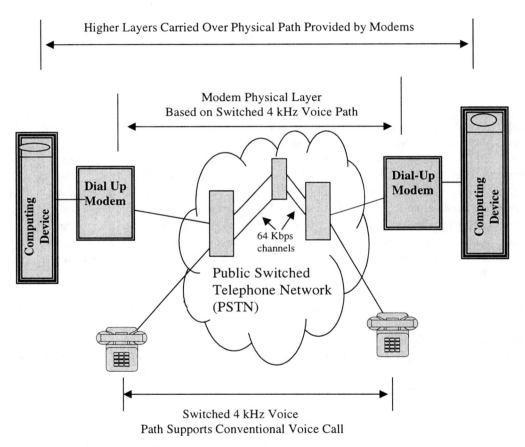

Figure 14.1   The telephone carrier provides physical layer connectivity.

Figure 14.2 illustrates the situation where a carrier provides DSL access. There must be a DSL modem at the user's premises and also at the carrier's site where the copper is terminated. The carrier must provide a network specialized for carrying data to efficiently connect the modem to the various remote sites that a user might need to reach.

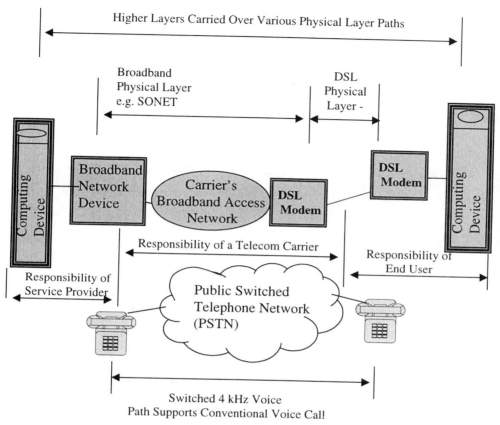

Figure 14.2   Schematic of a DSL end-to-end environment.

In this end-to-end DSL network, there are three natural "realms of responsibility." The equipment at the site of the user of DSL is one realm. The carrier's network, which provides the DSL service to the user and connects the user traffic to one or more service providers, is a second. The service provider, such as Internet service providers (ISP), or corporate environments whose networks actually support the servers and applications desired by the users, is the third realm.

Each of these realms may need to be managed separately from the others. In many cases, they will be owned by different entities. In any case, the demarcation of responsibility is natural. Both the user realm and the service provider realms are likely to exist before the DSL service is implemented, while the carrier has the responsibility of serving many customers. For example, the user already owns a PC, or the LAN at the user's small business is already deployed before DSL service is installed. The service provider might be an existing ISP, such as America Online, or could be an extensive corporate network. Similarly, The DSL service gives the user access to the service provider at a higher bandwidth, but the service must fit into an environment where the service provider uses access methods, such as dial-up modems or private lines, which already exist and are being operated today. The carrier is providing DSL, a new technology, to give the user access to the service provider. However, the existing model where the Telco hides the details, operations, and management of its network from its customers is still desirable. The users of DSL are unlikely to want the Telco to be involved in the management of their internal networks and the Telco must manage this new service with existing personnel and familiar methods.

14.1   An Overview of a Generic DSL Architecture

Figure 14.3 illustrates the structure of a generic DSL access between users and service providers.

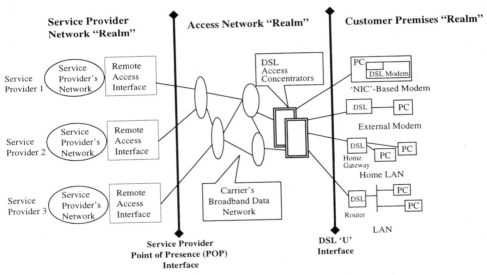

Figure 14.3   Generic DSL access architecture.

This architecture can be divided into several components:

14.1.1   The Customer's Premises

The DSL customer's computing environment must be connected to the DSL modem. The requirements for this interface depend upon the customers computing environment and the applications that they are supporting over the ADSL interface. Some of the likely configurations are:

- The user has a single PC with an internal DSL modem card. Applications on the PC terminate the data communications stack and control the modem. Other applications allow the user to interact with the network to select connections with the appropriate service providers. This configuration is likely in a residential environment to support either Internet entertainment or work-at-home.

- The user has a single PC with an external DSL modem. The ADSL data communications stack and its control may occur in the external modem. An application on the PC allows the customer to interact with the network to select connections with service providers. This configuration is likely in a residential environment to support either Internet entertainment or work-at-home.

- The DSL modem resides in a small router. The LAN connections are probably based upon Ethernet and support multiple hosts, all of which can use the DSL connection to access a wide area network using the IP protocol. The DSL link serves the same function as the private DS0/T1 line that currently is used to link small routers to their central sites. A likely user of this configuration would be a small branch office of a larger organization or a small business.

- The DSL modem is connected to an existing router or network element. The DSL link appears to the router as if it was supporting a conventional private line. The DSL link and the back-end network connect to the user's data network, such as a frame relay network. The DSL line serves the same function as private line services, such as T1 or fractional T1 used to connect a user's equipment to an existing data network.

- The DSL modem supports a home LAN. A home may now support several computing devices. These may need to be linked on a local network that is optimized for installation and operation by a relatively unsophisticated user. The technologies to provide this home LAN are still under development. By providing a single DSL link to a device supporting this LAN, all users in the home can access remote sites over the DSL link.

14.1.2 The DSL Loop

This is the copper loop pair that carries the traffic from the customer's site to the central office or remote termination. The characteristics and requirements for an ADSL loop are discussed extensively in other sections of this book.

14.1.3 Termination of DSL in the Carrier's Central Office or Remote Site

In a DSL implementation, anywhere from several dozen to many thousands of DSL loops will need to be terminated at a Central Office. A typical CO may terminate anywhere from 5,000 to 50,000 copper pairs. The largest offices will terminate as many as 160,000 lines. Future DSL service penetration is unknown at this time. However, some estimates are that as many as 20% of customers will be served by ADSL within five years. In this case, a typical office will need to terminate between 1,000 and 10,000 ADSL users, and the largest offices will terminate 32,000 ADSL users.

The terminating equipment must be able to terminate the DSL lines, support interfaces to the back-end data network, and multiplex the user's traffic efficiently onto the back-end network. The termination must have interfaces to allow the carrier to perform administrative functions on the DSL network such as bringing into service new customers, terminating customers who have ceased to receive the service, and test and resolve problems. The term DSLAM (digital subscriber line access multiplexor) has become somewhat of generic term for the devices that will terminate ADSL service in the Central Offices.

The devices that will terminate DSL in remote environments where the telephone service is supported by remote carrier systems have special requirements. In these remote sites the copper loop terminates remotely from the Central Office. Telephone service is provided by a digital loop carrier system that terminates the copper and multiplexes the POTS traffic over either copper or fiber trunks. If DSL service is to be offered to customers served by these remote sites, DSL multiplexors must be built to meet special requirements. They must operate in environmentally difficult conditions, in some cases without air conditioning or heat. The Central Office has a main distribution frame that allows any loop to be connected easily to any piece of equipment; remote sites often lack this flexible connectivity. Remote sites also vary in size from extremely small installations with 100 lines to remote carrier systems supporting 2,000 lines. DSL systems must exist to support as few as 20 and as many as 400 customers economically from these remote installations.

In the case of ADSL, the carrier must separate the POTS signal on the loop and deliver it to the switched voice network; ADSL signal is sent to the DSLAM.

14.1.4 The Carrier's Back-End Data Network

The carrier is responsible for terminating the DSL physical layer at the site where its copper loops are terminated and placing the user's traffic on a back-end data network. In the case of ADSL, a single DSLAM will terminate between 200 and 500 ADSL lines and concentrate the traffic on a DS3 (45 Mbps) trunk for delivery to the back-end data network. The back-end network will connect all the Central Offices that serve DSL customers to all the service providers who wish to be reached by users served by DSL.

This data network for any large deployment of DSL must be separate from the existing voice network because the existing telephone networks are very inefficient for carrying data. A large deployment of ADSL would be uneconomical using the existing infrastructure as a back-end.

The switches are designed assuming that voice calls average about three minutes in length. However, data sessions are longer. An Internet "surfing" session that lasts less than a half-hour is a rarity. Many desirable applications, such as connections between a branch office and a central site, require permanent connections. Resources in the voice network are allocated assuming that the statistics of calls (the likelihood that a user will make a call at any particular time and the average length of a call) are like those found for voice calls. These resources include potential call paths within the switches and potential paths on the trunks between switches. When the statistics of calls change dramatically, the voice network becomes overloaded as these resources become exhausted at times of maximum use. The carrier must incur the considerable expense of adding additional resources to their network.

Already before the deployment of DSL, the existence of "long holding time" data calls, either to reach ISPs for Internet recreation or, corporate LANs for employees working at home has had serious effects on the quality service that some phone companies can provide their customers. The phone customers, both those placing data calls and those placing voice calls are denied service, either by waiting to receive a dial tone or, by receiving an "all circuits busy" signal instead of a connection. Usually it is the nondata customer who sees the degraded service. The telephone carrier must enhance their network. However, tariff rules, also designed for voice, often prevent the telephone company from recovering any of the additional expenses due to the network enhancements required to deal with these longer calls. Removal of the long holding time switched data calls from the voice network and placing them on a separate network with different economics therefore becomes an imperative for telephone companies.

Much of the economic advantage to a carrier of providing a DSL-based access service comes from this need to remove long data calls from the PSTN as from potential revenue from the new DSL service itself. The DSL access and its new back-end network avoid the expenses of adding additional non–revenue-producing resources to the PSTN. However, this benefit is seen only if the data traffic sent over ADSL is transported over a network designed for the efficient transport of data.

The existing voice network architecture assumes that a dedicated connection is required between any two points that communicate. The resources of this circuit are then dedicated to the communications even if no data traffic is currently being sent. Data traffic is typically very bursty; that is, there are many periods during any communication where no data is being sent at all between the two end points. For example, an Internet connection will be nearly idle while a user types a query to a web server, followed by a flurry of activity when the web pages are downloaded. While the user is looking at the content of the page on their PC, there may be no transfer of data at all for a long period. Even private lines provided by telephone carriers to connect remote sites on corporate networks are used inefficiently because of the same bursty nature of the data communications carried on the lines.

Data network architectures take advantage of this burstyness by allowing concentration of the data over the networks. Connectionless packet architectures such as those supporting IP (Internet protocol) do not allocate connections to user sessions. Instead, traffic for numerous users is multiplexed over connections allocated between routers without allocating dedicated resources to any particular user. The trunks between routers can then be filled efficiently. ATM (asynchronous transfer mode) does allocate connections to individual sessions. However, unlike the voice network that has a granularity of bandwidth of 64 Kbps per connection, ATM has no such minimum granularity. Additionally, the ATM architectures allow a trunk to be allocated connections whose total bandwidth might exceed the capacity of the trunk. The bursty statistical nature of the data and ATM's use of small 53-byte cells to encapsulate the data allows the ATM switches to concentrate the traffic on the trunks between ATM elements. Each user session appears to get a connection with a specific bandwidth dedicated to it. However, the data are really being multiplexed with other data on the trunk.

The voice network, both its switches and its transmission facilities, does not have the capacity to carry the volume of high-bandwidth traffic that will be generated by a DSL-based access network. An ADSL loop may have bandwidths as high as 6 Mbps per user in the downstream direction and 1 Mbps upstream. A voice channel is 64 Kbps; a T1 private line supports only 1.5 Mbps. Even a DS3 link, with a capacity of 45 Mbps, would support only seven such ADSL connections if architecture based upon dedicated circuits were constructed. A data network architecture with ability to both multiplex and concentrate traffic is therefore required if there is to be any significant deployment of DSL-based services.

14.1.5 The Interface to the Service Provider's Network

The service provider must be able to terminate the traffic delivered by the carrier over the back-end network. The service provider already operates a network. If they are an ISP they typically terminate dial-up modem calls from users at a data rate of between 28.8 and 56 Kbps. As ADSL service begins, the service provider will need to terminate customers who can now communicate at rates as much as 200 times faster.

The carriers back-end network will multiplex traffic from all DSL customers in an area that wish to connect to a particular service provider. In a large metropolitan area such as Chicago with 2,000,000 telephone subscribers, a 20% penetration of ADSL would mean that as many as 400,000 customers would need to be terminated by the service providers.

Service providers will vary in size. A major national ISP may need to terminate tens or hundreds of thousands of customers in a metro area, while smaller specialized services (e.g., certain corporate networks) might terminate only several hundred customers served by ADSL. The physical interface between the carrier's back-end network and the service providers may thus need to be as large as multiple OC-3 SONET optical interfaces operating at 155 Mbps, or as small as a T1 interface operating at 1.55 Mbps.

14.2 Potential ADSL Services and the Service Requirements

A DSL access network can support transport of data traffic for many different applications between the user and service provider. Among the possible applications that have been identified are:

- *Internet Access (residential "entertainment")* — ADSL access supports a faster version of today's Web "surfing" for home users. No service guarantees are made due to a need for low cost to the subscriber.

- *Internet Access (premium service)* — An enhanced service is offered to consumers and business. Service guarantees are made for a fee. The user is a "power user" at a residence accessing the public Internet.

- *IP Telephony and Derived Voice Lines* — IP or ATM-based telephony *is* enhanced by DSL access. Appropriate functionality in the access network allows these new forms of voice service to have similar performance to existing voice services.

- *Videoconferencing (two or more users videoconference)* — DSL's higher bandwidth allows enhanced quality.

- *Digital Compressed Video (clips and movies)* — The user views a digital movie in real time delivered from a server owned by a service provider

- *Radio and Audio Feed* — The user listens to digital audio delivered in real time from a server owned by a service provider.

- *Banking Transaction (e.g., ATM machine or branch to main office connection)* — A bank branch or automatic teller is connected to the bank's network.

- *Telecommuting and Remote Office (high-end service; low-end is same as residential Internet access)* — The user connects to their work site from home, or a branch office is connected to the corporate network. Applications may be very data intensive (e.g., remote CAD/CAM or high-end graphics production).

- *Multiplayer Interactive Games* — "Twitch" games supporting multiple players have extremely rigorous requirements with respect to transmission delay and variation (jitter).

Each of these applications will have different sensitivities to transmission delay and jitter, data loss, or corruption. Some require a minimum guaranteed bandwidth or constant bandwidth through the carrier's network, while a few services may only require a best effort and will not require any bandwidth commitment at all. The carrier does not necessarily control which applications need to be supported over their network. The customers, users, and service providers will determine the application mix.

Table 14.1 lists some of the qualities required of a carrier's network to provide access for the services described above.

14.3 Specific Architectures for Deploying ADSL in Different Business Models

ADSL and other DSL technologies are being deployed in a number of different business environments. Each of these situations places different requirements with respect to scale, opera-

Table 14.1   End-to-End Application Service Requirements

Application	Acceptable Average Delay	Sensitivity to Data Loss	Sensitivity to Data Corruption	Qualities of the Protocols Supporting the Service
Internet access (residential "entertainment")	~100 ms	Moderately sensitive	Moderately sensitive	No committed bandwidth required
Internet access (premium service)	~100 ms	Moderately sensitive	Moderately sensitive	Minimum guaranteed bandwidth required
IP telephony	~ 50 ms	Very sensitive	Tolerant of data loss	Minimum guaranteed bandwidth and ability to support a real time services required
Videoconferencing	~ 20 ms	Moderately sensitive	Moderately sensitive	Constant bit rate service required
Digital compressed video (clips and movies)	Delay tolerant (but jitter intolerant)	Sensitive	Sensitive	Constant bit rate service required
Radio and audio feed	Delay tolerant, (but jitter intolerant)	Moderately sensitive	Moderately sensitive	Minimum guaranteed bandwidth and ability to support a real time services required
Banking transaction (e.g., ATM machine or branch to main office connection)	>200 ms	Sensitive	Sensitive	Dedicated virtual circuits required for security; Minimum guaranteed bandwidth required
Telecommuting (high-end service; low-end is same as residential Internet access)	~100 ms	Sensitive	Sensitive	Minimum guaranteed bandwidth required
Multiplayer interactive games	Very Intolerant	Sensitive	Sensitive	Minimum guaranteed bandwidth and ability to support a real time services required

tional requirements, services supported, and business model. All of these issues affect the choice of architecture deployed to implement ADSL-based access.

Three important classes of business models for ADSL are:

1. The Large Local Telco: This category includes the regional Bell companies (RBOC), the larger independent American Telcos (e.g., GTE and United/Sprint), and Telecom administrations outside the United States. These providers are common carriers who

own the copper loop that serves the users. Until recently they held monopolies on local access. As the governments change regulation these monopolies are ending. These carriers see ADSL services as both a source of new revenue and a way of protecting their existing customer base from inroads from emerging competitors. Millions of potential ADSL customers are currently customers of the incumbent large Telco.

2. The Competitive Provider: In the United States, this category includes long distance companies offering broad ranges of competitive local service sometimes using facilities leased from the incumbent local phone company (e.g., AT&T), specialized companies offering only data services over leased facilities and companies such as Teleport and MFS that own their own local facilities. In Europe, telecommunications competition is less well developed than in the United States. However, it appears that a similar structure, where competitors are able to support local customers, is emerging as the European Commission changes the regulatory environment.

3. The Campus and Niche Provider: This category of provider of ADSL service includes many different situations. For example, the owner of a condominium or rental building may offer ADSL service to tenants. A university campus may use DSL technologies to provide high-bandwidth connectivity to outlying buildings. A small rural telephone company may provide DSL services to customers in the center of town. A military installation might use ADSL to provide high-bandwidth connections to outlying sites.

Each of these categories of DSL service provider has different requirements, and a different place within the ADSL architecture. Some of the most important differences are:

1. Regulation.
2. Scale of Service: How many customers will be supported by a mature access service?
3. Common Carrier vs. Integrated Service: Will a provider of ADSL support access many different and independent providers of data services, or will access and the provisioning of service to the DSL customer be packaged as an integrated unity.

Table 14.2 Comparison of Service Requirements

	Telco	Competitive Telecommunications Provider	"Niche"
Regulation	Heavily regulated	Nonregulated	Nonregulated
Scale of Service	>10^5 customers in a metro region	~10^3–10^5 customers in a metro region	~10^2–10^3 customers in an implementation
Common Carrier vs. Integrated Service	Common Carrier. The Telco provides an access network to multiple independent services.	Either as a common carrier or integrated provider.	Integrated Service: The access network provided by ADSL is integrated as seemless whole with a data service.

14.4   Several ADSL Architectures

14.4.1   A Packet-Based Architecture for Small Deployments

The first ADSL trials by carriers employed a router-based architecture. The niche environment, as defined above, continues to have many of the same characteristics as these early deployments as they serve relatively few customers from a single point.

The simplest version of this architecture, shown in Figure 14.4, locates at each Central Office (CO) an arrangement of ADSL modems (ATU-Cs) connected to an IP router through an Ethernet hub or switch performing MAC layer concentration. As each hub is supported by one router port, the concentrator enables a much larger number of subscribers to use the network than would be implied by the number of router ports implemented. A separate Ethernet link connected each CO modem to the switch or hub. By combining the multiplexing function within the central office environment, the CO modems connect directly to a shared high-speed backplane. The router may be located in the CO or at a central site. In some cases it may be integrated into the Central Office equipment supporting the ADSL ATU-Cs. Figure 14.4 also illustrates the protocol stack supported in this end-to-end architecture.

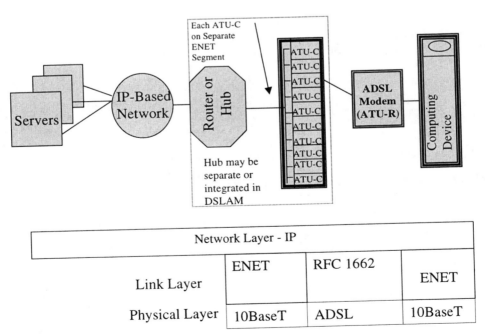

Network Layer - IP			
Link Layer	ENET	RFC 1662	ENET
Physical Layer	10BaseT	ADSL	10BaseT

Figure 14.4   Router-based access network.

In this architecture the customer appears to the IP router in the central site as a host device connected to the network over an Ethernet LAN segment.

The use of ADSL allows this LAN segment to be extended to the user's premises from the Central Office. The use of a packet-based transmission convergence and link layer (as described in Chapter 12) on the ADSL physical layer allows IP to be carried from the Central Office to the users site. The user's PC behaves as if it is connected directly to the router over a LAN. Multiple connections into a single PC are supported TCP/IP as with any host connected to a conventional LAN/WAN network.

This configuration integrates PC into the LAN/WAN as administered by the owner of the router. This is acceptable and even desirable for those environments where the ADSL access and access to a corporate LAN or the Internet are completely integrated. This occurs in niche environments such as college campuses using ADSL to equip remote offices and professors' or students' residences with high-speed access to a campus network; or apartment complexes offering high-speed Internet access to their tenants. It was also a desirable architecture for the earliest deployment in ADSL trials by common carrier, as it could be implemented with equipment available in 1995. Since protocol conversion is not required in the network and it uses the well understood Internet suite of protocols, it is simple to implement and operate for these environments.

However, this router-based access architecture has a number of disadvantages for other environments. It does not scale well and therefore will not be able to support the broad market for residential broadband. It does not support a separation with respect to either management or services between the access provided by a carrier and the network services provided by a service provider such as an ISP or corporate network. Therefore, it is not particularly suitable for deployments by telephone companies or other common carriers. This architecture does not easily support multiple services to one user. Access to both an ISP and a corporate LAN by a user is not possible. Services requiring guarantees of service quality (delivery of high-quality videoconferences or high-quality voice connections, for example) are difficult to support. For these reasons, most carriers who plan a wide deployment of ADSL are planning a service based upon using ATM to connect the ADSL users to service providers.

14.4.2 ATM Access Networks

As illustrated in Figure 14.5, ATM architectures for ADSL connect a user's site to multiple service providers through a network based upon ATM transport. In these architectures the link level on the ADSL loop is based upon ATM and an ATM network connects the ADSL equipment in the CO to the service provider's network. ATM virtual circuits connect the user to the service provider's sites. Multiple service providers may be connected to simultaneously from a single user. As different virtual circuits may have different ATM "service contracts," the virtual connections may have guarantees of service quality to support application requirements. Since the ADSL loop is dedicated to a particular user, as is the ATM virtual circuit between the Central Office and the service provider, the ATM connection has additional security compared to that provided by router-based ADSL access or alternative methods of broadband access such as cable modems.

Currently, a carrier offering an ADSL architecture based on ATM end-to-end will use ATM permanent virtual circuits (PVC) to provide ATM circuits between a user's terminal and a service provider such as an ISP or corporate LAN gateway. ATM PVCs must be configured by a network administrator and cannot be altered by the user. This requires the carrier to expend considerable resources to configure its customers as they are activated or change their services. If the virtual circuit connects to the Internet (via an ISP, for example) or to a corporate "Intranet," the ATM circuit will carry IP traffic between a router in the service provider's network and the user's environment. A user can access multiple service providers over multiple virtual circuits. If traffic other then IP is carried (e.g., compressed video) a virtual circuit can be configured with qualities optimized for that traffic. Once preconfigured, a user's computer can communicate with the appropriate service provider by communicating over the appropriate ATM PVC. However, as the access network grows in size and capabilities, it will become impossible to preconfigure PVCs to connect users to the many possible service providers that may be supported.

Implementation of ATM switched virtual circuits (SVC) between the user's site and the service provider allows the connection to be established in real time. Signaling from the user allows a connection to be requested with the specific qualities required for the service. SVCs will greatly reduce the effort to provision service to a new customer and would also permit customer to freely roam among service providers. However, the development of an SVC-based ADSL access network requires the development of the following functions that do not exist yet in current ATM switching elements or ADSL equipment:

- ADSL equipment that supports the ADSL signaling standard Q.2981
- ATM switches with the capacity to support the call setup demands from ADSL access networks serving an entire metropolitan area
- Equipment at the service provider site that connects the service provider's existing network to the carriers ATM access network

These functions are currently under development by vendors of ADSL and ATM equipment.

The end-to-end ATM architecture in Figure 14.5 shows a simple model of an ATM access network. All traffic is switched through an ATM backbone. This gives high flexibility for connecting multiple services and provides support for all the traffic requirements defined earlier in this chapter. Each Central Office is equipped with a DSL access multiplexer (DSLAM). The DSLAM contains integrated ADSL modems for each subscriber, extending to subscriber premises over existing copper lines. The ATM access network is under control of the carrier. At the customer premises, the ADSL modem may be integrated into the PC, be connected to the PC over a serial connection such as the Universal Serial Bus (USB), be integrated by the ADSL modem into the PC, or implement some form of premises network, either Ethernet or ATM.

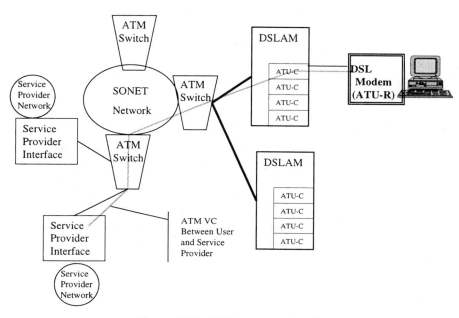

Figure 14.5 ATM access network.

Regardless of whether PVCs or SVC are used through the access network, applications must exist which link the user's applications to the broadband access provided by ADSL. In the case of a user on a PC accessing the Internet or a corporate LAN, it is desirable that the user can use the existing tools provided on the PC which support dial-up access. In the case where the PC is connected to a LAN, which uses ADSL access to reach remote sites, the user should not need to be aware of the fact that a router on the LAN uses ADSL. In either case, the service provider should require a minimal change to their existing practice to support ADSL-based customers.

Several solutions to these issues are being implemented as ADSL access networks are deployed.

- RFC 1483
- PPP over ATM
- The Level 2 Forwarding Gateway
- PPP-Terminated Aggregation

14.4.3 RFC 1483

RFC 1483 issued by the Internet Engineering Task Force (IETF) [1] defines several methods for encapsulation of other data protocols over ATM virtual circuits. It allows the definition interfaces that allow the ATM circuit to be used by a bridge to transport link level protocols (such as Ethernet) to the remote site. It also allows the definition of interfaces that allow routed protocols such as IP to be transported over the ATM virtual circuits. RFC 1483 has been used in most early implementations of ATM over ADSL architectures. Figure 14.6 illustrates access architecture based on RFC 1483.

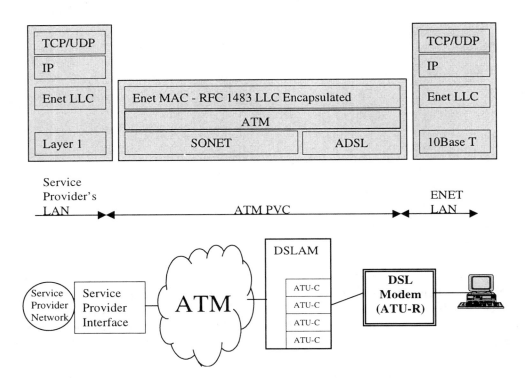

RFC 1483 with LLC encapsulation

Figure 14.6   ATM access based upon RFC 1483.

RFC 1483 has been useful in early implementation as the standard is mature and is supported by existing IP routers used by service providers. Its disadvantage is that it does not support negotiation end to end of the capabilities of the protocol carried over it. Facilities that are important in defining a connection over IP, such as assignment of the IP address to the remote device, or providing for security of the connection cannot be negotiated via RFC 1483. These parameters must be preconfigured. In the case of an implementation where the user uses the ADSL connection to permanently connect a router to a remote site, this may not be a problem, as the configuration of the IP parameters will occur when this router is first set up. RFC 1483 is therefore acceptable when preconfigured PVCs are used in the network. However, when a user needs to connect a computer to a service provider over a transient connection, the negotiation of the IP parameters will need to occur in real time and use of RFC 1483 alone becomes unacceptable.

14.4.4 PPP over ATM

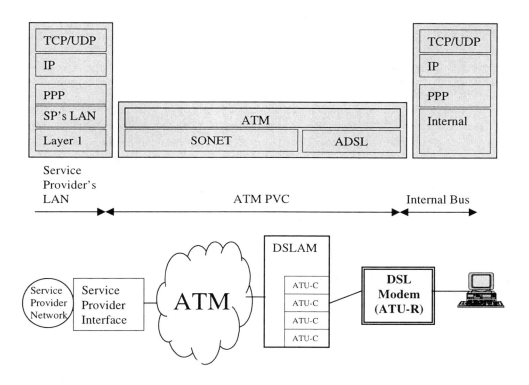

PPP Over ATM carried on a PVC

Figure 14.7 PPP over ATM architecture.

The IETF has defined point-to-point protocol (RFC 1662, RFC 1661). It is used over dial-up connections to provide a link layer for IP. It supports real-time assignment of the IP address and negotiation of authorization and authentication of the user. Point-to-point protocol (PPP) over ATM is being defined by the IETF to allow the transfer of the PPP protocol over an ATM virtual circuit [2]. It has already been implemented by several ADSL vendors. Since PPP is already supported by all Internet service providers and on almost all PCs, the use of PPP over ATM is very compatible with existing equipment, software, and the procedures of the existing Internet. For example, ISPs utilize AAA (authentication, authorization, and accounting) security servers in their access configurations. The security servers are dependent on PPP session establishment between the ISP and the end user. By adopting the same service access protocol as the lower-speed dial-up users do today, service convergence is enhanced for the ISP. The ADSL subscriber also benefits by ready access to the more mature service platform available via PPP already in place for narrowband both from the service performance perspective and the service extension perspective. Figure 14.7 illustrates an ADSL access architecture based upon PPP over ATM.

14.4.5 Tunneled Gateway Architecture

Although PPP over ATM provides a solution to many of the issues involving the interworking between ADSL, ATM and IP interfaces must be provided which allow the service provider to terminate the PPP sessions delivered over the access network. Reducing the number of virtual circuits delivered by the carrier to the service provider will simplify the network for both the service provider and the carrier. The tunneled gateway architecture concentrates many PPP sessions over one virtual circuit between the Central Office and the service provider. It may be implemented with a L2TP (layer 2 tunneling protocol) access concentrator (LAC), which maps PPP sessions within individual PVCs from access network to PPP sessions in an L2TP tunnel. In the case of a DSL access to the tunnel is carried within a single PVC to the selected service provider. L2TP is used to convey a PPP session through the IP network. Figure 14.8 illustrates this architecture.

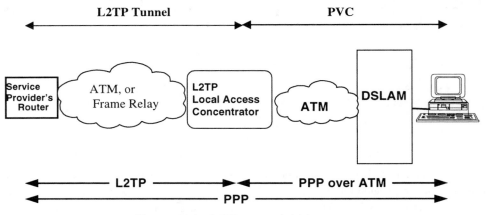

Figure 14.8 L2TP access architecture.

Service provider selection is possible because the PPP header contains identification of the chosen service provider. The domain name of the service provider in this header is inspected and the traffic from a user can be placed on the virtual circuit supporting the tunnel to the correct service provider.

14.4.6 PPP Terminated Aggregation

When PPP-terminated aggregation is implemented, the PPP over ATM session is terminated in a broadband access server that is placed within the carriers network. The IP traffic carried within the PPP connection is then placed on a routed IP network. Since the broadband access server terminates the PPP session, it is responsible for assigning the user their IP address, and ensuring the user is authorized to access the network. Figure 14.9 illustrates this architecture.

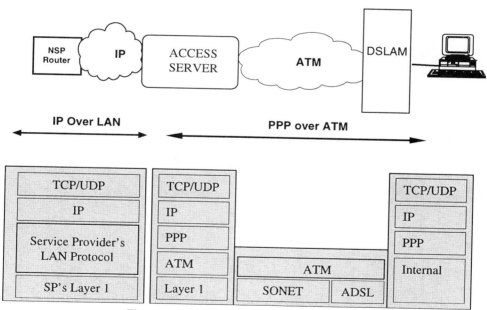

Figure 14.9   PPP-terminated aggregation.

Since it uses an ATM link layer to reach the end user, the carrier can easily evolve the access network to take advantage of the flexibility provided by ATM to support complex future services. At the same time this architecture allows the carrier to easily integrate their access into the service providers existing IP networks.

References

[1]   J. Heinanen. Multiprotocol Encapsulation over ATM Adaptation Layer 5î, RFC 1483, Telecom Finland, July 1993.

[2]   T. Kwok et al., *An Interoperable End-to-End Broadband Service Architecture over ADSL Systems (Version 3.0), ADSL Forum Contribution 97.215.*

Network Architecture and Regulation

15.1   Private Line

The dedicated private-line network is the simplest DSL network architecture. A permanent point-to-point circuit connects two DSLs together. The circuit is provisioned manually upon receiving an order from the customer and remains in place until the customer requests disconnection of service. This architecture is used for most HDSL systems.

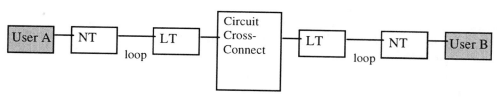

NT: DSL transceiver at user site, LT: DSL transceiver at network end

Figure 15.1   Private-line architecture.

15.2   Circuit Switched

With the circuit-switched network architecture, real-time circuit switching permits basic rate ISDN DSLs to establish 64 kb/s dedicated connections to any other DSL. This permits the user to immediately set up a connection to any other ISDN user. However, BRI permits only two connections per user, and the bandwidth is dedicated for the duration of each connection. Thus, network resources are poorly used during idle portions of the connection time. Note that the DSL transceiver at the network end of the loop (LT) is typically integrated within a Class 5 local digital switch.

Circuit switching is also used for conventional switched voice calls. A circuit-switched connection provides relatively low and constant signal delay and guarantees full throughput for the duration of the call. Call processing is performed during call set-up and tear-down. Virtually no processing is involved during the call. Call set-up may be blocked (denied) if the traffic load on the switch is too great. Circuit switches are engineered to assure a very low blocking probability.

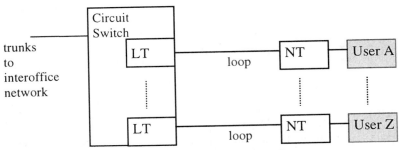

Figure 15.2   Circuit-switched architecture.

15.3 Packet Switched

The first data-oriented ADSL systems employed an architecture with external ADSL LT connected via 10baseT Ethernet connections to an Ethernet switch, which then connected to a packet router. With packet switching, the router examines each packet to determine its source and destination address and then forwards the packet to the next router in the path towards the destination. Each successive router repeats this function. This architecture permits efficient statistical multiplexing of bursty traffic. However, packet switching has been poorly suited to applications demanding assured throughput or low/constant delay. Such applications include video, audio, and real-time process control. No bandwidth is reserved or guaranteed with conventional packet switching. In the late 1990s, work was underway to improve packet switching to provide better performance for high-priority traffic.

The integrated end-to-end asynchronous transfer mode (ATM) network architecture permits efficient support of data and video applications, which include:

- TCP/IP-based Internet access without guaranteed throughput (unspecified bit rate, UBR)
- Assured throughput rate for high-performance packet service (variable bit rate, VBR)
- Circuit emulation service for video (constant bit rate, CBR)

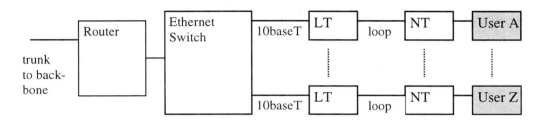

Figure 15.3 Packed-routed architecture.

15.4 ATM

Asynchronous transfer mode is a broadband switching technology that transports fixed-length information cells via connection-oriented virtual circuits. At one time, it appeared that ATM would be relegated to backbone and large business applications. ATM's best hope for reaching the mass market is the digital subscriber line access multiplexer (DSLAM), which performs low-cost statistical multiplexing to aggregate the traffic to and from many DSLs. DSLAM architecture use of ATM permits it to manage the traffic for several grades of service efficiently and fairly. The DSLAM is a multiservice platform that accommodates a growing demand for service. The DSLAM may contain line cards that terminate ADSLs and other types of DSL. The integration of the ADSL transceiver within the DSLAM reduces hardware costs, eases the support of effective traffic management, and simplifies system operation, administration, maintenance, and provisioning (OAM&P). To keep cost low, the DSLAM performs a minimum of protocol processing and does not perform local switching or routing of traffic between lines in a DSLAM. Switching and routing is performed by ATM switches and IP routers deeper in the network. Since the original DSLAM concept, the term *DSLAM* has also been used to describe DSL access systems that contain routing, switching, and protocol functions such as user authentication.

ATM connections may consist of permanent virtual circuits (PVCs) and switched virtual circuits (SVCs). Upon the receipt of a customer service order, PVC connections are established by the entry of connection information into network management systems by a network administrator. SVCs are set up, modified, and torn down in real time by a network connection management system in response to ATM signaling messages (ITU-T Q.2931) sent by the customer's equipment. SVCs are more complex to implement in DSLAMs and switches, but reduce the ongoing cost to manage service. SVCs allow the customer to roam among service providers, such as connecting their corporate office to work-at-home and later connecting to an ISP for Internet access.

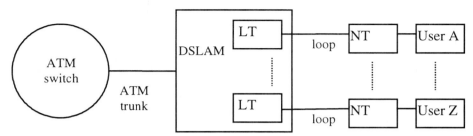

Figure 15.4   Integrated ATM architecture.

15.5   Remote Terminal

In addition to the CO-based architectures described previously, the subscriber loop may be served from a remote service node. The use of remote service nodes is increasing, and the placement of the nodes is progressing closer to the customer site. For digital loop carrier (DLC) applications, dedicated fibers connect the DLC multiplexer (typically serving hundreds of lines) to the CO. The fiber-to-the-curb (FTTC) architecture places optical network units (ONUs) less than 1,000 feet from the customer site. An ONU serves up to 8 to 16 lines. The ONUs are connected via fiber (either dedicated or a daisy chain) to a host terminal. Another alternative is a coaxial cable bus architecture that can be inexpensive due to a minimal amount of remote electronics and the cost sharing of one network-end termination by hundreds of customers. However, fiber-in-the-loop (FITL) architectures have the following advantages of a star-type architecture: higher reliability, faults usually affect very few customers, faults are more easily localized, more bandwidth can be guaranteed per customer, greater assurance of privacy, ability to block services to nonsubscribers, and the customer termination unit is less complex.

15.6   Competitive Data Access Alternatives

In some countries, competitive local exchange carriers (CLECs) may locate equipment in the Central Office (CO) of the incumbent local exchange carrier (ILEC, the traditional phone company), and lease copper loops from ILEC. For example, with *physical collocation*, the CLEC may place a DSLAM in the CO and provide ADSL service over the *unbundled* copper loops. In the physical collocation scenario, the DSLAM is owned and maintained by the CLEC. In the full-service example shown in Figure 15.5, the server, switching, and DSLAM are all provided by the CLEC.

The *virtual collocation* scenario is identical to physical collocation, except that the DSLAM is maintained by ILEC technicians instead of the CLEC. In some cases, virtual collocated equipment may be placed in the same equipment line-up as the ILEC equipment, whereas physical collocated equipment is often segregated to a separate space within the CO.

Since space within the CO may be expensive or unavailable, the CLEC or *alternative service provider* may choose to place the DSLAM outside of the CO. The DSLAM site should be

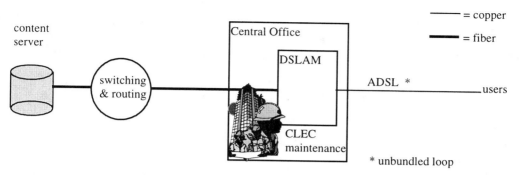

Figure 15.5   Physical collocation.

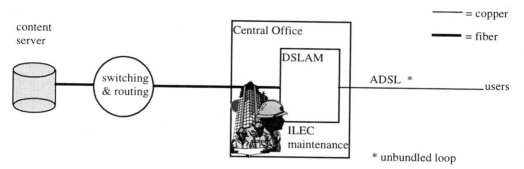

Figure 15.6   Virtual collocation.

as close as possible to the CO to minimize the loss of loop reach due to the additional segment of cable linking the DSLAM to the CO. To avoid excessive crosstalk in the case of ADSL, the DSLAM-CO cable may be a private cable dedicated to only serving the alternative carrier's DSLAM.

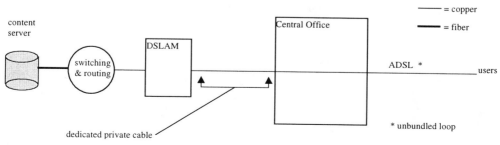

Figure 15.7   CO proximity location.

A network service provider (NSP, which may also provide content) may choose to access its customers via an access service provided by the ILEC. In this case, the ILEC provides the access portion of the service, and the NSP could provide switching/routing and content functions. The ILEC may perform some switching to permit the connection of multiple NSPs to the access network.

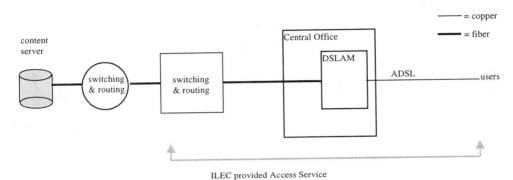

Figure 15.8   NSP with ILEC access service.

For private use, an enterprise-owned DSLAM may be placed at a campus location to serve users within a private enterprise via wire loops within the campus.

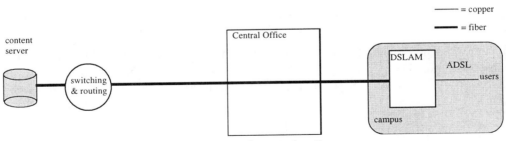

Figure 15.9   Campus location.

Many other competitive data access alternatives are possible, including cable modems and wireless.

15.7   Regulation

Until recently, each nation's telecommunications services were provide by one source, either by a department of the government (known as a PTT) or a private monopoly (such as AT&T prior to 1984). The sole provider was heavily regulated by the government and typically earned a fixed rate of return. This structure provided the users a simple and uniform service environment. However, it hindered innovation and price reductions.

The regulatory environment in many countries is changing to service provided by many competitive telecommunications firms. An example of this transition is the privatization of many PTTs, the divestiture of AT&T, and the U.S. Telecommunications Act of 1996 that requires incumbent local exchange carriers (ILECs) to provide *unbundled* network components to competing service providers. Certified/competitive local exchange carriers (CLECs) are permitted open and equal access to loops, switches, phone numbers, and operational support systems. This permits competitive service providers to lease elements instead of the bundled end-to-end service. In this regulatory environment, *price cap* regulation is sometimes applied to services.

In the United States, interstate and foreign telecommunications are regulated by the U.S. Federal Communications Commission (FCC), and intrastate telecommunications are regulated by state Public Utilities Commissions (PUCs). ILEC services are subject to tariff approval by the regulators. Typically, a tariff describing the proposed service and providing a network interface disclosure must be filed with the regulators six months prior to tariff approval and service turn-up. ILECs provide for the placement of CLEC equipment within the ILEC's Central Office (CO) if space is available. With *physical collocation*, the ILEC equipment in the CO is maintained by CLEC technicians. With *virtual collocation*, the IELC equipment in the CO is maintained by ILEC technicians. Collocated equipment must comply with CO equipment requirements, including Bellcore FR-2063 (NEBS). A CLEC may place their DSL terminating equipment at a CO, lease unbundled loops to their customer, and directly provide DSL-based service. What is less clear is the feasibility of unbundled access to customers served via digital loop carrier (DLC); especially integrated digital loop carrier (IDLC), which has a multiplexed interface to the local digital switch. With more than 10% of customers served by DLC in many areas, unbundled access via DLC is an important open issue.

Since the unbundled wire pairs leased by a CLEC are in the same cable binder groups as the wire pairs used by other CLECs and the ILEC, it is imperative that the signals transmitted on unbundled loops be spectrally compatible with the transmission systems used by the other carriers. The transmitted power spectral density (PSD) at both ends of the loop must remain below certain limits during start-up and during operation. Standards Working Group T1E1.4 is developing a standard on spectral compatibility to aid in the management of loops transmission systems. Another consideration for unbundled loops is the occasional rehabilitation and upgrade of the loop plant, including the transfer of service from a CO copper loop to a DLC-fed loop.

With certain exceptions for older types of equipment, U.S. regulators require that end customer be able to own all telecommunications equipment located at the customer premises; such equipment is called *customer premises equipment* (CPE). The CPE must comply with the technical requirements stated in the service tariff. The customer may purchase or lease the compliant

CPE from any available source. For example, the customer may obtain the CPE from a consumer electronics retailer, a value-added reseller (VAR), or the phone company.

The regulatory environment has a pervasive impact on telecommunications services, products, and architecture. Due to the frequent changes in regulatory policy, telecommunications companies must keep abreast of recent developments.

Standards

The development of standards was much simpler when telecommunications networks were solely provided by state-run phone companies (PTTs) and regulated monopolies such as the pre-divestiture AT&T. There was one central planning organization that dictated the standards. When a standard is developed privately by one or a few companies who have sufficient dominance to control market adoption, it is called a de facto standard. Today, the standards are developed by a collection of committees consisting of hundreds of companies with diverse interests. The standards developed by recognized open and fair standards bodies are called de jure standards.

The standards arena contains international organizations such as the International Telecommunications Union (ITU) and International Standards Organization and International Electrotechnical Commission (ISO/IEC), which address the harmonization of standards for nearly all countries. The European Telecommunications Standards Institute (ETSI) addresses standards for its member countries. The IEEE, originally a United States–dominated organization, is now becoming more internationally focused. National standards for the United States are developed by the Electronics Industry Association/Telecommunications Industry Association (EIA/TIA) and the American National Standards Institute (ANSI)–accredited Committees T1 and X3. TTC develops standards for Japan. Implementation forums have been formed to quickly address specific topics. The many standards organizations that do not have direct bearing on DSLs are not shown in Figure 16.1.

Not shown in Figure 16.1 is the UAWG (Universal ADSL Working Group), a consortium formed in late 1997 by leading telephone companies and computer companies to accelerate the progress on splitterless ADSL (ADSL-Lite). The UAWG is not considered a standards body, but its work contributed to the T1E1.4 and ITU-T standards bodies.

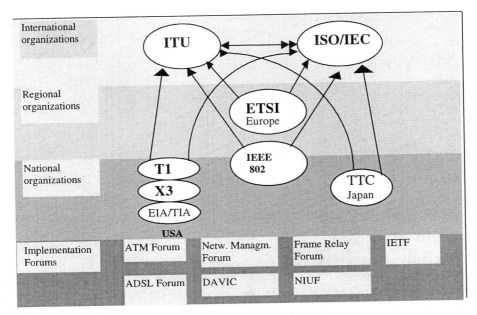

Figure 16.1   DSI related standards organizations.

16.1   ITU

Founded in 1865 as the International Telegraph Union, the ITU changed it name to the International Telecommunications Union in 1934 and became an agency of the United Nations in 1947, with its headquarters in Geneva, Switzerland. One hundred and eighty-seven countries were members of the ITU in 1996. The ITU consists of:

- The General Secretariat, which provides administrative and financial functions
- ITU-D, the Development Sector, which coordinates and assists the development of new telecommunications infrastructure, primarily in developing countries
- ITU-R, the Radio-Communications Sector, which ensures the fair, efficient and economical use of radio-frequency spectrum for all uses
- ITU-T, the Telecommunications Standardization Sector, which studies technical, operating, and policy questions. Previously know as the CCITT, the ITU-T issues Recommendations (nonbinding standards) for the purpose of international standardization of systems and services. In 1997, the ITU-T consisted of:
  - World Telecommunications Standardization Conference (WTSC), which defines general ITU-T policy and approves "questions" (i.e., projects)
  - Telecommunications Standardization Advisory Group (TSAG), which provides advice regarding work priorities and strategies

- Telecommunications Standardization Bureau (TSB), which provides secretariat (administrative) and coordination functions
- Study Group 1 — Service definition
- Study Group 2 — Network operation: network management, numbering, routing, service quality
- Study Group 3 — Tariff and accounting principles
- Study Group 4 — Network maintenance, TMN
- Study Group 5 — Protection against electromagnetic environment effects: protection of equipment and humans from electromagnetic hazards
- Study Group 6 — Outside plant: cables and associated structures
- Study Group 7 — Data network and open system communications: data communications network aspects, frame relay
- Study Group 8 — Terminals for telematic services: fax
- Study Group 9 — Television and sound transmission
- Study Group 10 — Languages for telecommunications applications
- Study Group 11 — Switching and signaling: intelligent network
- Study Group 12 — End-to-end transmission performance of networks and terminals
- Study Group 13 — General network aspects: consideration new concepts, IP aspects, BISDN, Global Information Infrastructure (GII)
- Study Group 14 — Modems and transmission techniques for data, telegraph, and telematic services: modems, ISDN terminal adaptors
- Study Group 15 — Transmission systems and equipment: subscriber access systems, DSLs, fiber optic transmission
- Study Group 16 — Multimedia

Membership classes consist of:

- Administrations: countries
- Recognized Operating Agencies (ROAs): network and service providers
- Manufacturers, Scientific, and Industrial Organizations (SIOs)
- International and regional organizations

The ITU-T may be reached at ITU-T, Place des Nations, CH1211 Geneva 20 Switzerland; http://www.itu.ch.

16.2   Committee T1

Committee T1, formed at the time of AT&T's divestiture, is the telecommunications standards body for the United States. Committee T1 is sponsored by the Alliance for Telecommunications Information Solutions (ATIS) and is accredited by the American National Standards Institute (ANSI). There were 67 voting members of Committee T1 in 1997.

In addition to Committee T1, which determines policy and direction, the committee contains an Advisory Group and the following technical subcommittees (TSCs):

- T1A1: Performance and Signal Processing
  - T1A1.2: Network Survivability
  - T1A1.3: Performance of Digital Networks and Services
  - T1A1.5: Multimedia Communications Coding and Performance
  - T1A1.7: Signal Processing and Network Performance for Voiceband Services
- T1E1: Interfaces, Power and Protection for Networks
  - T1E1.1: Physical Interfaces and Analog Access
  - T1E1.2: Wideband Access
  - T1E1.4: Digital Subscriber Loop Access
  - T1E1.5: Power Systems - Power Interfaces
  - T1E1.6: Power Systems - Human and Machine Interfaces
  - T1E1.7: Electrical Protection
  - T1E1.8: Physical Protection and Design
- T1M1: Internetwork Operations, Administrations, Maintenance and Provisioning
  - T1M1.3: Internetwork Operations, Testing, Operations Systems and Protocol
  - T1M1.5: OAM&P Architecture, Interface and Protocols
- T1P1: Systems Engineering, Standards Planning and Program Management
  - T1P1.1: NII/GII
  - T1P1.2: Personal Communications Service Descriptions and Network Arch.
  - T1P1.3: Personal Advanced Communications Systems
  - T1P1.5: PCS 1900
  - T1P1.6: CDMA/TDMA
  - T1P1.7: Wideband-CDMA
- T1S1: Services, Architectures, and Signaling
  - T1S1.1: Architecture and Services
  - T1S1.3: Common Channel Signaling
  - T1S1.5: Broadband ISDN
- T1X1: Digital Hierarchy and Synchronization
  - T1X1.3: Synchronization and Tributary Analysis Interfaces
  - T1X1.4: Metallic Hierarchical Interfaces
  - T1X1.5: Optical Hierarchical Interfaces

The TSC and its subtending working groups normally meet during the same week four times per year. Additional interim meetings may be held when needed. All meetings are open to nonmembers.

The name of technical subcommittee T1E1 has often been confused with the committee's work on transmissions systems related to T1 and E1 carrier. As can be seen from the list of committee names above, the committee's name has nothing to do with T1 and E1 carrier.

In addition to developing US standards and technical reports, committee T1 develops US positions to international standards bodies (ITU, ISO, IEC).

Committee T1's offices are at 1200 G Street, NW Suite 500, Washington, DC, 20005 USA, Phone: 202-434-8845, Fax: 202-347-7125, Web: http://www.t1.org.

16.3 ETSI

The European Telecommunications Standards Institute develops standards and reports for its member countries. ETSI was formed in 1988 to carry on the work which had previously been done by CEPT. In 1997, ETSI membership consisted of 41 administrations, 220 manufacturers, 70 public network operators, 60 private service providers and others, 27 users, and 25 associate members.

ETSI Technical Committees consist of:

- TM — Transmission and Multiplexing
  - TM1 — Transmission equipment, fibers and cables
  - TM2 — Transmission network management and performance
  - TM3 — Interfaces, architecture and functional requirements of transport networks
  - TM4 — Radio relay systems
  - TM6 — Access transmission systems on metallic cables (DSLs)
- NA — Network Aspects
  - NA1 — User interfaces, services and charging
  - NA2 — Numbering, addressing, routing & internetworking
  - NA4 — Network architecture, operations, maintenance, principle, and performance
  - NA5 — Broadband networks
  - NA6 — Intelligent networks
  - NA ECTM — European Coordination of TMN
- EE — Environmental Engineering
- ERM — EMC and Radio Spectrum
- BRAN — Broadband Radio Access Networks
- HF — Human Factors
- ICC — Integrated Circuit Cards
- CN — Corporate Networks
- CTM — Cordless Terminal Mobility
- MTS — Methods for Testing and Specification
- DECT — Digital Enhanced Cordless Telecommunications

- DTA — Digital Terminals and Access
- RES — Radio Equipment and Systems
- SEC — Security
- MTA — Multimedia Terminals and Applications
- PTS — Pay Terminals and Systems
- SPS — Signaling Protocols and Switching
- STQ — Speech Processing, Transmission, and Quality
- TMN — Telecommunications Management Networks

ETSI may be reached at http://www.etsi.fr.

16.4 ADSL Forum

The ADSL Forum, founded in 1995, is an international body that promotes the use of ADSL technology and develops technical specifications related to ADSL's use. The ADSL Forum had 300 member companies in 1998. The ADSL Forum consists of a marketing committee that performs industry education and public relations, and a technical committee that develops technical reports regarding the implementation of ADSL systems. In 1997, the technical committee had the following working groups:

- Access network (also addressing end-to-end system architecture)
- ATM mode
- Packet mode
- Customer premises (wiring, splitters, customer equipment)
- Network migration
- Network management
- Testing and interoperability

The ADSL Forum may be reached at 39355 California St., Suite 307, Fremont, CA. 94538; 1-510-608-5905; http://www.adsl.com.

16.5 ATM Forum

The ATM Forum is an international forum founded in 1991 with the purpose of assisting the rapid and widespread implementation of ATM technology. The ATM Forum had more than 900 member companies in 1997. The ATM Forum consists of a managing Board, a Technical Committee, Market Awareness Committees for Asia Pacific, Europe, and North America. The Technical Committee consists of the following Working Groups:

- Broadband Intercarrier Interface
- LAN Emulation
- Multiprotocol
- Network Management
- Physical Layer
- Private Network-Network Interface
- Residential Broadband
- Security
- Service Aspects and Applications
- Signaling
- Testing
- Traffic Management
- Voice and Telephony over ATM
- Wireless ATM

The ATM Forum may be reached at 2570 West El Camino Real, Suite 304, Mountain View, CA 94040-1313; 1-415-949-6700; http://www.atmforum.

16.6 DAVIC

The Digital Audio-Visual Council develops technical specifications to assure the end-to-end interoperability for digital audio-visual applications. DAVIC is an international forum that primarily addresses service provider and CPE aspects for video-on-demand and broadcast video services. DAVIC had over 200 member companies in 1997. In addition to the Forum Management committees, DAVIC consists of working groups addressing Applications, Physical Layers, Information Representation, Security, Subsystems, and Systems Integration. As of 1997, DAVIC's principal product was the DAVIC 1.0 specification for TV distribution, near-video on demand, video on demand, and basic forms of teleshopping. DAVIC may be reached at http://www.davic.org.

16.7 IETF

The Internet Engineering Task Force (IETF) is comprised of users, researchers, service providers, and vendors of hardware and software, it focuses on Internet protocol (IP)–based solutions. The IETF is chartered by the Internet Society (ISOC) and receives guidance from the Internet Engineering Steering Group (IES) and the Internet Architecture Board (IAB). In 1996 the IETF consisted of eight areas which contained about 90 working groups. The IETF develops specifications called "Request for Comments" (RFCs), which proceed in stages from less formal to more formal (proposed, draft, and full standard). To progress forward, evidence of multiple interoperable implementations must be provided. The IETF procedures are documented in RFC 2026. The IETF may be reached at 703-620-8990, http://www.ietf.org.

16.8   EIA/TIA

The Electronics Industry Association and Telecommunications Industry Association (EIT/TIA) is ANSI accredited. EIA/TIA develops U.S. standards and technical reports pertaining to equipment, and develops U.S. positions to international standards bodies (ITU, ISO, IEC). EIA/TIA T41.9 has developed technical recommendations that have assisted the FCC with Part 68 of the U.S. telecommunications rules. The EIA/TIA may be reached at Suite 300, 2500 Wilson Blvd., Arlington, VA, 22201; 703-907-7703; http://www.tiaonline.org.

16.9   IEEE

The Institute of Electrical and Electronic Engineers (IEEE) sponsors many standards committees. The most relevant to DSLs is IEEE Committee 802, founded in 1980 by the IEEE Computer Society. IEEE Committee 802 addresses ISO layer 1 and 2 aspects of local area network (LANs) and metropolitan area networks (MANs) and works closely with IEC/ISO JTC 1 SC6. In 1997, IEEE Committee 802 consisted of the following active working groups:

- P802.1: Overview, architecture, bridging, management
- P802.2: Logical link control
- P802.3: CSMA/CD (Ethernet)
- P802.4: Token bus
- P802.5: Token ring
- P802.6: DQDB (Dual Queue Dual Bus)
- P802.8: Broadband and fiber optics
- P802.9: Integrated services
- P802.10: Security
- P802.11: Wireless
- P802.12: Demand priority
- P802.14: Cable TV

IEEE P743 has developed techniques for measuring the analog characteristics of DSL signals. The IEEE may be reached at 445 Hoes Lane, Piscataway, NJ 08855; 908-562-3820; http://www.ieee.org.

16.10   The Value of Standards and Participation in Their Development

Each type of DSL was nurtured and defined in an innovation incubator: standards committees. Members of the standards committees know that the standards process accomplishes much more than just writing the standard. Some managers fail to understand the value of sending top members of their technical staff to frequent standards meetings. Trying to explain the creative synergy and learning opportunities for participating in the standards process can be rather like trying to explain why a rose is beautiful. It is hard to explain the obvious: standards set the future course of the industry.

Standardization enables the provision of interoperable elements and systems from multiple independent sources. This helps to ensure that compatible elements/systems will continue to be available even if the original source of supply is terminated. The feasibility of replacing one source of supply with another enables a competitive environment that stimulates low prices, good support, and continuing improvements. One may buy a mixture of interchangeable elements from different suppliers. Standards specify performance, reliability, safety characteristics, and methods for measurement and test. The buyer has the assurance that the standard elements comply with a comprehensive set of requirements. Thus, even if the buyer is not an expert, they can be assured that the standard element has the necessary performance and features. Standard systems provide the user (who may be a network technician) a common look and feel; this makes the system quick to learn. Standard elements may be reused elsewhere. Standards improve manufacturing and vendor efficiencies by reducing the number of unique versions of product that must be developed and supported.

What is less obvious is that the standardization process is a stimulus to innovation. Vendors are more likely to introduce new products when they see industry-wide buy-in to the new products. Through interactions in the standards committees, the vendors see the interest of the potential buyers. To assure success, the vendors and operators determine that the associated systems will be in place. Through the personal relationships developed among the representatives of many industry sectors, the committee members explore what could be possible, each other's intentions, needs, and concerns. To develop a full system view, an open interchange of ideas is needed among the system vendors, unit vendors, IC vendors, software providers, customer equipment vendors, test equipment vendors, network operators, content providers, end users, and consultants. The lively review and criticism of proposals by peers in the same and other industry sectors is a rapid and effective means to find the best overall solution and ensure that nothing is overlooked. A well-managed standards committee exhibits both competition and cooperation. The members compete with each other in an effort to find a better solution. Yet the members also know that they must cooperate to produce results in a timely manner. Given member companies who provide their top technical talent and strong committee leadership, extraordinary results can be attained by the multidisciplinary interactions. Standards committees are often criticized for being slow, but it does take time to develop industrywide buy-in and to develop a high-quality system definition. Once the standard is in place, one can proceed quickly and with confidence.

The best standards specify only what is necessary to assure interoperability, performance, reliability, and safety, while not overly constraining implementation optimization and improvement. A well managed standards process does not hinder competition, innovation, availability, and cost reduction. Suppliers can differentiate their products with supplemental features, cost efficiencies, versatility, ease of use, customer service, and other aspects.

Without a standard, a market leader may delay the entry of competitors into a market. However, if the market is attractive, competitors will eventually find a way to produce similar and perhaps compatible products. In most cases, the market leader would have been more successful from the larger market stimulated by a standard.

Involvement in standards development is an excellent means to learn what other companies (customers and competitors) have done and plan to do. Members of the standards committees are in the ideal position to learn of the latest developments in technology and business plans. Standards set the future course of the industry, and those present during the development of the standard understand its implications and implementation far better than an outside reader of the final document. There are opportunities to help steer the directions dictated by the standards to benefit one's own company. Lastly, participation in the committees can help establish a company's credibility, and the personal relationships with fellow committee members can be most useful building business relationships. The members who are willing to share some information, provide high-quality contributions, and compromise when necessary can have a far larger influence over the final standard than other members. The chairperson of the standards committee can have significant influence over the outcome, however, there are limits.

A company cannot participate in every standards committee. Each company must individually select those committees that have direct impact on their company's interests. Furthermore, the opportunity to stimulate progress, the ability to influence the outcome, the prognosis for timely results, the value of the knowledge gained, and the relationships built must be considered for each committee. Once a company decides which committees it should attend, it must then decide if it will only monitor the meetings or actively participate and contribute. Attending the meetings and only taking notes (monitoring) has nearly the same cost to the company as active participation. Only active participation provides the benefits of speeding progress, establishing company credibility, influencing the outcome to benefit the company, and learning the lessons that only come from being fully engaged in the process. Members must work to understand the needs and concerns of the other members and be flexible whenever possible.

16.11 Standards Process

The typical standards process begins with the proposal for a new standard or technical report. This may first require the approval of a new project and possibly a new committee or working group. Project approval authorizes a specific committee to perform certain activities. Contributions (technical papers that may make a proposal) are provided to the working group by its members. The working group chairperson will appoint one or more editors to prepare a working document based on the working group's agreements and contributions that have gained the working group's agreement. Most standards committees use a consensus process whereby extensive efforts are made to devise a solution to which all (or nearly all) members can agree. The consensus process can take longer than a simple voting process where the majority wins, however, the consensus process often generates superior industry buy-in. Also, in-depth consideration of minority opinions can often add value to the end result. When a final draft proposed standard is ready, it is sent with a voting ballot to all voting members of the committee. Members may vote YES, NO, or ABSTAIN. Members may also attached detailed comments to their ballot response. The committee reviews the ballot comments (regardless of whether the comment was on a YES or NO vote) and attempts to resolve all comments by making revisions to the draft document. If necessary the revised document is sent out for a second ballot (sometimes called a default ballot). For most committees, the standard is approved when a two-thirds majority of the votes are YES. Upon approval, the document is published. Many committees will review an approved standard 5 years after its issue. To expedite the issuance of a standard, it may be issued with some portions to be addressed in a future issue of the standard. In this case, work on the Issue 2 standard will begin upon completion of the Issue 1 standard.

In the United States, Section 273 of the Telecommunication Act of 1996 requires nonaccredited standards developments organizations (NASDOs) to follow rules of due process which assure open and fair operation. Further, U.S. antitrust laws require standard development organizations to avoid anticompetitive behavior. As a result, standard activities must avoid discussion of cost, price, product availability, market allocation, and other topics that might lead to anticompetitive behavior. Most standards development organizations require their members to identify any intellectual property (e.g., patents) that may be necessary for implementation of a proposed standard and to agree to license the intellectual property at "fair and reasonable" terms. This may be interpreted as offering a license to all parties at a cost that does not prevent them from offering products that are truly competitive.

Adoption of certain portions of a proposed standard, or the entire proposed standard may be delayed if members of the standards committee believe that the use of a patented technique is required and the terms and conditions for the patent are not acceptable. For this reason, most standards bodies recommend that companies holding applicable patents submit a statement that the company will license the patents on a nondiscriminatory basis for a reasonable fee.

16.11.1 When to Develop a Standard

The success of a standard depends on timing. If work starts too soon, the committee will waste time due to poor requirements and constantly changing technology fundamentals. If work starts too late, much of the market window may be lost or proprietary products may form one or more de facto standards. Work on a standard may begin while a technology is in the laboratory stage. However, before completion of the standard, it is best to have completed some field trials of prestandard systems to provide the benefit of practical experience learned from operating in the field. The committee leadership must keep a close watch on the interests of the committee members, the marketplace, and technology trends. New projects should be started quickly once several committee members express a strong interest in contributing work to a new project. In the case of ISDN, strong arguments can be made that standards were too late to address the market and too early for maturity of the technology.

16.11.2 Is a Standard Needed?

There is no definitive rule for when to start work on a standard or if a standard should be developed at all. Standards are necessary when equipment from multiple vendors must interwork. In many cases, a standard can help create a market that attracts many competitive suppliers. Often, the end user's interests drive the need for standards. End users desire standards to assure service uniformity across many service providers. Network operators desire standards to enable multiple sources for the systems they will buy. Equipment vendors desire standards to avoid the need to develop many redundant versions of their products. Regulatory and legal developments can create the need for new standards.

Compliance with consensus standards is voluntary in most countries. Market dynamics may drive suppliers to comply with the standards or to provide proprietary products for differentiation. A lack of standards can stimulate the introduction of government regulation where noncompliance results in penalties. Even the harshest critic of the standards process will admit that the regulatory process is inflexible and much slower.

So, when a new standard is needed, where will it be developed? Selection of the most suitable committee should be based largely on the expertise of the committee membership and the committee charter. Several committees may have an interest in the new topic. Sometimes certain companies will attempt to place the development of a standard in a committee where they have strong influence. In the event that more than one committee must be involved in the standard's development, it is vital that the committees establish and maintain a cooperative and coordinated relationship. This may be accomplished by one committee having lead responsibility and the other committee(s) providing input in defined areas. Another method is for two committees to jointly develop the standard as peers; this often involves a series of joint meetings. It is essential that conflicts between committees be quickly resolved to avoid an intercommittee war, which can waste an enormous amount of time.

In some cases, there is no existing committee with an interest in developing the new standard. The creation of a new committee may be necessary. The creation of a new subcommittee within an existing standards development organization requires the approval of the parent organization and the election or appointment of subcommittee leadership. The creation of a new standards development organization that is not a subpart of an existing standards development organization requires a massive administrative effort to begin and also to maintain. Thus, an exhaustive effort should first be made to find a way for the standard to be developed within an existing standards development organization.

16.11.3 Standard or Standards?

Some have suggested that multiple standards for the same item can help stimulate competition. However, the fundamental value of a standard lies in the uniformity it creates. All systems complying to the standard should interwork, and as future enhancements are introduced there is single base for which compatibility must be maintained. Lively competition is often seen with standardized products (e.g., V.34 modems). A market with multiple standards represents a high-risk venture, which may discourage the entry of suppliers and confuse customers. There are some cases where multiple standards are warranted. Different standards may be needed to address different applications, for example, a low-cost, low-performance consumer-grade version, and a higher-cost, higher-performance professional grade. For political reasons different standards may apply for different areas, for example, different standards for different countries. International standards bodies such as the ITU-T try to minimize the extent of country-specific standards. In a voluntary standards environment, a company or an alliance of companies can create and publish its own specification. Thus, it is not necessary to produce redundant standards.

Glossary

2B+D: Two 64 kb/s B channels and one 16 kb/s D channel are transported by Basic Rate ISDN.

2B1Q: 2-Binary 1-Quaternary line code is used for BRI and HDSL. Binary information is represented by four amplitude levels.

3B2T: A baseband line code where three binary bits are encoded into two ternary symbols.

4B3T: A baseband line code where four binary bits are encoded into three ternary symbols.

ABCD Parameters: The transfer characteristics of a two-port network describing the input voltage and current to the output voltage and current.

ADSL: Asymmetric Digital Subscriber Line simultaneously transports high-bit-rate digital information towards the subscriber, lower rate data from the customer, and analog voice via one twisted-wire-pair.

AFE: Analog Front End, functions including the analog-digital conversion, analog filter, and line driver.

AGC: Automatic Gain Control, receiver adaptation to the received signal level so as to reduce dynamic range of the signal input to the analog-to-digital converter.

AMI: Alternate Mark Inversion line code, also known as "Bipolar," is used for T1 transmission systems. Binary information is represented by pulses with three possible amplitudes.

ANSI: American National Standards Institute accredits standards bodies, for example, Committee T1 for telecommunications.

ATIS: Alliance for Telecommunications Industry Solutions sponsors Standards Committee T1.

ATM: Asynchronous Transfer Mode.

Attenuation: The signal loss resulting from traversing the transmission medium.

ATU-C: ADSL Transmission Unit at the Central Office side of the subscriber line. The ATU-C performs transceiver functions including modulation, coding, and equalization.

ATU-R: ADSL Transmission Unit at the Remote side of the subscriber line. The ATU-R performs transceiver functions including modulation, coding, and equalization.

AWG: American Wire Gauge measure of conductor diameter for wires.

AWGN: Additive White Gaussian Noise.

B8ZS: Bipolar with 8 Zero substitution line code use for T1 transmission. Ones are encoded as pulses of alternating polarity and eight consecutive zeros are represented by a pulse of the same polarity as the previous pulse.

Baseband: Transmission using a technique that places signal energy in a frequency band starting at approximately zero.

Baud: A symbol or element modulated for transmission.

B Channel: A 64 kb/s bearer channel used for ISDN.

BER: Bit Error Rate. The ratio of errored bits to transmitted bits.

Biphase: A baseband line code, also known as the Manchester line code.

Bit stuffing: Extra bit(s) that are conditionally inserted into the frame to adjust the transmitted bit rate.

BRI: Basic Rate ISDN provides 2B+D transport. See 2B+D, ISDN.

Bridged Tap: A wire stub connected to the transmission path.

Byte: A group of eight bits.

CAP: Carrierless AM/PM is a line code similar to QAM used for some DSL systems.

Cell: A fixed-length protocol unit used in a data link layer protocol. Since it is fixed length, a cell requires no special delimiting symbols.

CO: Central Office, also known as Local Exchange. A building where telephone lines terminate and connect to transmission and switching equipment.

Companding: Compressing the dynamic range of a signal prior to transmission, with matching expansion at the receiver to regain the original signal.

Convolutional Code: A code that depends on the current bit sequence and the encoder state.

CRC: Cyclic Redundancy Code. Used to detect bit errors.

Crosstalk (XTLK): The unintentional electromagnetic coupling of a signal to other wire pairs in a cable. See NEXT and FEXT.

CSA: Carrier Serving Area loop design rules specify the characteristics of loops served by Digital Loop Carrier sites.

Cyclic Code: An error correcting code implemented with a feedback shift regiser.

dBrnc: The logarithmic power ratio of a C-message weighted filtered signal with respect to 1 nano-watt.

D Channel: a 16 kb/s channel used by Basic Rate ISDN lines to carry signaling and packet information.

DA: Distribution Area is a loop serving area for a Feeder Distribution Interface (FDI).

DDS: Digital Data Service. A private line data service using AMI-type transmission of symmetric rates from 1.2 to 64 kb/s.

DFE: Decision Feedback Equalizer. An adaptive filter used to compensate for the frequency response of the channel.

DFT: Discrete Fourier Transform. A signal transformation that is often implemented as a fast Fourier transformation on a digital signal processor.

Discrete Time Domain: Signal values that are defined at periodic time intervals.

Distribution Cable: The portion of the telephone loop plant that connects the feeder cable to the drop wires.

DLC: Digital Loop Carrier. A system that is often located remotely from the Central Office to multiplex the service from many customer lines on to a high-speed line between the Central Office and the DLC remote terminal.

DMT: Discrete Multi-Tone is a multicarrier transmission technique that uses a Fast Fourier Transform (FFT) and Inverse FFT to allocate the transmitted bits among many narrowband QAM modulated tones depending on the transport capacity of each tone.

Downstream: Information flowing from the network to the customer.

Drop Wire: The section of the local loop connecting the distribution cable to the customer premises.

DS0:   A 56 or 64 Kb/s full-duplex service provided by telephone service providers.

DS3:   A 45 Mb/s full-duplex service provided by telephone service providers.

DSL: Digital Subscriber Line may be used to indicate the Basic Rate ISDN transceivers and the local loop to which they are connected. May also be used to indicate all types of DSL.

DSLAM: Digital Subscriber Line Access Multiplexor.

DSP: Digital Signal Processing performs filtering and other signal modifications in the digital domain by first converting analog signals to a digital representation.

DSX-1:   DS1 Cross-connect. A 1.544 Mb/s AMI signal used for short distances to interconnect equipment within a CO.

DWMT: Discrete Wavelet Multitone. A version of DMT that uses a wavelet type transform in place of the FFT. See DMT.

E1:   2.048 Mb/s symmetric transmission.

EC: Echo Cancellation is a DSP time domain technique for removing echos.

ECH: Echo-Canceled Hybrid is a 2-to-4 wire conversion with Echo Cancellation. A hybrid transformer is often used to interface to the line.

EMC: Electromagnetic Compatibility prevents unintended Radio Frequency Interference (see RFI).

EMI: Electromagnetic Interference.

ETSI: European Telecommunications Standards Institute.

FCC: Federal Communications Commission is the U.S. government agency, which regulates the radio, television, and telecommunications industries.

FDD: Frequency Division Duplex, two-way transmission via Frequency Division Multiplexing (FDM).

FDI: Feeder Distribution Interface is a cabinet that contains a cross-connect field used to connect feeder cables to distribution cables.

FDM: Frequency Division Multiplexing permits more than one information stream to be sent over a line by subdividing the line into separate frequency bands.

FEC: Forward Error Control. Errors are corrected by the receiver using redundant information sent by the transmitter.

Feeder Cable: The portion of the telephone loop plant that connects the Central Office to the distribution cable.

FEXT: Far-End Crosstalk results from transmitted signals being coupled to another wire pair interfering with the reception at the far end of the line.

FFT: Fast Fourier Transform. An algorithm for efficiently implementing via digital signal processors the conversion from the time-domain to the frequency-domain.

FIR: Finite Impulse Response. A FIR filter utilizes a limited number of delay and multiplication elements.

FM: Frequency Modulation uses changes in frequency of a carrier signal to represent information.

Fourier Transform: Using a sinusoidal expansion to represent a signal.

Fractionally Spaced Equalizer: An equalizer using multiples of the symbol rate.

Frame: A variable-length unit used in a data link protocol. Since it can vary in length, a frame requires special delimiting symbols, flags, to begin and end it.

Full Duplex: Simultaneous transmission in upstream and downstream directions.

FUNI: Frame-based ATM UNI (User Network Interface).

Galois Field: A closed algebra field used to describe an encoder or decoder.

Gaussian Distribution: A bell shaped distribution function, also know as a normal distribution.

Half Duplex: Transmission that alternates the upstream and downstream information flow.

Hamming Distance: The number of bits having a different value for a pair of codewords.

HDLC: High-level Data Link Control. A layer 2 communications protocol.

HDSL: High-bit-rate Digital Subscriber Line permits 1.544 or 2.048 Mb/s symmetric transmission over two or three wire pairs, which meet CSA design rules.

HDSL2: Second-generation HDSL uses one pair of wires for 1.544 Mb/s symmetric transmission over CSA loops.

HEC: Header Error Check. Used in ATM cells.

HTU-C: HDSL Transmission Unit at the Central Office side of the line. The HTU-C performs transceiver functions including modulation, coding, and equalization.

HTU-R: HDSL Transmission Unit at the Remote side of the line. The HTU-R performs transceiver functions including modulation, coding, and equalization.

IDFT: Inverse Discrete Fourier Transform. A discrete form of the IFFT.

IDSL: A DSL that uses BRI transceivers to convey 128 kb/s packet-mode transport via a local loop.

IFFT: Inverse Fast Fourier Transform. An algorithm for efficiently implementing via digital signal processors the conversion from the frequency-domain to the time-domain.

Impedance: The relationship between the voltage applied to the resulting current.

Impulse Noise: An unwanted signal of short duration often resulting from the coupling of energy from an electrical transient from a nearby source.

Impulse Response: The time-domain response of a network to a impulse input.

ISDN: Integrated Services Digital Network provides end-to-end digital circuit switched and packet switched transport. *(See BRI)*

ISI: Inter-Symbol Interference. Signal energy of a symbol that transfers into nearby symbols due to channel distortion.

IP: Internet Protocol.

ISP: Internet Service Provider.

ITU: The International Telecommunications Union develops international standards.

kft: Kilofeet, a thousand feet.

LAN: Local Area Network.

LMS: Least Mean Square. An algorithm used for adaptive filters.

Loaded Loop: A loaded loop contains series inductors, typically spaced every 6 kft, for the purpose of improving the voice-band performance of long loops. However, DSL operation over loaded loops is not possible due to excessive loss at higher frequencies.

Loopback: A diagnostic test method whereby the signal received by an element is replicated and retransmitted to the source of the signal. In some cases, the system is taken out of service to perform the loopback.

LT: Line Termination. The LT is the DSL transceiver at the network end of the line.

MDB: Modified DuoBinary. A baseband line code.

MDF: Main Distributing Frame. A wiring cross-connect field in the Central Office used to connect outside lines to CO equipment.

MPEG: Motion Picture Experts Group of ISO develops video compression standards.

NEXT: Near-End Crosstalk results in the signal transmitted from a wire pair coupling into another wire pair and interfering with the reception of signals at a receiver located at the same end of the line as the disturbing transmitter.

NIC: Network Interface Card, a communications interface circuit card for a PC.

NID: Network Interface Device, a connection point located at the demarcation between the network and the customer installation.

NT: Network Termination is the DSL transceiver at the network end of the line.

OCn: Optical Carrier of n times 51.84 Mb/s.

OSI: Open Systems Interconnection. A protocol reference model.

Passband: The modulation of signals, which permits the placement of the modulated signal in a specified frequency band. For example, CAP, DMT, QAM.

PC: Personal Computer.

PLL: Phase Locked Loop. A feedback loop with narrow bandwidth used to recover signal timing.

PM: Phase Modulation uses changes in the phase of a carrier signal to represent information.

PMD: Physical Media Dependent. A protocol sublayer.

POTS: Plain Old Telephone Service, circuit switched analog voice.

PPP: Point-to-Point Protocol.

PSD: Power Spectral Density. Signal power at a function of frequency.

PSTN: Public Switched Telephone Network. The traditional circuit switched voiceband network.

PVC: Permanent Virtual Circuit. A provisioned ATM or Frame Relay connection.

QAM: Quadrature Amplitude Modulation is a passband modulation technique which represents information as changes in carrier phase and amplitude.

QoS: Quality of Service. Virtual circuit characterization in terms of throughput, delay, and delay variation.

QPSK: Quadrature Phase Shift Keying. A passband modulation method.

Quad Cable: Cables where four wires are twisted as a unit. High crosstalk is experienced among the wires within a quad unit.

Quantization Noise: The noise resulting from analog to digital conversion.

Quat: A quaternary symbol representing two bits with four-level symbol.

RADSL: Rate Adaptive Digital Subscriber Line is an ADSL that adjusts its transmission rate to the capacity of the line.

RFC: Request For Comment. A term used for an IETF specification.

RFI: Radio Frequency Interference.

RS Code: Reed Solomon code. A code used for error detection or correction.

SDSL: Variously used to describe symmetric DSLs and single-pair DSLs. May or may not carry POTS (voice) on the same loop.

SNMP: Simple Network Management Protocol. A network management protocol using TCP/IP to access Management Information Base (MIB) information.

SNR: Signal to Noise Ratio. A signal quality measure.

STP: Shielded Twisted Pair. A metallic sheath is provided to reduce RFI.

SVC: Switched Virtual Circuit. An ATM or Frame Relay connection established in real-time via signaling messages.

T1 (Committee): Telecommunications standards committee for the United States.

T1 (Carrier): 1.544 Mb/s symmetric AMI transmission.

T1E1.4: The US standards Working Group responsible for DSL standards. T1E1.4 is one of several Working Groups within Technical Subcommittee T1E1.

TCM: Trellis Coded Modulation. A convolutional code that provides coding gain without increasing bandwidth.

TCM: Time Compression Multiplexing, also known as Ping-Pong, permits two-way transmission by the use of alternating short one-way transmission bursts.

TDM: Time Division Multiplexing.

TCP: Transmission Control Protocol.

Throughput: The effective rate of information flow though a system.

Upstream: Information flowing from the customer to the network.

UDP: Unacknowledged Datagram Protocol.

UTP: Unshielded Twisted Pair

VC: Virtual Circuit.

VDSL: Very-high bit rate Digital Subscriber Line systems transmit information at rates from 12 to 52 Mb/s.

Viterbi: An algorithm used for reception of trellis coded modulation.

VLSI: Very Large Scale Integrated circuit used to package a large amount of circuitry on a silicon chip.

VTU-O: VDSL transmission unit at the ONU.

VTU-R: VDSL transmission unit remote, the VDSL transceiver at the customer's premises.

WAN: Wide Area Network, covering more than one city.

White Noise: Noise with equal power at all frequencies.

XDSL: A generic term which applies to all DSL technologies.

Selected Standards and Specifications

International Telecommunication Union (ITU, formerly CCITT)

G.703: Primary rate (T1/E1) Transmission Systems

G.960: ISDN-BRI Digital Section (T to V interfaces)

G.961: ISDN-BRI Digital System (DSL line format)

G.729: Annex A defines 8 kb/s voice coding

G.991.1: HDSL (first generation)

G.991.2: reserved for future HDSL (second generation)

G.992.1: full rate adsl (similar to ANSI T1.413 Issue 2)

G.922.2: splillerless adsl ("G.lite")

G.993: reserved for future VDSL Recommendation

G.994.1: handshake for DSL

G.995.1: overview of DSL Recommendations

G.996.1: test procedures of DSL

G.997.1: physical layer management tool for DSL

H.320: Desktop Video

H.321: ATM desktop Video

H.322: Ethernet LAN Video

H.323: 802.3 digital video at rates of 28.8 kb/s and above

H.324: Multimedia Terminals for switched telephone service

I.121: BISDN

I.150: ATM

I.321: BISDN Protocol Reference Model

I.327: BISDN Architecture

I.356: ATM Performance
I.361: ATM Layer Specification
I.362: ATM Adaptation Layer specification
I.364: BISDN Connnectionless Data
I.371: BISDN Traffic and Congestion Control
I.413: BISDN UNI
I.430: ISDN BRI S/T Interface
I.431: ISDN-PRI User-Network Interface
I.432: BISDN UNI Physical Layer
I.610: BISDN OAM
V.21: voiceband modem, 300 b/s FSK
V.23: voiceband modem, 1200 b/s FSK
V.26bis: voiceband modem, 1200 b/s PSK
V.27ter: voiceband modem, 4.8 kb/s
V.32: voiceband modem, 9.6 kb/s
V.32bis: voiceband modem, 14.4 kb/s
V.34: voiceband modem, 28.8 kb/s QAM
V.42: data compression
V.61: simultaneous voice and data based on V.32bis
V.70: simultaneous voice and data via PSTN based on voice-band modems (DSVD)
V.75: control of parameters for simultaneous voice and data voice-band modems
V.76: voice and data multiplexing for voice-band modems
V.90: voice-band modem, 56 kbps/28.8 kbps asymmetric

ANSI - Committee T1

T1.101: Synchronization of Digital Networks
T1.210: OAM&P Principles of Functions, Architecture, and Protocol
T1.214: OAM&P Generic Model
T1.215: OAM&P Fault Management Messages
T1.216: ISDN Management - BRI
T1.217: ISDN Management - PRI
T1.218: ISDN Management - Data Link and Network Layers
T1.219: ISDN Management - Overview
T1.231: Digital Hierarchy - Layer 1 In-Service Digital Transmission Performance Monitoring
T1.304: Central Office Temperature and Humidity
T1.307: Fire Resistance for Equipment
T1.308: Central Office Equipment - ESD
T1.311: DC Power
T1.313: Electrical Protection of CO Equipment

T1.315: Voltages for DC Power
T1.316: Electrical Protection of Outside Equipment
T1.318: Electrical Protection Equipment at Entrance to Customer Building
T1.319: Fire Propagation Testing
T1.328: Protection from Physical Stress and Radiation
T1.329: Earthquake Resistance
T1.401: Analog Voice-grade Switched Access Lines Using Loop-Start
T1.403: DS1 Metallic Interface
T1.404: DS3 Metallic Interface
T1.408: ISDN Primary Rate - Layer 1
T1.410: Carrier to Customer Metallic Interface Digital Data at 64 kbits/s and Subrates
T1.413: ADSL
T1.601: ISDN Basic Rate Network Side of NT, Layer 1
T1.604: Minimal Set of Bearer Services for ISDN
T1.605: ISDN Basic Rate at S & T Reference Points
T1.606: Frame Relaying Bearer Service
TR28: HDSL

Other ANSI Standards

NFPA 70: National Electrical Code
TIA/EIA-568A: premises wiring

Bellcore Specifications

TR-EOP-000001, *Lightning, Radio Frequency, and 60-Hz Disturbances at the BOC Network Interface*
TR-NWT-00057, *Digital Loop Carrier*
FR-78, *Physical Design for Reliability Family of Requirements*
TR-TSY-000228, *Generic Human Factors Requirements for Network Terminal Equipment*
TR-TSY-000303, *Integrated Digital Loop Carrier*
FR-357, *Component Design for Reliability Family of Requirements*
TR-STS-000383, *Common Language® Bar Code Labels*
FR-440, *Transport Systems Generic Requirements (TSGR) Family of Requirements*
GR-342-CORE, *DS3 Interface*
TR-TSY-000454, *Supplier Documentation for Network Elements*
GR-485-CORE, *COMMON LANGUAGE® Equipment Coding Processes and Guidelines*
GR-499-CORE, *Transport System Generic Requirements (TSGR)*
TR-TSY-000509, *LSSGR*
TR-NWT-000870 *Electrostatic Discharge Control in the Manufacture of Telecommunications Equipment*

GR-929, *Reliability and Quality Measurements for Telecommunications Systems (RQMS)*
TR-NWT-000930, *Hybrid Microcircuits Used in Telecommunications Equipment*
GR-1089-CORE, *Electromagnetic Compatibility and Electrical Safety*
TA-NWT-001210, *HDSL*
FA-NWT-001211, *HDSL Operations*
GR-1248, *ATM Operations*
GR-1421-CORE, *Generic Requirements for ESD-Protective Circuit Packet Containers*
FR-2063, *Network Equipment - Building System (NEBS) Family of Requirements (NEB-SFR)*

IEEE Standards

IEEE 110: Powering and Grounding
IEEE 241: Electric Power Systems
IEEE 802.3: CSMA/CD MAC for LAN
IEEE 802.4: Token Bus MAC for LAN
IEEE 802.5: Token Ring MAC for LAN
IEEE 802.6: DQDB Metropolitan Area Network
IEEE 820-1984: Standard Loop Performance Characteristics
IEEE 1394: Firewire serial bus for interconnection of computer peripherals

Underwriter's Lab

UL 497: Overvoltage Protection
UL 567: Grounding and Bonding
UL 1283: EMI Filters
UL 1449: Transient Voltage Surge Protection
UL 1459: Telephone Equipment
UL 1863: Communications Circuit Accessories
UL 1950: Information Technology Equipment

FCC

FCC Part 15: Electromagnetic Interference
FCC Part 68: Telephone Equipment

ATM Forum

ATM Forum User-Network Interface Specification - versions 3.1 & 4.0
ATM Forum - Physical Interface Specification for 25.6 Mb/s over Twisted Pair Cable
   AF-PHY-0040.000, November 1995

XDSL CD-ROM Contents

Notice: The contents of this CD-ROM are Copyrighted by Prentice Hall, 1999.

T1E1.4 Meeting Reports

89-018 July 88
89-019 Oct 88
89-114 March 89
89-161 June 89
89-186R1 August 89
89-241 Sept 89
89-285 Dec 89
90-077 March 90
90-135 June 90
90-203 Sept 90
90-240 Dec 90
91-051 Feb 91
91-097 May 91
91-156 August 91
91-183 Nov 91
92-056 Feb 92
92-128A May 92
92-195 August 92
92-253 Dec 92
93-103A March 93

93-103B April 93
93-147 May 93
93-244 August 93
93-277 Oct 93
93-330 Nov 93
94-040 Jan 94
94-077 Feb 94
94-109 April 94
94-147 June 94
94-172 Sept 94
94-193 Dec 94
95-040 Feb 95
95-083 June 95
95-120 August 95
95-168 Nov 95
96-056 Jan 96
96-149 April 96
96-280 July 96
96-370 Nov 96
97-134 Feb 97
97-216 May 97
97-261 June 97
97-362 Sept 97
97-463 Dec 97
98-068 Jan 98
98-135 March 98
98-136 April 98
98-226 May 98
98-230 June 98

Annual Log of T1E1.4 documents

89-000R5
90-000R4
91-000R3
92-000R2 (first half of 1992)
92-000R3 (last half of 1992)
93-000R5
94-000R6final
95-000R4final
96-000R2

97-000R4
98-000Q3

Draft of XDSL Standards and Technical Reports

91-006R3 – T1.410 DDS interface
91-005 – T1.605 Basic Rate ISDN S/T interface
92-004 – T1.601 Basic Rate ISDN U interface
92-002R2 – TR28 HDSL
96-006 – Draft Version 2 TR28 HDSL
94-007R7 – Final draft version of Issue 1 ADSL Standard T1.413
97-007R6 – Issue 2 ADSL Standard T1.413 as provided for default ballot July 1998
98-002R1 – Early Draft of Spectral Compatibility Standard
T1E1/97-104R2a – Draft Proposed RADSL CAP/QAM Standard as provided for ballot October 1997
T1E1.1/98-003R1 – T1.401.03 Analog Phone interface
T1E1.2/94-003R1 – T1.403 DS1 "T1 Carrier" interface
T1E1.2/98-006R2 – T1.403 Robbed Bit Signaling
T1E1.1/98-028 – TR-005 T1E1 Connector Catalog

Selected T1E1.4 Documents

1992

92-026 – Bellcore, ADSL DMT and QAM Performance
92-027 – Cambridge Univ., ADSL, HDSL, ISDN Spectral Compatibility
92-037 – UCLA, QAM ADSL Performance
92-143 – Bellcore, Impulse Noise
92-144 – Bellcore, ADSL CAP and QAM Complexity
92-147 – Bellcore, ADSL Testing and Selection
92-148 – Bellcore, ADSL Crosstalk Testing
92-149 – Bellcore, CAP ADSL Design
92-152 – Bellcore, QAM ADSL Performance
92-153 – Bellcore, ADSL Coding
92-154 – Bellcore, ADSL Impulse Noise Cancellation
92-155 – Bellcore, ADSL Echo Canceller
92-161 – Bellcore, HDSL Sealing Current
92-164 – Bellcore, ADSL Maintenance
92-165 – SBC, ADSL Deployment
92-166 – SBC, ADSL Overhead

92-167 – Adtran, HDSL Transmit Power and Pulse Shape
92-169 – Nortel, ADSL, HDSL, ISDN Spectral Compatibility
92-197 – Amati, ADSL Market Requirements
92-198 – Amati, ADSL DMT Design
92-200 – Amati, ADSL Programmable Trellis Coder
92-204 – Amati, ADSL Migration
92-205 – Amati, ADSL Performance
92-210 – Amati, ADSL Impulses from POTS
92-212 – GTE, ADSL Standards Process
92-214 – BT, ADSL Impulse Noise
92-219 – Bellcore, ADSL Test Plan
92-221 – Bellcore, Insertion Loss of HDSL Test Loops
92-227 – Bellcore, Loop and Inside Wire Background Noise
92-233 – AT&T, HDSL Spectral Compatibility

1993

93-007 – November 1993Draft of T1.413 ADSL Standard
93-014 – Bell Atlantic, Focus for ADSL Work
93-015 – Ameritech, ADSL Requirements
93-018 – Amati, Why DMT for ADSL
93-020 – Amati, Echo Cancelled ADSL
93-026 –Amati, DMT ADSL Spectral Compatibility
93-029 – Bellcore, ADSL Test Method
93-030 – Bellcore, QAM ADSL Measured Performance
93-031 – Bellcore, DMT ADSL Measured Performance
93-032 – Bellcore, CAP ADSL Measured Performance
93-033 – Bellcore, Impact of ADSL on T1
93-034 – Bellcore, Impulse noise testing for ADSL
93-035 – Bellcore, RS code for ADSL
93-037 – Bellcore, DMT, CAP, QAM ADSL Performance
93-038 – Bell Atlantic, VDSL
93- 040 – Bell Atlantic, ADSL Capabilities
93-047 – Bellcore, Sealing Current
93-048 – GTE, DMT ADSL Measured Performance
93-054 – AT&T, CAP ADSL Units Tested
93-059 – AT&T, CAP vs. DMT ADSL
93-067 – Aware, DWMT ADSL
93-077R1 – Chair, ADSL Working Agreement
93-079 – Chair, ADSL Consensus
93-083R3 – Amati, DMT Specification Overview

93-084 – Amati, DMT Transmitter

93-086R1 – Amati, TEQ for ADSL

93-087 – Amati, ADSL Training

93-088 – Amati, ADSL DMT Loading Algorithm

93-090 – Amati, ADSL PSD

93-091 – Amati, ADSL FEC

93-095 – Nynex, ADSL Impulse Noise Performance: DMT, CAP, QAM

93-109 – AT&T, ISDN Sealing Current

93-113R1 – Amati, ADSL Activation

93-114 – Amati, ADSL Reveille Sequence

93-116 – Amati, ADSL Echo Cancelled Mode

93-117 – Amati, ADSL Revised FEC and Interleaving

93-118 – Amati, ADSL Trellis Coding & Tone Ordering

93-119R2 – Amati, ADSL Framing and Synchronization

93-122 – Amati, ADSL Channel Analysis

93-123 – Amati, ADSL Coding and Echo Cancellation

93-136 – Ameritech, ADSL System Reference Model

93-127 – Ameritech, ADSL in the Customer Premises

93-128 – Amati, ADSL Scrambler

93-129 – Alcatel, ADSL – Impact of Jitter

93-130 – Alcatel, ATM for ADSL

93-131 – Alcatel, ADSL FDM vs. Echo Cancelled

93-132 – Consltronics, Test Loops

93-149 – Amati, ADSL FDM/EC and Coding

93-150 – Bell Atlantic, ADSL Loop Reach, Rates, PSD

93-151 – GTE, ADSL Testing

93-160 – Alcatel, ADSL FDM vs. EC

93-177 – Amati, ADSL Coding

93-178 – Amati, ADSL & T1 Crosstalk

93-179 – Amati, ADSL Trellis Coding

93-180 – Amati, ADSL Trellis Coding

93-182 – Amati, ADSL Scrambler

93-184 – Amati, ADSL Bit Swap

93-185 – Amati, ADSL Echo Canceler Performance

93-197 – Ameritech, ADSL POTS Splitter Location

93-199 – Ameritech, T1 Carrier Loops

93-206 – Bell Atlantic, ADSL Loop Reach, Rates, PSD

93-220 – Bellcore, ADSL Next & Fext

93-225 – BT, ADSL Operations and Maintenance

93-229 – NTI, ADSL Premises Wiring

93-237 – Amati, ADSL Coding and Echo Cancellation

93-238 – Bellcore, ADSL Operations

93-239 – Bell Atlantic, ADSL Loop Reach, Rates, PSD

93-240 – GTE ADSL Impairments, Noise, Testing

93-246 – Amati, Multidrop ADSL

93-247 – Amati, ADSL Coding

93-248 – Amati, ADSL & T1 Crosstalk

93-258 – Bellcore, ADSL Premises Wiring

93-262 – Bellcore, ADSL Transmit Power

93-263 – Bellocore, ADSL RFI

93-264 – Orckit, ADSL Trellis Code

93-268 – NTI, ADSL & T1 Crosstalk

93-272 – Aware, ADSL Trellis Codes

93-274 – Aware, ADSL Trellis Code Recommendation

93-276 – Aware, Amati, ADSL Activation & Training

1994

94-019 – Bellcore, ADSL performance in the presence of T1 carrier

94-021 – Bellcore, HDSL performance in the presence of ADSL

94-026 – NTI, Distribution with the customer premises

94-032 – Amati, Collocation of POTS splitter for ADSL

94-033 – Amati, Customer premises configurations

94-040 – Minutes of January 1994 meeting

94-041 – Minutes of ad-hoc conference call on testing

94-042 – Minutes o ad-hoc conference call on POTS splitter location

94-043 – BT, POTS Splitters

94-068 – Telecom Australia, Impulse noise from ring trip

94-077 – Minutes of February meeting

94-087 – Amati, Improved performance from doubling the number of ADSL carrier tones

94-088 – Amati, 15 Mb/s ADSL via doubling the number of ADSL carrier tones

94-109 – Minutes of interim meeting

94-117 – Intellectual Property Rights Notice

94-126 – Amati, Multidrop ADSL

94-129 – Aware, 1-bit DMT constellation

94-131 – Bellcore, POTS splitter filter characteristics

94-132 – Bellcore, Downstream ADSL power spectral density

94-133 – ADI, Specification of analog components

94-134 – ADI, D/A converter aspects

94-135 – Bellsouth, POTS splitter characteristics

94-147 – Minutes of June meeting

94-153 – Results of T1E1 LB 94-05 (ADSL)

94-155 – ADC, Overhead bits for HDSL repeaters to report CRC errors

94-156 – Telecom Australia, Extended impulse noise test

94-161R1 – ECI, Method for ADSL bit swap

94-165 – Motorola, Announcement of ADSL Forum meeting

94-166 – Westell, Return loss simulation results

94-172 – Minutes of September meeting

94-176 – Results of ADSL default letter ballot

94-182 – Ameritech, Proposal for new Spectral Compatibility project

94-183 – Amati, Proposal for VDSL project

94-185 – Westell, ADSL rates for ATM

94-190 – ECI, Ameritech, Spectral Compatibility project proposal

94-193 – Minutes of December meeting

1995

95-007R2 – Final Draft ANSI T1.413 ADSL Standard (Issue 1)

95-013 – Ameritech, Starting text for Spectral Compatibility report

95-015 – Alcatel, ATM over ADSL

95-018 – Telstra, Impulse noise testing

95-020 – BT, RFI for VDSL

95-021 – Tut Systems, high speed transmission methods

95-022 – Amati, Line code complexity

95-023 – Amati, DMT VDSL performance with RFI

95-024 – ADC, HDSL repeater start-up

95-027 – IBM, Reverse ADSL

95-028 – Comments from ADSL letter ballot

95-029 – Aware, DMT and DWMT RFI performance

95-030 – Ericsson, POTS Requirements

95-031 – Orckit, QAM and Multitone VDSL performance

95-044 – Ameritech, HDSL2

95-045R5 – Working Group Mission and Scope

95-048 – Tellabs, Rate vs. Reach for symmetric DSL

95-049 – Tellabs, VDSL performance estimates

95-057 – Aware, VDSL performance with RFI

95-058 – AT&T, CAP HDSL Specification

95-060 – Amati, VDSL performance and complexity

95-061 – GTE, DMT VDSL performance

95-062 – GTE, CAP VDSL performance

95-063 – GTE, Impact of digital broadcast standard on VDSL

95-064 – ETSI recommended revisions to BRI standards

95-065 – Amati, high rate ADSL frame format

95-066 – Results of T1 letter ballot 95-01 (ADSL)

95-067 – Metalink, HDSL2 proposal

95-068 – Metalink, HDSL2 performance at 1.5 Mb/s

95-069 – Metalink, HDSL2 performance at 2 Mb/s

95-070 – Metalink, HDSL2 spectral compatibility with DMT ADSL

95-071 – Metalink, HDSL2 spectral compatibility with HDSL and BRI

95-073 – Orckit, VDSL complexity

95-075 – Orckit, ADSL rates above 8 Mb/s

95-076 – AT&T, CAP HDSL2 performance

95-077 – Bellcore, RFI model for ADSL

95-083 – minutes of June meeting

95-085 – Charles Ind., comparison of HDSL specifications

95-086 – ADC, HDSL start-up recommendation

95-087 – Detuche Telecom, ADSL over BRI

95-088 – results of T1E1 LB-495 (VDSL project)

95-089 – Amati, DMT HDSL2 performance estimate

95-090 – Amati, VDSL architecture

95-092 – GTE, HDSL2 performance

95-095 – GTE, TCP/IP performance over ADSL

95-097 – BT, Measured RFI into loops

95-100 – Metalink, HDSL2 performance

95-101 – Motorola, ADSL fine gain adjustment

95-102 – Motorola, C-RATES1 message for large interleave depth (ADSL)

95-103 – Tut Systems, VDSL transmission method

95-104 – Tut Systems, RFI for VDSL

95-105 – Tut Systems, VDSL Complexity

95-106 – AT&T, CAP and PAM HDSL2 performance

95-107 – Adtran, CAP and PAM HDSL2 performance

95-108 – Bellcore, HDSL2 transmission methods

95-109 – Racal Datacom, BRI standards issues from ETSI ETR-080

95-113 – BT, ETSI work on VDSL

95-114 – Bellcore, Ad Hoc recommended updates for T1.610 BRI standard

95-117R4 – Amati, VDSL system requirements

95-120 – minutes of August meeting

95-121 – ECI, Medium rate DSL

95-125 – Telstra, Impulse noise survey

95-127 – Bellcore, Echo canceled hybrid HDSL2 performance and spectral compatibility

95-128 – Bellcore, FDM HDSL2 performance and spectral compatibility

95-130 – results of default T1E1 LB-495 (VDSL project)

95-131 – AT&T-NS, VDSL system requirements

95-132 – Bellcore, RFI for VDSL

95-133 – Bellcore, Rationale for 6 dB margin design (ref. T1D1.3/85-241)

95-134 – US West, ADSL crosstalk into T1

95-135 – US West, ADSL performance with remotely located ATU-C

95-137 – Adtran, HDSL2 crosstalk into T1, ADSL, HDSL, and ISDN

95-138 – Adtran, HDSL2 performance

95-139 – Orckit, ADSL rates above 8 Mb/s

95-140 – Orckit, Coding for ADSL rates above 8 Mb/s

95-141 – BT, Far-end crosstalk

95-142 – Nortel, Balance measurements of BT cables

95-143 – Nortel, Impact of amateur radio on VDSL

95-144 – Nortel, RFI model for VDSL

95-145 – Nortel, Impact of RFI on DMT VDSL

95-146 – Nortel, Demographics aspects of amateur radio

95-147 – Metalink, HDSL2 performance with POTS

95-148 – Metalink, HDSL2 with Reed Soloman coding

95-149 – Revised ETSI ETR-080, BRI

95-149A – Living list for ETSI ETR-080, BRI

95-150 – ETSI Liaison regarding HDSL repeater start-up

95-151 – Independent Editions, VDSL Applications

95-152 – Independent Editions, ADSL data rates

95-154 – Level One, HDSL2 performance

95-155 – Davis 2.0 Specification Part 8

95-168 – Minutes of November meeting

95-172 – ETSI VDSL Technical Report DTR/TM-03068

1996

96-013 – BellSouth, Characterization of RFI in Two Metropolitian Areas

96-014 – BT, VDSL Splitter

96-015 – BT, VDSL Cable Characteristics

96-016 – Amati, Tutorial Review of DMT for ADSL Issue 2

96-020 – USWEST, Proposed VDSL Rates

96-021 – USWEST, VDSL Payload Mapping

96-024 – Bellcore, HDSL2 Transmission Methods

96-025 – Bellcore, RFI for VDSL

96-026 – Impact of ADSL Delay on TCP Performance

96-028 – GTE, VDSL Node Size vs. Line Rate

96-031 – Nortel, RFI for VDSL

96-032 – ADC, HDSL2 Framing

96-033 – ADC, HDSL2 Line Code

96-034 – Adtran, HDSL2 Line Powering

96-035 – Adtran Effect of Periodic Non-Gaussian Noise on HDSL2

96-036 – Adtran, HDSL2 Next Models

96-037 – Adtran, HDSL2 Proposal using SC-PAM

96-038 – AT&T, HDSL2 Spectral Compatibility

96-040 – Independent Editions, VDSL Service Requirements

96-041 – Independent Editons, VDSL System Requirements

96-045 – Level One, HDSL2 Noise Models

96-046 – Level One, Performance of Alternative HDSL2 Transmission Methods

96-047 – Metalink, Performance of NML for HDSL2

96-048 – Deutsche Telecom, VDSL Requirements

96-050 – ETSI, ETSI Reorganization

96-057 – Nordix/CDT, High Performance Twisted Pair Cable

96-059 – T1E1.4 Chair, Voltage Safety Limits

96-063 – Pairgain, HDSL2 Transmission Methods

96-064 – Pairgain, Veterbi Decoder Complexity

96-065 – Pairgain, Sequential Decoding Complexity and Gain

96-066 – Pairgain, Shaping Gain for HDSL2

96-067 – Pairgain, Turbo Codes

96-068 – Westell, TCP/IP Performance over ADSL

96-072 – Consultronics, Noise Generator Tutorial

96-073 – BT, Cable Models for VDSL

96-074 – BT, US 24 & 26 AWG Cable Models

96-075 – BT, Comparison of US and UK Drop Wire

96-076 – Bellcore, Working Draft of Spectrum Compatibility TR

96-077 – Bellcore, Cable Characteristics

96-078 – Bellcore, Performance of BRI-DSL Transceivers on Test Loops

96-081 – Amati, SDMT for VDSL

96-082 – Amati, RFI with VDSL

96-083 – Amati, SDMT for RFI with VDSL

96-084 – Amati, RF Cancellation with SDMT

96-085 – Amati, RF Egress with SDMT

96-086 – Amati, Eliminate 52 Mb/s VDSL Requirement

96-088 – Amati, SDMT for VDSL

96-089 – Amati, DMT ADSL Throttle

96-090 – Amati, 64 Tone Upstream for DMT ADSL

96-091 – Ameritech, Rate Adaptive ADSL

96-092 – Ameritech, Reduced ADSL Transmit Power at Remote Sites

96-093 – Ameritech, ADSL POTS Splitter

96-094 – Ameritech, HDSL2 Requirements

96-095 – Ameritech, HDSL Crosstalk Interferer Model

96-097 – BT, Primary Line Constants for US 24 & 26 AWG Cables

96-098 – BT, Measured RF Emission from VDSL Dropwire

96-099 – Harris, Impact of Peak Power on Line Driver

96-100 – Westell, FEXT Impact of RT-fed ATU-C on CO-fed ADSL in Same Cable

96-101 – Alcatel, Multipoint VDSL

96-102 – Alcatel, Spectral Shaping for VDSL

96-110 – ADC, Line power feeding voltages

96-113 – Motorola, ADSL Performance Enhancement

96-115 – Amati, VDSL RFI Egress & PSD

96-118 – Bell Atlantic, Need for CAP ADSL Standard

96-120 – Nortel, Amateur Radio RFI for UK

96-121 – Nortel, AM Broadcast RFI for UK

96-122 – Nortel, Whistler VDSL

96-123 – ADI/Aware, CAP/DWMT VDSL

96-126 – K. Maxwell, ADSL POTS splitter

96-132 – Bell South, VDSL bit rates

96-134 – TI, VDSL Channel Models

96-145 – K. Maxwell, VDSL Services Break Out Report

96-150 – ADC, HDSL2 Requirements Break Out Report

96-151 – Westell, ADSL POTS Splitter

96-152 – Westell, ADSL POTS Splitter

96-153R2 – J. Cioffi, VDSL System Requirements

96-159 – Alcatel, ATM over ADSL

96-161 – Broadcom, QAM for ADSL/VDSL

96-163 – BT, VDSL Power Back-off

96-164 – BT, VDSL FDM vs. TDD

96-166 – Bellcore, Estimate loop plant coverage

96-167 – Pulsecom, Distributed POTS Splitter

96-168 – Pulsecom, Premises wire crosstalk

96-169 – GTE, TCP/IP-WWW traffic

96-170 – Paradyne, CAP ADSL Specification

96-174 – Telstra, Impulse noise & ADSL

96-180 – Amati, VDSL FDD vs. TDD

96-183 – Amati, VDSL RFI Egress

96-185 – Stanford, Tutorial on Optimized Decision Feedback

96-186 – Amati, TDD for VDSL

96-188 – Amati, Reduction of upstream ADSL Overhead

96-191 – BT, Subjective effect of RFI on VDSL

96-194 – Lucent, VDSL Performance

95-195 – Nortel, Whistler VDSL

95-196 – Nortel, Rate Adaptive HDSL

96-207 – Bellcore, RFI in VDSL

96-208 – Adtran, HDSL2 line modeling

96-209 – Pulsecom, POTS Transient models

96-211 – Orckit, ADSL Clip Mitigation

96-212 – Orckit, ADSL rates above 8 Mb/s

96-219 – Bellsouth, Effect of Grounded Shield for VDSL Drop Wire

96-223 – Level One, HDSL2 Performance

96-224 – Level One, HDSL2 Performance

96-226 – Level One, HDSL2 Spectral Compatibility

96-227 – Pairgain, HDSL2 Sequential Decoding

96-228 – Consultronics, VDSL Testing

96-241 – GTE, Separate POTS Splitter at CO

96-242 – Bellcore, CAP & QAM interoperability

96-243 – Ameritech, ADSL Spectral Compatibility: CAP & DMT

96-245 – Ameritech, Need for Reduced ADSL Overhean

96-246 – Amati, VDSL SDMT Crosstalk canceler

96-247 – Amati, VDSL SDMT Add/Delete function

96-249 – Lucent, VDSL FDD vs. TDD

96-250 – GTE, Coexistence of VDSL and ADSL

96-251R1 – NEC, ATM over ADSL and VDSL

96-252 – NEC, ATM over ADSL and VDSL

96-254 – ADC and Level One, HDSL2 automatic power control

96-262 – Nortel, VDSL DMT Whistler performance

96-265 – Level One, Trellis and RS Coding for HDSL2

96-268 – Amati, SDMT VDSL Performance with foreign NEXT

96-269 – Amati, Mixing Symmetric and Asymmetric SDMT VDSL

96-270 – Pairgain, ADSL downstream NEXT into upstream

96-271 – Ameritech, Spectral Compatibility of VDSL with HDSL and ADSL

96-272 – OKI, The Case for Dual Standards

96-278 – GTE, Measured performance of DMT ADSL

96-284 – ADI, Perils of Dual Standards

96–288 – Motorola et. al., DMT is Best for ADSL

96–289 – Motorola et. al., The Case for One Standard

96-294 – Paradyne, CAP ADSL Performance

96-302 – Alcatel, Proposed method to compare VDSL systems

96-311 – Harris, ADSL Premises Wiring

96-316 – Harris, MTPR Clarification for ADSL

96-317 – BT, VDSL RFI measurements

96-318 – BBT, VDSL RFI for Aerial plant

96-320 – GTE, ADSL NEXT and Upstream Performance

96-321 – Lucent, Holmdel VDSL Test Facility

96-322 – Globespan, Turbo Coding for HDSL2

96-323 – Aware et. al., VDSL RFI Egress

96-324 – Lucent, VDSL System: FDD and TDD

96-326 – Lucent, VDSL DWMT performance

96-327 – Motorola, Rate Adaptive Start-Up for ADSL

96-329 –Amati, SDMT VDSL PMD Specification

96-330 – Amati, Mixing ADSL and VDSL

96-331 – Amati, Distributed ADSL Splitter

96-332 – Amati, Impact of RFI on DMT and single carrier

96-333 – Amati, SDMT VDSL Advantages

96-334 – Amati, Symmetric VDSL Applications

96-337 – Amati, VDSL FEC methods

96-338 – Amati, ADSL DMT vs. CAP performance

96-340 – PairGain Technologies, "A Modulation Strategy for HDSL2," 11/96.

96-361 – ADC & Ameritech, HDSL SNR field survey

96-364 – Alcatel, Network Timing Reference for ADSL

96-367 – ETSI, VDSL Draft Technical Reference

1997

97-016 – Telia, Zipper VDSL

97-019 – Draft TIA/EIA-711 Premises Architecture

97-021 – Harris, Placement of POTS HPF

97-022 – Harris, MTPR Clarification

97-024 – Globespan, HDSL2 Turbo Code

97-025 – TR41.9 FCC Part 68 for ADSL

97-038 – Alcatel, ADSL crosstalk within premises wiring

97-042R1 – Alcatel, ADSL EOC

97-043R1 – Alcatel, ADSL AOC

97-044 – Alcatel, ADSL Loss of Cell Delineation

97-047 – Alcatel, Reduced ADSL Overhead

97-053 – Pairgain, Spectral Compatibility Framework

97-054 – Pairgain, HDSL2 Sequential Coding

97-055 – Level One, HDSL2 Delay

97-057 – Level One, HDSL2 FEC Techniques

97-058 – Orckit, VDSL Interleaving

97-059 – Orckit, VDSL and ADSL Spectral Compatibility

97-065 – NEC, Effect of ADSL Dynamic Rate Change on ATM

97-066 – NEC, ADSL Clock for ATM

97-071 – ADTRAN, "Interference into T1: A Measured T1 Receiver Filter,"

97-073 – ADTRAN, "A Modulation Technique for CSA Range HDSL2," 2/97

97-074 – Pairgain, HDSL2 PSD and Coding

97-076 – Amati, Reduced ADSL Overhead

97-078 – Amati, VDSL FDD vs. TDD

97-083 – Nortel, VDSL RFI ingress for North America

97-094 – Aware, No bit swap for ADSL

97-095 – Aware, ADSL without POTS

97-099 – DSC, ADSL Rate Adaptation

97-100 – DSC, One-bit Constellation for ADSL

97-102 – DSC, ADSL Rate Adaptation

97-103 – DSC, ADSL Issue 2 Messages

97-104 – Lucent, VDSL Blind Training of DFE with RFI

97-106 – Lucent et al, Single Carrier VDSL Proposal

97-107 – Lucent, VDSL Latency & TCP/IP Throughput

97-109 – Pulsecom, ADSL POTS Splitter

97-118 – ADSL Rate Adaptation Break Out Report

97-124 – Siemens, Residential Wiring Practice – BICSI

97-129R1 – Alcatel, ADSL POTS LPF

97-130 – Gx, VDSL Requirements

97-131R1 – VDSL System Requirements

97-132 – BT, VDSL Latency

97-133 – BT, VDSL PSD

97-135 – GTE, VDSL Requirements

97-137 – Telia, VDSL Zipper DMT Scheme

97-138 – Telia, VDSL Zipper Performance

97-140 – Telia, Multimode VDSL

97-141 – Telia, VDSL from CO

97-142 – BT, VDSL Test Environment

97-143 – Harris, VDSL RFI

97-149 – BT, Measuring VDSL Impulse Noise

97-150 – BT, VDSL FEXT testing

97-154R1 – Alcatel, ADSL Indication of Options at Start Up

97-167 – ETSI VDSL Requirements

97-168R1 – Lucent, Single Carrier VDSL Draft Specification

97-169 – Lucent, Inside wire characteristics

97-170 – ADC Telecommunications, "Refined HDSL2 Transmission Masks: Performance
& Compatibility,"

97-176 – Globespan, HDSL2 Coding Method

97-177 – DSC, ADSL Rate Adaptation

97-179 – Pairgain Technologies, "Performance and Spectral Compatibility Comparison of POET PAM and OverCAPped Transmission for HDSL2," 5/97.

97-180R1 – Pairgain Technologies, "Normative Text for Spectral Compatibility Evaluations," 5/97.

97-181 – Pairgain Technologies, "On the Importance of Crosstalk from Mixed Sources,"

97-190 – Nokia, VDSL MQAM Line Code

97-191 – Level One, HDSL2 POET Performance

97-192 – Level One, HDSL2 Spectral Shaped Transmission

97-193 – Level One, HDSL2 Multistage Trellis Code

97-202 – Motorola, ADSL R-ACT3

97-203 – Motorola, ADSL EOC

97-205 – Motorola, ADSL C-QUIET1

97-207 – Alcatel et al, ADSL Spectral Copatibility

97-209 – ETSI VDSL Requirements

97-213 – ADI, ADSL Performance Measurments

97-214 – DSC, ADSL Rate Adaptation Break Out Report

97-217 – Alcatel, ADSL EOC/AOC Break Out Report

97-222 – Harris, MTPR Testing Primer

97-226R1 – Alcatel, ADSL Dynamic Clip Scaling

97-235 – Lucent, VDSL Power Boost

97-236 – Orckit, ADSL Spectral Compatibility

97-237 – Pairgain Technologies, "Performance and Spectral Compatibility of OPTIS HDSL2," 6/97.

97-239 – ADC, HDSL2 Performance and Spectral Compatibility

97-240 – Level One, HDSL2 OPTIS Performance

97-245 – Alcatel, ADSL Interoperability

97-246 – AFC, Spectral Compatibility Modeling

97-247 – Alcatel, ADSL POTS LPF

97-251 – Amati, ADSL Bit Ordering

07-252 – Amati, ADSL Dual Latency

97-257 – ADTRAN, "Principles of Agreement for HDSL2," 6/97.

97-259 – Nortel, ADSL Downstream PSD

97-265 – HDSL2 Break Out Report

97-266 – ADSL Dynamic Rate Adaptation Break Out Report

97-267R1 – T1.413 ADSL Issue 3 Living List

97-269 – HDSL2 Frame Format

97-270 – HDSL2 PAR Reduction

97-272 – ADSL CO Splitter location

97-273 – ADSL Fault Segmentation

97-274 – Harris, VDSL RFI Egress Measurement

97-279 – Bellcore, VDSL RFI

97-280 – NORDEX CDT, VDSL Aerial Drop Wire Characterization

97-281 – NORDEX CDT, VDSL Cat 5 Wire Characterization

97-283 – Telia, VDSL Synchronization via GPS

97-286 – Telia, VDSL & ADSL Coexistence

97-288 – Tellabs, VDSL Test Loop Simulation

97-289 – Tellabs, VDSL Impulse Response

97-294 – Bellcore, Spectral Compatibility Generic Specificaton

97-295 – Bellcore, Impulse Noise Survey Plan

97-296 – Bellcore, Primary and Secondary Cable Parameters

97-297 – Bellcore, Primary and Secondary Drop-Wire Paramters

97-300 – PairGain Technologies, "A 512-State PAM TCM for HDSL2", 9/97

97-301 – PairGain, HSL2 FEC

97-302 – Bellcore, Cable Crosstalk

97-303 – TI, Splitterless ADSL

97-307 – Cicada Semiconductor, "Performance and Spectral Compatibility of MONET PAM HDSL2 with Ideal Transmit Spectra—Prelim. Results," 9/97.

97-309 – Cicada, HDSL2 Worst Case Disturbers

97-311 – Alcatel, ADSL Dual Latency

97-312 – Alcatel, ADSL Bit Swap Command Set

97-315 – Alcatel, ADSL Ingress Measurements

97-317 – Alcatel, VDSL & ADSL Spectral Compatibility

97-320 – Level One, "OPTIS Template Specification for HDSL2," 9/97.

97-321 – Level One, HDSL2 OPTIS Start Up

97-322 – Alcatel, ADSL bi=0, gi=1

97-324 – Lucent, VDSL Bridged Taps

97-325 – Lucent, VDSL Blind Start Up

97-326 – Alcatel, ADSL & T1 Spectral Compatibility

97-327 – Pacific Bell, ADSL PSD for FDD

97-328 – PairGain, ADSL PSD

97-332R1 – VDSL SDMT Proposal

97-335 – Adtran, ADSL PSD Conference Call Report

97-336 – Adtran, Crosstalk from ADSL to HDSL

97-337 – ADTRAN, "Performance and Characteristics of One-Dimensional Codes for HDSL2," 9/97.

97-338 – Adtran, HDSL2 Programmable Trellis Encoder

97-339 – ADTRAN, "The Importance of Testing Widely Deployed Equipment in the Presence of Interference from New Line Codes," 9/97.

97-343 – Level One, HDSL2 Complexity Estimate

97-350 – ADSL PSD Break Out Report

97-353 – Amati, ADSL Clip Entropy

97-350 – 3com, ADSL Annex M – EOC

97-367 – Stanford, ADSL PAR Reduction

97-368 – SGS Thompson, VDSL Zipper FFT complexity

97-371 – DSC, ADSL Dynamic Rate Adaptation

97-372 – DSC, ADSL Rate Adaptation at Start Up

97-374 – ADSL Forum, ADSL EOC

97-378 – Level One, HDSL2 PSD & Crosstalk

97-380 – 3com, ADSL PSD

97-383 – Ameritech, Proposed HDSL Standards Project

97-384 – Ameritech, Proposed Spectral Compatibility Standards Project

97-388 – Committee T1 Letter Ballot Process

97-389 – Ameritech, Splitterless ADSL

97-390 – T1.413 Issue 2 LB 652 Ballot Results

97-391 – CAP/QAM RADSL LB 653 Ballot Results

97-395 – TI, Splitterless ADSL

97-397 – TI, ADSL Clip Mitigation

97-401 – Level One, HDSL2 Standard Structure

97-403 – ETSI Cable Models for VDSL

97-404 – Amati, VDSL & ADSL Spectral Compatibility

97-408 – AMD, ADSL Home Wiring Measurements

97-409 – PairGain, HDSL2 16 PAM TCM

97-411 – NORDX CDT, Cable Characteristics for CAT5 and Distribution Cable

97-412R1 – Cicada Semiconductor, "Performance and Spectral Compatibility of MONET(R1) HDSL2 with Ideal Transmit Spectra—Prelim. Results,"

97-415 – Cicada, HDSL2 Worst Case Disturber Models

97-417 – Centillium, G.lite Proposal

97-418 – Centillium, Splitterless ADSL

97-419 – Centillium, G.lite upstream

87-420 – Centillium, G.lite performance

97-421 – Tollgrade, ADSL Fault Location

97-431 – Westell, Splitterless ADSL

97-434 – ADC Telecommunications, "Measured Spectral Compatibility of HDSL2 with Deployed HDSL," 12/97.

97-435 – Level One Communications, ADC Telecommunications, PairGain Technologies, "Updated OPTIS PSD Mask and Power Specification for HDSL2," 12/97.

97-436 – ADC, HDSL2 Measured Spectral Compatibility

97-437 – BBT, VDSL EMC for Underground Plant

97-438 – Globespan, HDSL2 FEC

97-440R1 – ADTRAN, "Test Results of HDSL Units with HDSL2 Noise Generator," 12/ 97.

97-441 – ADTRAN, "HDSL2: Common Elements for Symmetric DSL Rates," 12/97.

97-442 – ADTRAN, "Motivation for a Programmable Encoder for HDSL2," 12/97.

97-443 – ADTRAN, "Proposal to Break the FEC Logjam for HDSL2," 12/97.

97-444 – ADTRAN, "Simulated Performance of HDSL2 Transceivers," 12/97.

97-446 – Rockwell, Splitterless ADSL & POTS interference

97-447 – Rockwell, Splittereless ADSL & POTS Test Configuration

97-450 – Paradyne, G.lite Characteristics

97-469 – ADC Telecommunications, ADTRAN, Level One Communications, PairGain Technologies, "Performance Requirements for HDSL2 Systems," 12/97.

97-471 – ADC Telecommunications, ADTRAN, Level One Communications, PairGain Technologies, Siemens, "Basis for HDSL2 Standard," 12/97.

ALL 1998 T1E1.4 CONTRIBUTIONS, through September

ADSL Forum Technical Reports

Approved ADSL Forum Technical Reports provided with the permission of the ADSL Forum.

TR-001: ADSL Forum System Reference Model

TR-002: ATM over ADSL Recommendations

TR-003: Framing and Encapsulation Standards for ADSL

TR-004: Network Migration

TR-005: ADSL Network Element Management

TR-006: SNMP-based ADSL LINE MIB

TR-007: Interfaces and System Configurations for ADSL

TR-008: Default VPI/VCI addresses for FUNI Mode Transport

TR-009: Channelization for DMT and CAP ADSL Line Codes

TR-010: Requirements and Reference Models for ADSL Access Networks: The "SNAG" Document

TR-011: An End-to-End Packet Mode Architecture with Tunneling and Service Selection

Index

2B1Q, 25, 26, 32, 155–159
3B1O, 159
4B3T, 26, 166–168
4B5B, 168, 170
10BaseT, 161

A

ABCD modeling/theory, 64–65
  definitions, 65
activation, 297–298
ADC (analog-to-digital converter) accuracy,
    299
ADSL, 7, 9, 13, 18, 41–49
  ADSL + ISDN, 46
  ADSL Forum, 424
  ADSL Lite, splitterless, 47
  AEX bytes, 376
  AS-X channels, 371, 372
  ATM-0 channel, 373
  ATM-1 channel, 373
  fast byte, 377
  fast channel, 373
  fast data, 375
  fast data buffer, 375
  fast path, 371
  Forum TR-001, 369

ADSL (cont.),
  Forum TR-003, 388, 389
  frame, 382
  indicator bits, 377, 378
  interleave data, 375
  interleave data buffer, 375, 376
  interleaved channels, 373
  interleaved path, 371
  LEX bytes, 376
  loop, 398
  LS-X channels, 371, 372
  reference model, 370–373
  showtime, 379
  TC, 382
  T-interface, 370, 372
  U-interface, 370
  V-interface, 370, 373
ADSL overhead control (AOC), 377, 378
Al-Dhahir, Naofal, 200, 230
AMI (alternate mark inversion), 25, 162–163
analog devices, ADI (hybrid circuit), 129
analytic equivalent, 178–180
ANSI, 419, 421, 445
ANSI T1.403, 140, 163
ANSI T1.404, 165
ANSI T1.413, (DS3 or T3), 144, 235–6, 369

ANSI T1.601, 155–156
applications, 16–21
asymmetric transmission, 9
asynchronous transmission, 9
ATM, 10, 46, 412–414, 373, 385, 397
   access, 403
   cell delineation, 381, 382
   cell rate decoupling, 381
   cells, 381, 382
   Forum str-saa-funi_01.01, 388
   header, 381
   HEC (header error correction), 381
   idle cells, 383
   permanent virtual circuit (PVC), 404, 405
   stack, 366
   switch virtual circuit (SVC), 404, 405
   transmission convergence (TC), 372, 381
   UNI, 388
   virtual circuit, 381
   virtual connection identifier (VCI), 389
   virtual path identifier (VPI), 389
ATM Forum, 424, 446
attenuation constant, α, 68
ATU-C, 41, 42, 370–372, 382, 383, 385, 402
ATU-R, 41, 42, 372, 382, 383, 385
authentication, authorization, and accounting
     (AAA), 408
autocorrelation function, channel, 185
automatic gain control (AGC), 302
AWG, 112, 55–57

B
B8ZS, BnZS, 165–166
back-end data network, 396
balance, 84–85, 123
   longitudinal measurement, 332
banking transactions, 399
Barker code, 348
baseband
   codes, 155
   equivalent, 178–180
Bell 103 modem, 3

Bell 202 modem, 3
Bell 212 modem, 4
Bell, Alexander, 1, 53
BER (bit error rate), 23, 35, 358
Berrout, C., 263
binary symmetric channel (BSC), 265
binder group, 25, 56
Bingham, John A.C., 227
Bingham's canceller, 227–228
biquadratic, 117–8
bit distribution/table, 207 (Equation 7.37)
   B-tightness, 216
   efficiency of, 214
   energy tightness, 215
   swapping, 323–324
bit-error-rate tester (BERT), 332
bit-reverse addressing, 242
BRI, (*see* ISDN, basic rate)
bridged tap, 57, 58
BRITE, 27
broadband ISDN, 365
burst (error), 274
Butterfly operations, 243
Butterworth (filters), 108–110, 160

C
cable modem, 135
Campello, Jorge, 214
campus and niche provider, 401
CAP (carrierless amplitude and phase modula-
     tion), 174–175
capacity (Shannon/Channel), 155, 205
carrier serving area (CSA), 31, 36, 326–328
Cauer (filters), 112–3
CCITT, (*see* ITU)
CEV, 6
channel identification, 303–311, 314
CHAP, 387
characteristic impedance, 70, 78
Chebychev (filters), 110–112
Cheong, Kok-Wui, 233
   Cheong's precoder, 233

Chow, Jacky S., 230, 312
Chow, Peter S., 213–214
Chow's eigen-updating, 312, 317
Cioffi, John M., xvii, 200, 324
circuit switching, 411
clip, clipping, 237
CMIP, 357
CO (Central Office), 5, 6, 8, 54, 395, 396,
  402, 414–416
coax, CATV, 2, 12
coding gain, γ_c, 256
  fundamental, 261
  shaping, 261
collocation, 414, 415
common carrier, 401
common-mode choke (CMC), 123–4
competitive provider, 401
compression, data, 5
concatenation (of codes), 277
constellation, signal, 148
  lattice, 261
  mapping, 262
  partitioning, 256–260
Cook, John (Cook Pulse), 95
  Cook/Foster splitters, 126–127
  Cook/Sheppard splitters, 125
crossover probability, 265
crosstalk, 13, 53, 86, 328–329
  ADSL, 102–104
  cyclostationarity, 91–92
  distribution, 91
  far-end (FEXT), 15, 16, 86, 90–91
  HDSL, 102
  ISDN, 101
  near-end (NEXT), 15, 86–90
  self-NEXT, 44, 99
CSA (carrier serving area), 31, 36, 326–328
customer premises, 394
  equipment, 417
customer premises wire, (*see* inside wire)
cyclic codes, 265
  encoders, 270

cyclic reconstruction, 234
cyclic redundancy check (CRC), 265,
  281–285, 371
  polynomials, 284

D

DAC (digital-to-analog converter) accuracy,
  299
DAML, 29
data link control (DLC), 385
data link layer, 385
DAVIC, 51, 425
Davis, J., 237
DDS, 22
decision regions, 151
decoding/decoder, 150, 273
  tabular, 269
demodulation, 150
detection, 150–152
  maximum-likelihood, 151, 202–205
  symbol-by-symbol, 151
differential encoding, 160, 177, 201
digital compressed video, 399
digital transmission, 21
dispersion, 69
dispersionless (transmission line), 69
distribution plant, 54–56
distribution service area (DSA), 328
DLC, NGDLC, 6, 27, 46, 54–56, 414
DMT (discrete multitone), 221–222, 235
driver circuit, 129
DS0/T1, 395
DS1, (*see* ANSI T1.403)
DS3, (*see also* ANSI T1.404), 396, 398
DSL, xv, 1, 2, 5, 6, 16
  bit rates, 7, 14, 24
  margin, 23, 35
  terminology, 12, 433–442
DSL loop, 395
DSL modem, 392
  external, 395
  internal, 395

DSLAM, 356, 396, 413
DSP, Digital Signal Processor, 16
D-transform, 185
dual latency, 278–279, 373
duobinary, 201
duplex transmission
    dual duplex, 33
    full duplex, 9
    half duplex, 8
    single duplex, 32, 33
duplexing, 139
DWMT (discrete wavelet multitone), 225
dying gasp, 360
dynamic clip scaling, 237–238

E

E1 carrier, 24
E1 or E3 (see ITU g.703)
ECH (echo-canceled hybrid), 9, 15, 44, 45
echo cancellation, 140–142, 378
    adaptive, 142, 248–250
    circular echo synthesis (CES), 248
    frame synchronous, 249
EIA/TIA, 426
Eigen-functions, 217
elliptic (filters), 112–113
embedded operations channel (EOC), 10,
        377
emissions (egress), 104
encoder, 147–148
    matrix, 262
    memoryless, 148
    sequential, 148
energy, average, 148
entropy, 236
    of clip, 237
equalization/equalizer, 183, 186–200
    adaptive algorithm, 315
    decision-feedback (DFE), 188, 204, 311
        block, 233
        generalized, (see GDFE)

equalization/equalizer (cont.),
    fractionally spaced, 187
    frequency, (see FEQ)
    linear (LE), 186–188, 193
    TEQ (time equalization), 228–230, 311
    transmit, 196–200
error
    burst, 274
    control, 264–270
    correction, 264
    detected, 268
        probability, 269
    detection, 264
    locator/location, 270
    propagation, 190–191
    undetected, 268
        probability, 269
errored second, 153
    percentage of, 153
Ethernet, 161, 403
    bridging, 388
    hub, 402
ETSI, 30, 419, 420, 423
excess bandwidth, 171
exchange, 314
Eyuboglu, V., 237

F

facsimile (FAX), 3, 19–21
Fan, John, 224
fast path, 235–6
FDM, 10, 37, 44, 45
feedback filter, 188
feeder plant, 54
feedforward filter, 188
FEQ (frequency equalizer), 223, 226,
        230–231, 319
    modified FEQ, 231
FEXT, (see crosstalk)
FFT (fast Fourier transform), 242–247
    inverse (IFFT), 245–247

fiber, 2, 16
fiber-to-the-home (FTTH), 133–134
Forney, G. David Jr., 200, 256, 324
Foster, Kevin, (Cook/Foster splitter), 126–127
four-wire duplexing, 139–140
frequency division duplexing/multiplexing
    (FDD/FDM), 378, 143–144
frequency scaling, 114–116
frequency transformations, 115
Friese, M., 237, 238
FSK, 3
FUNI (frame-based user network interface),
    385, 388, 389
    data frame, 388

G

G.729, 4
G.992.1, 42, 48
G.992.2, 42, 47–49
G.994.1, G.995.1, G.996.1, 42
G.997.1, 43
gain estimation, 304, 307–309
    swap, 324
    updating, 318
Galois field, 265
gap, 154–155
Gatherer, Alan, 237, 239
Gaussian (noise), 149
    period of, 336
GDF (generalized decision feedback), 231
    key equation, 232
generalized impedance conversion (GIC), 119
generator matrix, 267
generic DSL architecture, 394
G.lite (*see* G.992.2 and ADSL, splitterless)
granularity, 212
Gray coding, 156
group delay, τ_g, 69
group velocity, ν_g, 69
guard period, 219

H

H.261, 16
hamming
    distance, 267
    weight, 267
HDB (high-density binary coding), 166
HDLC (high level data link control), 386, 387,
    388
HDSL, 7, 13, 30–35
HDSL2, 35–41
Henkel, Werner, 200, 230
Hood, Robin, 214
Huber, J., 238
Hughes-Hartog, Dirk, 214
hybrid (circuit), 128–129
hybrid fiber coax (HFC), 135

I

IDSL, 30
IEEE, 420, 426, 446
IEEE 802.2 (standard), 389
IETF, 425
Imai, H., 237
impulse (noise), 94–97, 330–331
ingress, 92
initialization, 378, 379
    activation and acknowledgement, 378
    channel analysis, 378
    exchanged, 379
    transceiver training, 378
insertion loss, 64, 122, 126
inside wire, 47–49, 60–62
integrated service, 401
interleaving, 274–276
    block, 274
    convolutional, 275
    depth, 274
    Tong's, 275
interleave path, 235
intermodulation distortion, 217

International Organization for Standards
(ISO), 363
Internet, 16, 18
premium service, 399
residential entertainment, 399
Internet Engineering Task Force (IETF), 385,
405, 407
Internet protocol suite, (*see also* IP), 365
Internet service provider (ISP), 393, 397, 398,
403, 404, 408
intersymbol interference (ISI), 171, 183–6
mean-square, 185–6
IP (Internet protocol), 365, 395, 397, 404, 407,
408
address, 388, 409
router, 402, 406
telephony, 399
IPX, 388
Isaksson, Mikael, 227, 250
ISDN, (*see also* ANSI T1.601), 22,
155–163
basic rate, 7, 10, 13, 25–29
extended range, 27, 28
German, 167
Japanese, 163–164
repeater, 27
ISO 3309 (standard), 386
ITU, ITU-T, 4, 5, 25, 30, 42, 355, 419–421
ITU g.703, 140, 166
ITU g.dmt (g.922.1), 144
ITU g.lite (g.922.2), 144
ITU-T Recommendation I.432, 382, 383

J

Jedwab, J., 237

K

Karhunen-Loeve, 217
Kasturia, Sanjay, 233
Kschischang, F., 237

L

L2TP (Layer 2 Tunneling Protocol), 408
LAN, 393, 395, 403, 405
Lanczos algorithm, 312
large local TELCO, 400
latency, 34, 36
paths, 371
layered model, 363
leap-frog (circuits), 118
level-two forwarding and gateway, 405
line code, 148
line power, 63
linear phase, 69
link control protocol (LCP), 385, 387, 389
LMS (least mean square) algorithm, 315, 320
load coil, loaded loop, 1, 28, 58, 59
loading, 208–217
Campello, 214–217
Chow, 213–214
dual bit-mapping, 217
Greedy, 213
Hughes-Hartog, 214
margin-adaptive, 210–211, 213–4
rate-adaptive (RA), 209–210, 212–214
logical link control (LLC), 389
long holding time data calls, 397
loop,
length, 59, 60
plant, 1–3, 53–60
qualification, 360, 361
loop-back, 358, 359
loop timing, 303, 337

M

MAC layer concentration, 402
magnitude scaling, 114–116
management, 355–359
Manchester, modulation or encoding, 161–162
margin, (*see also* DSL), 155, 336
McDonald, Richard, 23

MDF, 54, 55
Medley sequence, 304
merge, 202
metric, 202
MIB, 357
minimum distance, 151–152
MMS43, (see 4B3T)
modal modulation, 217
modem, 392
    voice-band, 3–8
modified duobinary, 201
modulation, 147
    linear, 148
MPEG, 16, 43
MPEG 2, 371
MTU, Metallic Termination, 359
Muller, S.H., 238
multicarrier (transmission), 206
multichannel line code, 205
multiplayer interactive games, 399
multitone (transmission), 205

N

Narula, A., 237
nearest neighbors, 152
    average, 152
network control protocol (NCP), 385, 388
NEXT, (see crosstalk)
NID, 47, 56
noise
    crosstalk, 86–104
    enhancement, 187
    estimation, 304, 309–311
        updating, 318
    Gaussian (additive white), 149
    narrowband, 251
noise-equivalent channel, 186
numerically controlled oscillator (NCO), 350
Nyquist
    criterion, 171
    pulse, 171

O

OAM&P, 355–361
OC-3, 398
Ochiai, H., 237
ONU, 49–52
open systems interconnection (OSI), 363
OSS, 357

P

packet-based architecture, 402
packet switching, 412
PAM, 38
PAP, 387
parity
    matrix, 256, 268
partial response, 200–201
partial transmit sequences (PTS), 238
partitioning
    channel, 217
passband, 107–8
    codes, 156, 172–177
passive optical network (PON), 134
Patterson, K.G., 237
PC, 393, 403, 404, 405
PCM modem, 4, 5
peak-to-average ratio (PAR), 237
performance monitoring, 358
phase
    constant, β, 68
    delay, τ_p, 69
    detector, 337
    jitter, 338
        data-dependent, 342
    spectrum, 338
    velocity, v_p, 69
physical layer frame, 371, 372, 373, 374
physical layer super frame, 371, 372, 375
physical medium dependent (PMD) sublayer,
        367, 369
pilot, 343

ping-pong, (*see* TCM)
PLL (phase-lock loop), 337
    first order, 338
    second order, 338
POET, 37–39
point-to-multipoint transmission, 11
point to point protocol (PPP), 365, 385, 387,
        388, 408
    encapsulation, 387
    over ATM, 405, 407, 408, 409
    terminated aggregation, 405
point-to-point transmission, 11
Polley, M., 237, 239
POTS, 12, 41, 47–49, 53, 54, 56, 396
power, 72
    analog, 149
    control, 300–302
    digital, 149
power feeding, 32
precoding, 190, 234
    flexible, 192–3
    Laroia, Rajiv, 192
    Tomlinson-Harashima, 191–2
primitive polynomial, 271
private line, 411
probability of (error), 152–154
    bit error, 152–153
    errored second, 153
    error-free seconds, 153
    normalized error, 152
    symbol error, 152
propagation constant, γ, 68, 78
protocol stack, 365
PSD, 38
PSK, 4
PSTN, 3, 5, 22, 397
public switched telephone network (PSTN),
        (*see also* PSTN), 391, 392, 397
pulse amplitude modulation (PAM), 159–160
PVC, 413

Q

QAM (quadrature amplitude modulation),
        173–4, 176–7
    OQAM (offset QAM), 176–177
quad wire, 54
quat, 34

R

radio and audio feed, 399
radio noise (RFI, radio frequency interfer-
        ence), 92–94
    AM, 94, 329
    amateur (HAM), 93, 254, 330
    cancellation, 253–256
RADSL, 43
raised cosine pulses, 171
    square-root form, 171–172
Raleigh quotient, 312
random reencoding, 238
realms of responsibility, 393
reciprocal circuits, 105
Reed-Solomon codes, 265, 269
reflection coefficient, 73
regulation, telecommunications, 2, 16, 17,
        401, 414–418
repeater, 1, 7, 21, 25, 27, 28, 30
resistance design (RD), 56, 57, 325
return loss, 73–4, 123, 127
revised resistance design (RRD), 326
RFC 1331, 386
RFC 1332, 388
RFC 1334, 386
RFC 1483, 389, 405, 406, 407
RFC 1638, 386
RFC 1661, 385, 387, 407
RFC 1662, 386, 387, 407
RFC 1664, 386
RFC 1877, 386, 388
RFC 2331, 386
RLCG parameters, 67–68, 78–79

RLCG parameters *(cont.)*,
  admittance per unit length, 68
  impedance per unit length, 68
  measurement of, 78
  tables, 80–82
rotor, 341

S

SAI, 54
scale of service, 401
scrambler, 285
  self-synchronizing, 285
  synchronous, 285
SDSL, 33, 35
sealing current, 63, 64
selective mapping (SLM), 238
serial line interface protocol (SLIP), 385
service, 17
shaping, 262
  trellis, 262
shuffle
  matrix, 231
  tone, 235
signal power loss, 13
signal-to-noise ratio (SNR), *(see also* SNR),
  geometric, 208
Silverman, Peter, xviii
simplex transmission, 8
small router, 395
SNMP, 357
SNR (signal noise ratio), 379
SONET, 398
spectral compatibility, 13, 36, 37, 97–104
spectral estimation, 318–319
splice, wire, 57
splitter (POTS or ISDN), 41, 47, 48, 120–127
splitterless ADSL, *(see* ADSL Lite)
Spruyt, Paul, 227
standards, 419–431
  process, 428
  value, 427

Starr, Thomas, xvii
stopband, 107–8
survivor, 202
SVC, 413
SVD (singular value decomposition), 220
switched capacitor, 118–9
symbol, data, 147
symbol period   148
synchronization, 302–303
  frame, 348
syndrome, 268

T

T1, *(see also* ANSI T1.403), 22–25, 30–32
T1.413, 42
T1.601, 64
T1 circuit, 398
T1 Committee, 419–423, 444
T1E1.4, xvi, 23, 369, 419, 422
T3, *(see* ANSI T1.404)
TCM (time compression multiplexing), 9, 25
TCP/IP, 46, 365
telecommuting, 399
telephone line, *(see* loop plant)
Tellado, Jose, 237
testing, 325–337
three-ports, 119–129
time division duplexing/multiplexing
      (TDD/TDM), 10, 142–143
  Japanese ISDN, 143
  VDSL, 143, 247–248
timing recovery, 341
  add/delete, 346
  all digital, 341
  band-edge, 343
  decision-directed, 344
  open-loop, 341–342
  pilot, 343
  pointers, 346–7
  square-law, 342
TL1, 357

TMN, 355–357
tone injection, 241–2
tone reduction, 240
Tong, Po T., 276
transfer time for image, 21
transmission convergence (TC) sublayer, 367
trellis codes, 40, 256
turbo codes, 263–264
twisted-pair wire, 53
two-port modeling, 64–67, 105–6
     bridged-tap, 74
     input impedance, 67
     insertion loss, 64, 83–84
     load coil, 75
     lossless, 107
          realization, 113–4
     return loss, 64, 73
     transfer function, 65–66, 75–77, 83–84

U

UAWG, 419
UDP, 365
unbundled loop, 417
Unger model, 36
Ungerboeck, Gottfried, 256
universal discrete symmetric channel (UDSC), 265–6
universal serial bus (USB), 404

V

V.21, 3
V.22, V.22bis, V.34, V.42, V.61, V.70 V.90, 4
VA3400 modem, 4
VC multiplexing, 389
VDSL 7, 13, 49–52
vector modulation (vector coding), 219–220
Verbin, R. (Verbin's method), 239
video, 16, 19–21

videoconferencing, 399
Viterbi, Andrew, 202
Viterbi decoding, 40
Viterbi detection/algorithm, 202–204
voice channel, 398
voltage-controlled oscillator (VCO), 337
     discrete, 348
     oversampled control of, 351–352
     VCXO (VC crystal O), 351
VSB (vestigial sideband modulation), 175–6

W

Wander, 338
water-filling, 198–199, 209
wavelength, λ, 68–69
wavelet, 224
Wei, L.F., 258
Werner, Jean-Jacques, 64, 174
white noise, 150
whitening filter, 189
Wiese, Brian, 231, 251
Wiese's Canceler, 251–253
windowing
     in channel identification, 309
     receiver, 228
     transmitter, 226
wireless, 2
wireless local loop (WLL), 136
Wulich, D., 237

X

XDSL, 2

Z

zero-forcing, 223, 313, 320
Zimmerman, George, 336
zipper, 250–251
Zogakis, T. Nicholas, 263

http://www.phptr.com/

| What's New? | What's Cool? | Destinations | Net Search | People | Software |

PRENTICE HALL

Professional Technical Reference

Tomorrow's Solutions for Today's Professionals.

Keep Up-to-Date with
PH PTR Online!

We strive to stay on the cutting-edge of what's happening in professional computer science and engineering. Here's a bit of what you'll find when you stop by **www.phptr.com**:

@ **Special interest areas** offering our latest books, book series, software, features of the month, related links and other useful information to help you get the job done.

@ **Deals, deals, deals!** Come to our promotions section for the latest bargains offered to you exclusively from our retailers.

@ **Need to find a bookstore?** Chances are, there's a bookseller near you that carries a broad selection of PTR titles. Locate a Magnet bookstore near you at www.phptr.com.

@ **What's New at PH PTR?** We don't just publish books for the professional community, we're a part of it. Check out our convention schedule, join an author chat, get the latest reviews and press releases on topics of interest to you.

@ **Subscribe Today!** **Join PH PTR's monthly email newsletter!**

Want to be kept up-to-date on your area of interest? Choose a targeted category on our website, and we'll keep you informed of the latest PH PTR products, author events, reviews and conferences in your interest area.

Visit our mailroom to subscribe today! **http://www.phptr.com/mail_lists**

LICENSE AGREEMENT AND LIMITED WARRANTY

READ THE FOLLOWING TERMS AND CONDITIONS CAREFULLY BEFORE OPENING THIS CD PACKAGE. THIS LEGAL DOCUMENT IS AN AGREEMENT BETWEEN YOU AND PRENTICE-HALL, INC. (THE "COMPANY"). BY OPENING THIS SEALED CD PACKAGE, YOU ARE AGREEING TO BE BOUND BY THESE TERMS AND CONDITIONS. IF YOU DO NOT AGREE WITH THESE TERMS AND CONDITIONS, DO NOT OPEN THE CD PACKAGE. PROMPTLY RETURN THE UNOPENED CD PACKAGE AND ALL ACCOMPANYING ITEMS TO THE PLACE YOU OBTAINED THEM FOR A FULL REFUND OF ANY SUMS YOU HAVE PAID.

1. **GRANT OF LICENSE:** In consideration of your purchase of this book, and your agreement to abide by the terms and conditions of this Agreement, the Company grants to you a nonexclusive right to use and display the copy of the enclosed software program (hereinafter the "SOFTWARE") on a single computer (i.e., with a single CPU) at a single location so long as you comply with the terms of this Agreement. The Company reserves all rights not expressly granted to you under this Agreement.

2. **OWNERSHIP OF SOFTWARE:** You own only the magnetic or physical media (the enclosed CD) on which the SOFTWARE is recorded or fixed, but the Company and the software developers retain all the rights, title, and ownership to the SOFTWARE recorded on the original CD copy(ies) and all subsequent copies of the SOFTWARE, regardless of the form or media on which the original or other copies may exist. This license is not a sale of the original SOFTWARE or any copy to you.

3. **COPY RESTRICTIONS:** This SOFTWARE and the accompanying printed materials and user manual (the "Documentation") are the subject of copyright. The individual programs on the CD are copyrighted by the authors of each program. Some of the programs on the CD include separate licensing agreements. If you intend to use one of these programs, you must read and follow its accompanying license agreement. You may not copy the Documentation or the SOFTWARE, except that you may make a single copy of the SOFTWARE for backup or archival purposes only. You may be held legally responsible for any copying or copyright infringement which is caused or encouraged by your failure to abide by the terms of this restriction.

4. **USE RESTRICTIONS:** You may not network the SOFTWARE or otherwise use it on more than one computer or computer terminal at the same time. You may physically transfer the SOFT-WARE from one computer to another provided that the SOFTWARE is used on only one computer at a time. You may not distribute copies of the SOFTWARE or Documentation to others. You may not reverse engineer, disassemble, decompile, modify, adapt, translate, or create derivative works based on the SOFTWARE or the Documentation without the prior written consent of the Company.

5. **TRANSFER RESTRICTIONS:** The enclosed SOFTWARE is licensed only to you and may not be transferred to any one else without the prior written consent of the Company. Any unauthorized transfer of the SOFTWARE shall result in the immediate termination of this Agreement.

6. **TERMINATION:** This license is effective until terminated. This license will terminate automatically without notice from the Company and become null and void if you fail to comply with any provisions or limitations of this license. Upon termination, you shall destroy the Documentation and all copies of the SOFTWARE. All provisions of this Agreement as to warranties, limitation of liability, remedies or damages, and our ownership rights shall survive termination.

7. **MISCELLANEOUS:** This Agreement shall be construed in accordance with the laws of the United States of America and the State of New York and shall benefit the Company, its affiliates, and assignees.

8. **LIMITED WARRANTY AND DISCLAIMER OF WARRANTY:** The Company warrants that the SOFTWARE, when properly used in accordance with the Documentation, will operate in substantial conformity with the description of the SOFTWARE set forth in the Documentation. The Company does not warrant that the SOFTWARE will meet your requirements or that the operation

of the SOFTWARE will be uninterrupted or error-free. The Company warrants that the media on which the SOFTWARE is delivered shall be free from defects in materials and workmanship under normal use for a period of thirty (30) days from the date of your purchase. Your only remedy and the Company's only obligation under these limited warranties is, at the Company's option, return of the warranted item for a refund of any amounts paid by you or replacement of the item. Any replacement of SOFTWARE or media under the warranties shall not extend the original warranty period. The limited warranty set forth above shall not apply to any SOFTWARE which the Company determines in good faith has been subject to misuse, neglect, improper installation, repair, alteration, or damage by you. EXCEPT FOR THE EXPRESSED WARRANTIES SET FORTH ABOVE, THE COMPANY DISCLAIMS ALL WARRANTIES, EXPRESS OR IMPLIED, INCLUDING WITHOUT LIMITATION, THE IMPLIED WARRANTIES OF MERCHANTABILITY AND FITNESS FOR A PARTICULAR PURPOSE. EXCEPT FOR THE EXPRESS WARRANTY SET FORTH ABOVE, THE COMPANY DOES NOT WARRANT, GUARANTEE, OR MAKE ANY REPRESENTATION REGARDING THE USE OR THE RESULTS OF THE USE OF THE SOFTWARE IN TERMS OF ITS CORRECTNESS, ACCURACY, RELIABILITY, CURRENTNESS, OR OTHERWISE.

IN NO EVENT, SHALL THE COMPANY OR ITS EMPLOYEES, AGENTS, SUPPLIERS, OR CONTRACTORS BE LIABLE FOR ANY INCIDENTAL, INDIRECT, SPECIAL, OR CONSEQUENTIAL DAMAGES ARISING OUT OF OR IN CONNECTION WITH THE LICENSE GRANTED UNDER THIS AGREEMENT, OR FOR LOSS OF USE, LOSS OF DATA, LOSS OF INCOME OR PROFIT, OR OTHER LOSSES, SUSTAINED AS A RESULT OF INJURY TO ANY PERSON, OR LOSS OF OR DAMAGE TO PROPERTY, OR CLAIMS OF THIRD PARTIES, EVEN IF THE COMPANY OR AN AUTHORIZED REPRESENTATIVE OF THE COMPANY HAS BEEN ADVISED OF THE POSSIBILITY OF SUCH DAMAGES. IN NO EVENT SHALL LIABILITY OF THE COMPANY FOR DAMAGES WITH RESPECT TO THE SOFTWARE EXCEED THE AMOUNTS ACTUALLY PAID BY YOU, IF ANY, FOR THE SOFTWARE.

SOME JURISDICTIONS DO NOT ALLOW THE LIMITATION OF IMPLIED WARRANTIES OR LIABILITY FOR INCIDENTAL, INDIRECT, SPECIAL, OR CONSEQUENTIAL DAMAGES, SO THE ABOVE LIMITATIONS MAY NOT ALWAYS APPLY. THE WARRANTIES IN THIS AGREEMENT GIVE YOU SPECIFIC LEGAL RIGHTS AND YOU MAY ALSO HAVE OTHER RIGHTS WHICH VARY IN ACCORDANCE WITH LOCAL LAW.

ACKNOWLEDGMENT

YOU ACKNOWLEDGE THAT YOU HAVE READ THIS AGREEMENT, UNDERSTAND IT, AND AGREE TO BE BOUND BY ITS TERMS AND CONDITIONS. YOU ALSO AGREE THAT THIS AGREEMENT IS THE COMPLETE AND EXCLUSIVE STATEMENT OF THE AGREEMENT BETWEEN YOU AND THE COMPANY AND SUPERSEDES ALL PROPOSALS OR PRIOR AGREEMENTS, ORAL, OR WRITTEN, AND ANY OTHER COMMUNICATIONS BETWEEN YOU AND THE COMPANY OR ANY REPRESENTATIVE OF THE COMPANY RELATING TO THE SUBJECT MATTER OF THIS AGREEMENT.

Should you have any questions concerning this Agreement or if you wish to contact the Company for any reason, please contact in writing at the address below.

Robin Short

Prentice Hall PTR

One Lake Street

Upper Saddle River, New Jersey 07458

About the CD-ROM

As a convenience to persons who wish to conduct their own research into the origins of the XDSL standards, a CD-ROM is provided with this book that contains the equivalent of a three-drawer filling cabinet of the original standards committee documents. Standards specify *what* is required, but seldom explain *why* it is required. The documents on the CD-ROM provide extensive discussions of why the requirements are in the standards, and discuss many interesting concepts that were not adopted for the standards.

The CD ROM contains more than 900 T1E1.4 Standards Committee documents including:

- official reports of all T1E1.4 meeting from 1989 to 1998
- annual lists of all T1E1.4 documents from 1989 to 1998
- draft versions of many XDSL Standards and Technical Reports
- full text of the top papers presented at the T1E1.4 meetings from 1992 to 1998

The Standards Committee T1 may be reached at www.t1.org or by phone at 202-434-8845.

The CD-ROM also contains ADSL Forum Technical Reports 1 through 11. The ADSL Forum may be reached at www.adsl.com or by phone at 510-508-5905.

The CD-ROM also includes the Adobe Acrobat reader program that may be used to read to files on the CD-ROM. If you have not already the Adobe Acrobat Reader program installed on your computer, then it will be necessary to install the reader program from the CD-ROM. The reader program may be used for search for a word or phrase by typing **CTRL F** and then entering the word or phrase.

Technical Support

The CD does not include software other than the Adobe Acrobat Reader. Therefore, technical support is not provided by Prentice Hall nor the author. If, however, you feel that your CD is damaged, please contact Prentice Hall for a replacement: disc_exchange@prenhall.com.